A CONTEMPORARY GEOGRAPHY OF UGANDA

A CONTEMPORARY GEOGRAPHY OF UGANDA

Editor
Bakama B. BakamaNume
Prairie View A&M University

MKUKI NA NYOTA
DAR-ES-SALAAM

Published by:
Mkuki na Nyota Publishers Ltd.
Nyerere Road, Quality Plaza Building
P. O. Box 4246
Dar es Salaam, Tanzania
www.mkukinanyota.com

TOURISM UGANDA
You're welcome

ISBN 978-9987-08-036-6

Contents

Acknowledgements

The authors are grateful to numerous individuals in various government agencies and private institutions who generously shared their time, expertise and knowledge about Uganda. These people include; students who worked as assistants in some of the research projects reported and government employees who provided data used here. None of these individuals is in any way responsible for the work of the authors, however.

As the editor, I wish to thank those who contributed directly to the preparation of the manuscript. These include; Dr. Daudi Basena, Margaret Kabamba, Richard Kazibwe, Dr. Joe Muwonge who reviewed all textual and graphic materials and the complete manuscript; Auvin Burnem and Patrick Elelwa who typeset the chapters and helped prepare the manuscript for publication; and Bernard Muhwesi, who prepared some of the maps for the text.

I acknowledge the generosity of the individuals, public and private agencies who allowed their photographs to be used in this study. We are indebted especially to those who contributed work not previously published.

The idea was developed and most of research for the book was done during three summer visits to Uganda. I was in Uganda on another research project which was funded by a grant from the National Institute of Health with Dr Raymond Sis as the Principal Investigator (PI). I would like to thank my PI for that indirect support. I would also like to extend thanks to the editorial board of Mkuki na Nyota Publishers, Ndugu Walter Bgoya, Deogratias Simba, Tapiwa Muchechemera, and others. Asante.

Finally, I would like to extend my sincere thanks to my wife Ann and my children Babirekere (Babi) and Waibi, for being very understanding and supportive when I was busy working on the book.

Bakama B. BakamaNume
Houston, Texas, USA.

Acknowledgement

Introduction

The last text on the geography of Uganda was written in 1975 by Professor Brian Langlands. Since the last publication, Uganda has undergone numerous changes. The population has more than tripled from less than 10 million to almost 30 million. The district boundaries have changed and the number of districts increases every year. New districts are created every year. Economic productivity has also shifted over the years. Furthermore, new and emerging diseases have surfaced in Uganda. This textbook addresses the need for an updated document on the geography of Uganda. This book was written by a joint group of Ugandan geographers. The contributors authored chapters in their areas of specialization. There are a total of twelve chapters in the book. These chapters are based on the most current data available.

Chapter 1 discusses the climate of Uganda. Ndyabahika Matete and Bakama BakamaNume, examine Uganda's climatic controls, zones, characteristics, trends and the relative importance of climate in Uganda's economic development. They propose future directions mainly for practical applications in agriculture, human settlement, health and industry and in planning short, medium and long-term economic benefits. They acknowledge that data on climatic conditions is now more readily available than before.

Chapter 2 deals with the geomorphology of Uganda. Yazidhi Bamutaze discusses the diversity of Uganda's landforms and the geological structural changes caused by both internal and external forces. Their emphasis is on surfacial forces (erosion) or river action. The chapter points to the fact that climate changes have resulted in the melting of most of the ice on Mt. Rwenzori, leaving behind glacial features. Mt. Elgon also depicts features of glacial erosion and morainic deposition. In the Lake Victoria and Lake Albert regions, the immediate coastal shores show lacustrine sand deposits. Around Lake Victoria, a secession of raised beaches occur at approximately 3.5m, 15m and 20m and on part of the shores of Lake Mobutu, bars and cuspate fire lands have been created by the prevailing winds from the west.

Chapter 3 is devoted to soils of Uganda. Bob Nakileza provides an analysis of the distribution of major soil types, soil degradation including the types, causes of soil degradation and trends and effects in Uganda. The chapter also discusses response strategies to the problem of soil degradation.

Chapter 4 deals with forestry in Uganda. Mukadasi Buyinza and Jockey Nyakaana examine forest resources in Uganda. They discuss the availability of forest resources, exploitation and the problem of deforestation.

Chapter 5 covers water and wetland resources. Bakama BakamaNume and Hannington Sengendo examine water and wetland resources in Uganda. The chapter is divided into two sections. The first section deals with the availability of water resources and problems of water resources. The second section examines wetland resources and related problems.

Chapter 6 deals with population issues in Uganda. Fredrick Tumwine examines population distribution, population growth, demographic characteristics, factors influencing fertility and possible interventions to reduce rapid population growth.

Chapter 7 is about urbanization in Uganda. Hannington Sengendo examines the evolution of urban societies. He presents a critical view of the internal structure of cities and processes that mold and shape the city, urbanization trends, causes and effects associated with urbanization and city growth and the environmental issues resulting from urbanization. He critically looks at trends in Uganda and what can be done to reduce some of the urbanization problems.

Chapter 8 analyzes medical geography. In this chapter, Bakama BakamaNume focuses on four health issues: disease distribution, mortality causes, distribution of medical facilities and HIV/AIDS and malaria prevalence. Furthermore, the analysis and discussion in this chapter demonstrates how health care issues can be examined and analyzed using geographic and cartographic techniques.

Chapter 9 is on political geography. Bakama BakamaNume presents an examination of spatial attributes of the Ugandan political process in the areal expression. The focus of this chapter is the evolution of administrative units in Uganda, the administrative system in Uganda and the electoral geography in Uganda.

Chapter 10 examines spatial variations in economic activities. Bakama BakamaNume examines economic activities – primary, secondary, tertiary, quaternary and quinary. Primary activities involve extraction and harvesting of resources. These activities include; agriculture, mining, forestry, hunting and quarrying. Secondary activities involve adding value to a raw material. Manufacturing and construction are secondary activities. This chapter focuses on primary and secondary sectors of the Ugandan economy. The emphasis is on agriculture, fisheries and manufacturing.

Chapter 11 is on development geography. Jockey Nyakaana provides an interpretation of and elements of sustainable development, as well as the relationship between geography and sustainable development in Uganda. The discussion gives a conceptual framework for linking the two. It also shows the linkage between the Millennium Development Goal 7 (MDG) for ensuring environmental sustainability and other MDGs.

And finally, Chapter 12 deals with Geographic Information Systems (GIS). Shaub Lwasa focuses on GIS as a new tool in Geography and geographic studies with some demonstrations of its capabilities in handling geographic data and information. The chapter is organized in three parts: the first part is an introduction on GIS in which definitions and theoretical underpinnings of GIS are elaborated; the second part is a discussion of GIS as a tool and method with some review of literature on methodology in Geography as well as the different fields where GIS is applied and used in Uganda; and the third part presents some case studies analyzed using GIS focusing on the procedures or algorithms and the uses of the outputs. The three case studies attest to the analysis of population dynamics and visualization and urban development. The chapter concludes with a brief overview on opportunities and limitations of using GIS in Uganda.

Districts Of Uganda

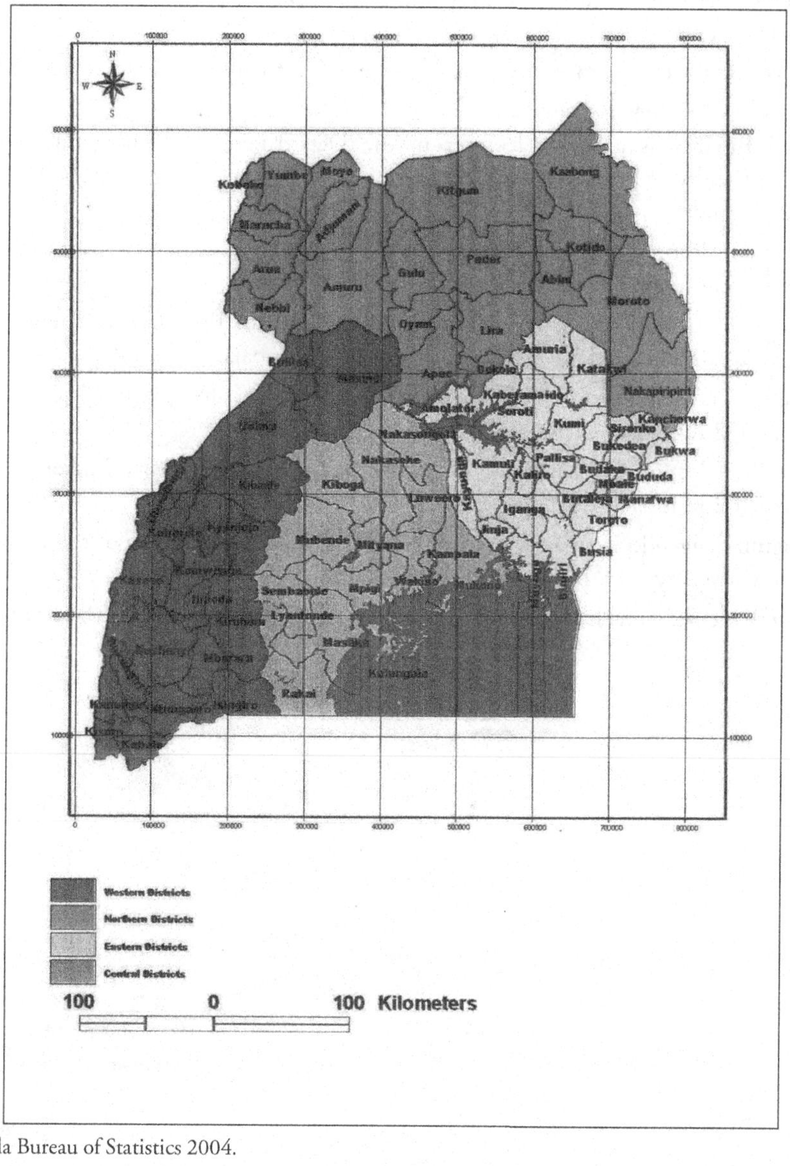

Source: Uganda Bureau of Statistics 2004.

The Uganda district map is continually changing. Districts are created every year. However, the official district data is based on the last National Census Data of 2002. For this reason, this book uses maps with 2004 district boundaries. County boundaries have not changed but there is not much data on the county level.

Contributors

1. Bakama BakamaNume is an associate professor of Geography at Prairie View A&M University, Prairie View, Texas
2. Yazidhi Bamutaze lecturer Department of Geography, Faculty of Arts, Makerere University, P.O. Box 7062 Kampala, Uganda
3. Mukadasi Buyinza is a senior lecturer in the Department of Community Forestry and Extension, Faculty of Forestry and Nature Conservation, Makerere University, P.O. Box 7062 Kampala, Uganda
4. Shuab Lwasa is lecturer in the Department of Geography, Faculty of Arts, Makerere University, P.O. Box 7062 Kampala, Uganda
5. Jockey B. Nyakaana is an Associate Professor and head of the Department of Geography, Faculty of Arts, Makerere University, P.O. Box 7062 Kampala, Uganda
6. Ndyabahika Matete is lecturer Department of Geography, Faculty of Arts, Makerere University, P.O. Box 7062 Kampala, Uganda
7. Bob Nakileza lecturer Department of Geography, Faculty of Arts, Makerere University, P.O. Box 7062 Kampala, Uganda
8. Hannington Sengendo is an associate professor of geography and dean of the Faculty of Arts, Makerere University, P.O. Box 7062 Kampala, Uganda
9. Fred Tumwine lecturer Department of Geography, Faculty of Arts, Makerere University, P.O. Box 7062 Kampala, Uganda

CHAPTER 1

CLIMATE OF UGANDA

Ndyabahika Matete and Bakama B. BakamaNume

1.1 Introduction

Climate is the average state of the atmosphere at a given point on the earth's surface (Beckinsale R.P, 1955). It consists of a mixture of average physical elements, which include; precipitation, temperature, wind speed and direction, atmospheric pressure, humidity, cloud cover and sunshine duration. Other elements include; soil moisture, soil temperature and evaporation, but the significance of these depends on particular situations (Robinson P.J and Sellers A. H, 1999). These averages are computed over a considerable time usually 30 to 35 years.

Climate is a major natural resource for a country and it benefits nearly all human activities. It is a major established factor in the resource and economic development of any country. On the other hand, climatic events such as cold or dry spells or floods can negatively influence human activities. This chapter examines Uganda's climatic controls, zones, characteristics, trends and the relative importance of climate in Uganda's economic development. It proposes future directions mainly for practical applications in agriculture, human settlement, health and industry and in planning short, medium and long-term economic benefits. At present, for the purpose of making better decisions, information on climate has become more important than ever before.

1.2 Types of Climate

The Climate of Uganda can be divided into five categories.

1.2.1 True Equatorial Climate

This is limited especially to islands of L. Victoria such a Buvuma, Ssese, Kalangala and Kome. It is characterized by:

a) Double maxima type of rainfall distribution centered around March and September.

b) Heavy and evenly distributed rainfall throughout the year, averaging between 2,000 - 2,500 mm per year.

c) Uniformly high mean monthly temperatures throughout the year, ranging between 24 ^{0}C and 27 ^{0}C.

d) Greatest insolation is received with a small diurnal range temperature of between 2 ^{0}C - 3 ^{0}C.

e) High relative humidity of 80%.

f) Extensive cloud cover and no distinct dry spell throughout the year.

1.2.2 Modified Equatorial Climate

This is experienced around L. Victoria and in parts of mountainous and hilly areas of Rwenzori and Kigezi. Uganda lies astride the equator but the large part of it is denied a true equatorial climate because of variations in altitude and other climatic controls. It is characterized by heavy rainfall of 1,500mm - 2,000 mm throughout the year; high temperatures and extensive cloud cover similar to that of true equatorial climate but with modifications.

1.2.3 Tropical savanna or Continental climate

This is a wet and dry tropical type of climate. It is transitional between an equatorial wet climate and a semi-arid or arid climate. Northern Uganda experiences this type of climate. It is characterized by:

a) Monomodal type of rainfall distribution centered around June.

b) Moderate and unreliable rainfall ranging from 750mm to 1500mm per annum.

c) High mean monthly temperature ranging from 24^0C to 30°C.

d) Low humidity.

e) Long dry season from September to April.

1.2.4 Semi arid or arid climate

North East Uganda including Karamoja, Moroto and Kotido districts and Ankole-Masaka corridor experience this type of climate. It is characterized by:

a) Little and unreliable rainfall of less than 750mm per annum.

b) Excessively high temperatures of over 30^0C.

c) High diurnal temperature ranges with very hot days and very cold nights.

d) Little cloud cover with clear skies.

e) Dry winds with low humidity.

1.2.5 Montane Climate

The mountainous and highland areas experience this type of climate. Varied types of climate are experienced ranging from tropical savanna to temperate climate, for example, Mt. Rwenzori. Below 1200m sea level, the climate is tropical savanna while from 1,200 m to 2,000 m above sea level the climate is equatorial. Above 2000m, the climate changes to temperate type and at 4500m, the snow line is reached.

1.3 Factors Influencing Climate in Uganda

The factors that influence climate in Uganda. include;; latitude, altitude and physical shape of the land, winds, water bodies, vegetation and the influence of human beings.

1.3.1 Latitudinal Effect

Uganda lies between 1^0S and 4^0N latitudes. This means that the sun is always overhead in the sky. Day time heating leads to the formation of an equatorial low pressure zone called Inter-tropical Convergence Zone (ITCZ). This zone migrates north and southwards following the apparent movement of the overhead sun. This results into regions north or south of the equator having one hot and wet season when ITCZ lies across the region, followed by a warm or hot dry season when it lies on the opposite side of the equator. Intense heating in this zone causes the air currents to rise which cool rapidly, condense and form clouds, which fall as rain.

ITCZ is responsible for the seasonal pattern of rainfall distribution in Uganda. This is because it influences the seasonal changes in the nature of Southeast and Northeast prevailing winds and gives rise to bimodal rainfall distribution in some parts and mono-modal distribution in other parts. Kalangala, Ssesse islands and Northwest shores of L. Victoria have bimodal distribution with rainfall seasons from March to May and September to November because the sun is along the equator in Uganda twice a year. Gulu, Kitgum and Lira districts have mono-modal rainfall distribution with a rainfall season from May to August.

However, in Uganda the local variations due to relief and presence of large water bodies introduce significant modifications such that the seasons cannot follow the classical ITCZ pattern with enough correlation to make this approach a practical one (Griffiths, J.F, 1972).

1.3.2 Altitude

Altitude is the height of the land above sea level. A rise in altitude leads to a fall in temperature at a lapse rate of 6.5 ^{0}C for every 1,000 m one ascends. This is because:

a) Gases like water vapour and carbon dioxide, particles of sand and dust, which absorb heat directly from the sun and also prevent heat from escaping rapidly in space, are found at a lower altitude. Mountain and hilltops have little of these and therefore heat escapes rapidly.

b) The amount of heat radiated by the earth's surface depends on the surface area. At lower altitudes, the surface area is wide hence more heat is radiated. At higher altitudes, the surface area is small hence less heat is radiated.

c) The atmosphere receives the greater part of its heat not directly from the sun but from heat absorbed by the earth. The heat warms the overlying lower layers of the atmosphere through radiation, convection and conduction.

Mountainous areas such as Rwenzori, Elgon and Muhavura tend to have lower temperatures and low humidity compared to areas at lower altitudes. The western rift valley constitutes the lowest areas in Uganda and is characterized by high temperature. Altitude has resulted in montane climate, which portrays various climatic conditions in high land and mountain environments. Although Mt. Rwenzori is along the equator, it has a permanent snow cover because of altitude, while Mt. Elgon experiences occasional frost conditions (Mawejje A.B, 2004). Altitude is largely responsible for the modification of the typical equatorial climate. Sometimes the higher you go the cooler it becomes. This is temperature inversion due to altitude effect. This is true in those areas where there are sharply marked mountains and valleys or hills and valleys such as Mbale and Kigezi highlands.

During the day, the valley bottom is warmed up more quickly than the hill tops. Cool air moves from tops of mountains down-wards, while warm air from the valley bottom rises. This makes the valley have low temperature during the day while mountain slopes have moderate temperature. This creates a situation where the valley floors are chilly because of cool air from hilltops. In the process of exchange of positions of cool and warm air, the two air masses meet above the valley. This results in fog conditions (Figure 1.1). At night, the earth loses heat rapidly and the atmosphere near the ground in the valley tends to be cooler than air higher above the surface of the valley. Thus greater heights are warmer than the valley floors.

The effects of temperature inversion include;:

a) Slowing of the development of conventional currents hence low precipitation.

b) Formation of fog, thus reduction of visibility.

c) Promotion of pollution in industrial locations.

d) Lowering of temperature in valleys, thereby, affecting crop growth and human settlement.

e) Reduction of morning hours' work especially in agricultural areas.

The Equator in Southwestern Uganda

Photography by Bakama B. BakamaNume (2004)

Figure 1.1: Temperature Inversion (Warm Air Raising, Fog, Valley, Cool Air descending

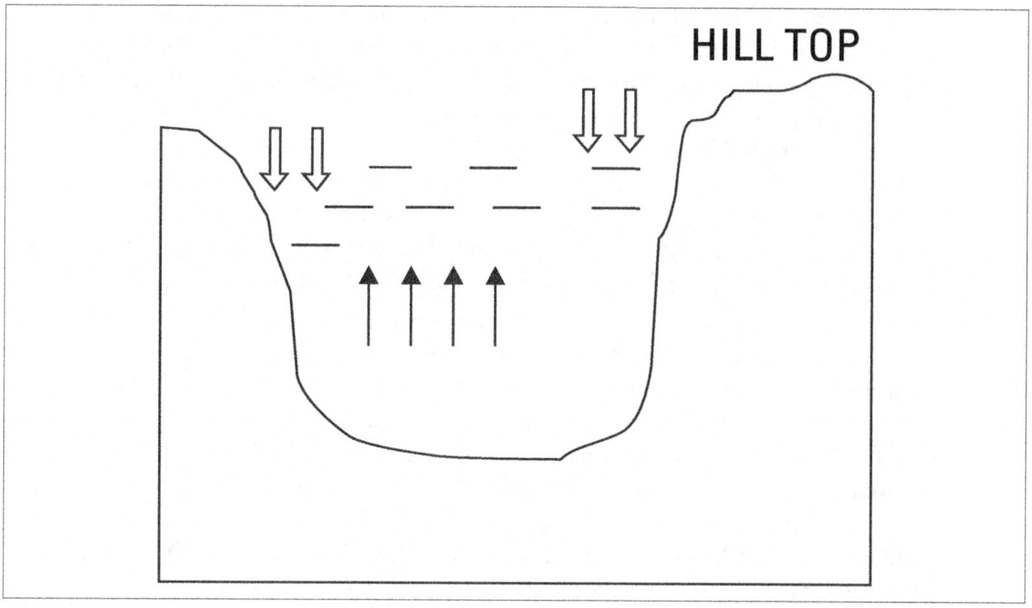

Altitude affects atmospheric pressure. Atmospheric pressure increases with decrease in altitude. The peaks of Mt. Elgon and Rwenzori have lower atmospheric pressure than the foothills. This is because of a bigger column of air above foothills and vice versa.

1.3.3 The general physical shape of the landscape

The country is characterized by various landforms which create local differences in climate. Mountains, Rwenzori, Elgon and Kigezi, as well as West Nile highlands, lead to the formation of relief rainfall on their leeward sides. The rainfall tends to increase up the slope until a zone of maximum rainfall is reached above which rainfall amounts decrease.

Tops of mountains tend to have low precipitation compared to middle slopes. Hills and mountains create a rain shadow effect resulting in inadequate rainfall. That is why low lying areas of L. Edward, George trough, Kasese, Mubende, parts of Bunyoro, Semuliki Valley and L. Albert flats in Western Uganda receive very low rainfall totals. Some of these areas are within the western rift valley where there are no relief barriers and so winds do not rise as they blow across. The lack of high areas means there is no rainfall formation in these areas.

Absence of highlands in Karamoja and Mbarara-Masaka corridor contribute to aridity conditions there. Karamoja has flat plains with a few isolated residual hills or inselbergs. Cumulo-nimbus clouds build up around these hills early in the morning leading to heavy rainfall (NEMA, 1996). Consequently, few areas of heavier rainfall, notably Mt. Moroto and around residual hills break the monotony of semi-desert climate. Much of Uganda is a plateau and it is characterized by undulating hills and valleys, lakes and swamps which also influence the climate.

1.3.4 The Effects of Winds

South East, North East and South westerly prevailing winds, South West monsoon, katabatic and anabatic winds systems influence the climate of Uganda. As the sun moves north or south of the equator, the Inter Tropical Convergence Zone (ITCZ) migrates in a similar pattern. The prevailing winds also move in a similar direction as the ITCZ. ITCZ attracts the NE trade winds from the Arabian Desert. These are dry winds which pass over the Ethiopian highland on their way to the low pressure of the ITCZ. Dry conditions are then extended to North Eastern Uganda which is in a rain shadow of Ethiopian highlands. The North Eastern and Northern slopes of Mt. Elgon from Ngora extending to Amuria are dry because of the effect of North East trade winds.

The South East trade winds are warm moist winds originating from the Indian Ocean in a bid to enter the ITCZ. As they move towards the equator they are warmed up by warm Mozambique Ocean currents and more moisture is picked from the ocean. They are forced to rise over the highlands in southern Tanzania and drop their moisture in the form of relief rainfall. They then become relatively dry but continue their North West movement until they pass over L. Victoria where more moisture is picked. This moisture is later dropped as rain in the Northern and North western shores of L. Victoria. That is why areas around Mabira Forest, Mukono and Mpigi have heavy rainfall. On reaching the equator the South East trade winds become deflected as South West monsoon winds and account for heavy rainfall on the south western sides of Mt. Elgon, Western Uganda hills, S.W Mufumbiro ranges, Mt. Rwenzori, Nebbi and Arua districts are influenced by Southwest monsoon winds which are moist laden from the Atlantic Ocean passing over the Congo basin between August and January. These areas are also influenced by the deflected South East trade winds, or south west monsoon winds.

Katabatic, or mountain winds, move down slope of a hill or mountain. They occur at night when the slope is maximally cooled. The cool denser air drains down slope under the influence of gravity. On the other hand, anabatic, or valley winds move upslope during the day when the slope is maximally heated by the sun's insolation.

The two types of winds cause:

a) Fog and chilly conditions
b) Precipitation up slope by anabatic, e.g. around Mt. Elgon
c) Precipitation down slope by katabatic, e.g. the northern shores of L. Victoria
d) Temperature inversion in valleys in the morning.

The meeting of NE and SE trade winds in the ITCZ triggers off the upward movement of conventional currents. This action leads to cloud formation through conduction and formation of frontal rainfall.

1.3.5 Water Bodies

L. Victoria and L. Kyoga are the main source of moisture which recharges the atmosphere through evaporation. As the water vapour rises, it condenses and falls as conventional rainfall. That is why Kalangala and Ssese Islands have average rainfall of over 2000mm per year.

Lakes also influence the climate of adjacent lands through land and lake breezes (Figure 1.2). These breezes occur in areas where land is lying in close proximity to water bodies.

Figure 1.2: Lake and Land Breezes

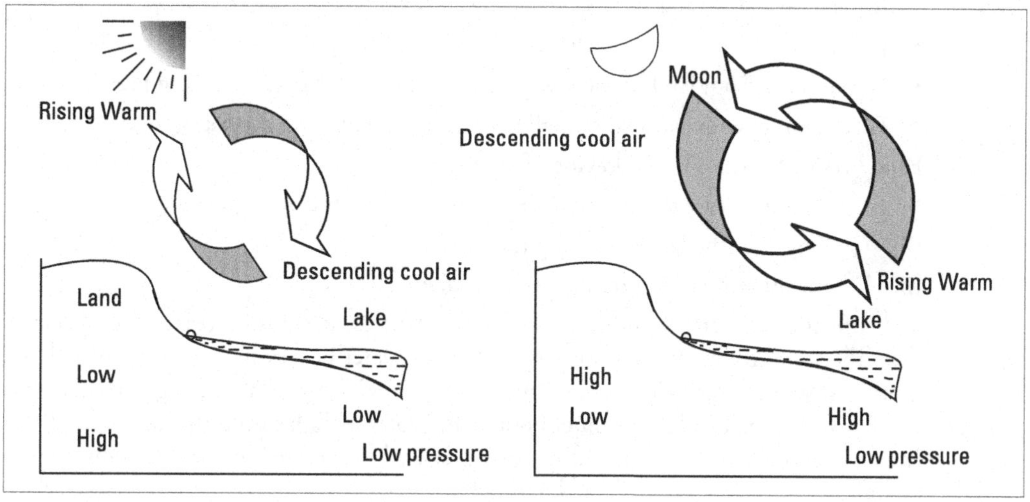

The movement of air between the land and lakes is caused by the following factors:

a) Differences in specific heat capacities of both land and water;
b) Differences in reflecting capacities of land and water;
c) Mobility of water compared to land which is solid;
d) Heat transmission through water as opposed to opaque land;
e) Differences in pressure on land and water.

Lake breeze processes are summarized as follows:

- There is intense heating of adjacent lands during the day;
- Land warms faster than the lake; therefore, temperatures are high over land and low over the lake;
- Convention currents of warm air rise over land and create a reduction in pressure over the land compared to the lake;
- Cool moist wind from the lake (high pressure zone) blows towards the land (low pressure zone) as a lake breeze to replace the rising air;
- Formation of conventional rainfall from risen air which falls on lake shores especially in the afternoons.

Lake breezes have got the following effects:

(i) Moist air can cause rainfall, or a drizzle especially on coastal areas
(ii) Cool temperature on land in late afternoon and evening such as in Kampala, Jinja and Entebbe urban centers.

Land breeze processes are summarized as follows:

- Rapid loss of terrestrial radiation on coastal lands at night.
- Land cools faster than the lake which leads to lower temperatures over the land than the lake, which retains most of its heat as it cools slowly.
- Low pressure is created over the warm lake and high pressure over the cold land.
- Wind blows from land to the low-pressure zone over the lake as a land breeze.
- Formation of conventional rainfall from risen air falls over the lake at dawn.

Land breezes have got the following effects:

(i) Cool temperature on the lake brought by the land breeze

(ii) Cooling over the lake causes fog formation.

(iii) Temperature inversion results over the lake.

The influence of water bodies on the climate of adjacent lands decreases with distance from the water body. Areas in proximity to large water bodies such as L. Victoria have their temperature modified by maritime conditions. North Eastern Uganda is partly dry due to the general lack of water bodies and from being far from the influence of the Indian Ocean. In general, the effects of water bodies on climate of surrounding areas include;:

(i) Moderation of temperatures of surrounding areas through land and sea breezes;

(ii) Raising humidity of surrounding atmosphere;

(iii) Creating their own wind system;

(iv) Creating differences in atmospheric pressure; and

(v) Creation of rainfall.

1.3.6 Vegetation

When vegetation is involved in the process of evaporation, transpiration occurs. The term "evapo-transpiration" is used to describe the process. This process influences temperature, humidity and rainfall of a given area. High evapo-transpiration rates from thick vegetation cover recharges the atmosphere with water vapour that cools, forming clouds and rain. According to National Environmental Management Authority (NEMA (1998), there are 1,276,370 hectares of forest reserves in Uganda. Uganda's wetlands are estimated to be 30,105km2 which represents 13% of the total area of Uganda according to NEMA (2001). Forests greatly contribute to rainfall formation in areas where they are found such as Mabira forest, Bundongo, Bwindi, Mt. Elgon, Rwenzori forests, Itwara and Mpanga. Swampy areas around L. Victoria, L. Kyoga, Kururuma valley wetlands, Kabale district valley swamps and swamps along Katonga and Rwizi rivers have a similar effect. Poor vegetation cover of acacia, scrub and thorny vegetation is partly responsible for dry conditions in Karamoja.

1.3.7 Human Influence

The physical environment has been interfered with by people through swamp reclamation, drilling of bore holes, industrialization, deforestation and poor farming methods. Swamp reclamation, which is intended to provide land for settlement, cultivation, dairy farming and industrialization, has resulted in the lowering of the water table. Swamp reclamation causes stagnant water to drain away and the depletion of vegetation. This means that the processes of evaporation and transpiration are interfered with resulting in low humidity and rainfall. Industrialization results in emission of gases, which absorb the sun's insolation causing a rise in temperature. This is why Jinja and Kampala industrial complexes experience high temperatures. Afforestation and re-afforestation programmes, on the other hand, have resulted in forests. From these forests, there is evapo-transpiration that results in water vapour that goes into the atmosphere, cools and forms clouds leading to precipitation.

1.4 Climatic Zones

Kakumirizi G W (1989) and NEMA (1999) identified five climatic zones in Uganda, based on precipitation. They include; Karamoja, Acholi-Kyoga, L. Victoria, Ankole-Southern Uganda and Western Uganda areas (See Figure 1.3).

Zone 1: Karamoja Zone

This zone experiences a semi- desert type of climate. Rainfall is less than 750mm. There is a single rainy season from May to August. An intense, dry, hot season is from November to March. When the overhead sun moves northwards towards the Tropical of Cancer in June, the rainfall belt shifts north so that areas in Karamoja experience a rainy season. Hills in this zone, together with the Southwest monsoon winds, play a significant role in rainfall development and distribution. Temperatures range from 300C and above with a high diurnal range of temperature especially during the dry season due to the absence of cloud cover.

Zone 2: Acholi -Kyoga Zone

This consists of parts of Northern and Eastern Uganda. This land is generally flat, embraced by L. Kyoga, Kwania, Opeta and Bisiria with their associated swamp and savanna vegetation immediately after the swamps. There are smaller climatic sub divisions in this zone largely due to the effect of L. Kyoga. This zone experiences one long wet season from April to October, but with two peaks, from April to June and August to October. These peaks are a result of the passage of the ITCZ. Annual rainfall ranges from 1000-1500mm per annum. Temperatures are relatively high; 240C is the minimum temperature during the rainy season.

Zone 3: Lake Victoria Zone

Lake Victoria, its Islands and the "fertile crescent" extending inland 80 km from the lakeshore, fall within this zone. It is characterized by flat topped hills separated by swamp valleys, swamp inlets of the lake, short grass on hill tops and forests in valleys giving rise to papyrus swamps. Some patches of natural vegetation in this zone have been cleared for crop production, settlement and industry giving rise to secondary vegetation in some cases. Rainfall received ranges from moderate to heavy. The Islands experience heavy rainfall of over 2000mm, while from the lakeshore; it ranges from 1500mm- 1750mm and progressively reduces towards inland. The zone has a double maxima rainfall pattern, March to May and September to November. A real dry season does not occur since rain is received even in the drier months of December to March and June to July. It experiences a high mean monthly temperature of 210-270C with a small diurnal temperature range and dense cloud cover.

Zone 4: Ankole- Southern Uganda

Most of East Mukono, Mpigi, Mubende and Western parts of Busoga, Masaka, Eastern Mbarara and Ntugamo districts are in this zone. The zone is characterized by flat topped hills in Ankole, Lakes Wamala and Kachira, Katonga and Mayanja rivers, together with their associated swamps, scrub and grass down lands of Ankole and scattered patches of forests. A modified equatorial type of climate with two rainfall peaks is experienced. March to May and September to November are the rainfall seasons. A pronounced dry season occurs in June and July; while a less severe and often-interrupted one is December to February. Rainfall ranges from 750mm per year in East Ankole to 1500 mm towards Zone 3.

Map 1.1: Major Climatic Zones in Uganda

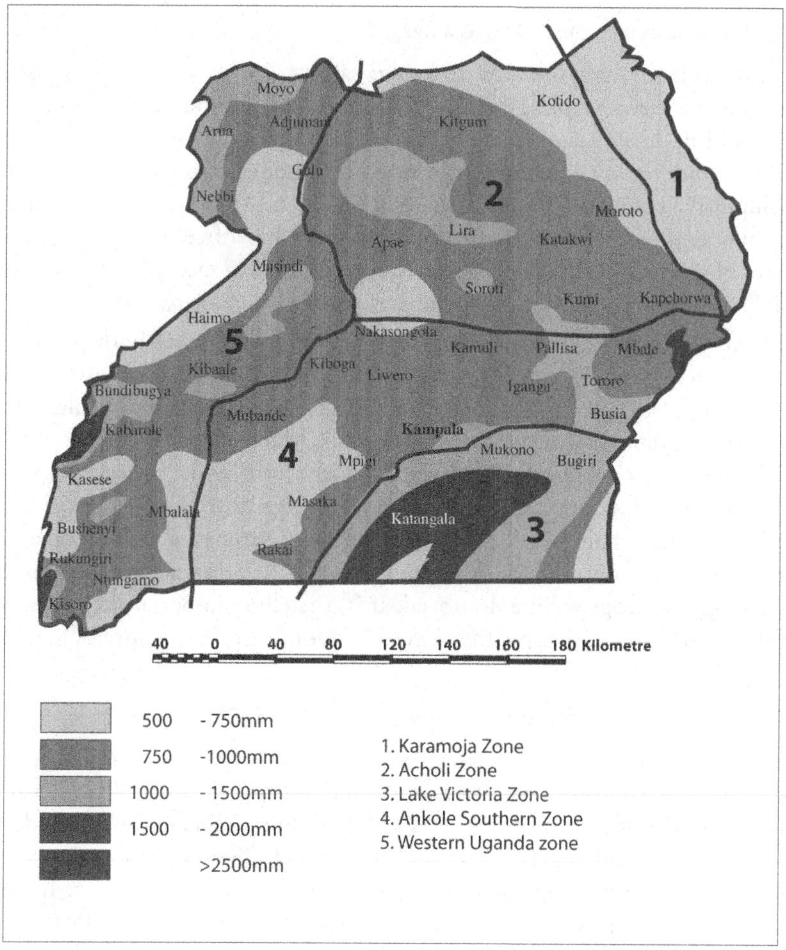

Source 1: Kakumirizi G W (1989) Physical Geography and Human Geography of Uganda. P. 69

2. NEMA (2001): State of Environment report for Uganda. P. 85

Zone 5: Western Uganda

This zone stretches along the Democratic Republic of Congo and the Uganda border from the Kigezi highlands in the South West through the western Rift Valley to West Nile. It encompasses Kigezi highlands, west Ankole hills, the Western Rift Valley, its escarpments and lakes, Mt. Rwenzori, Toro and West Nile hills. Other characteristic features of this zone include;: forests such as Kibale, Bundongo and Semiliki. It is a zone of transition between Congo basin equatorial climate and Uganda savanna climate. These highland areas receive more rain than the lower valley areas.

1.5 Climate and its relative importance to economic development

1.5.1 Climate as a resource and as a hazard

A resource is anything that human beings value. Climate is a resource since it influences human activities such as agriculture, transport and communication, forestry, recreation and tourism. It is a major factor in resource and economic development of every nation. Unlike other resources, climate as a resource is not easy to value, allocate, manage and manipulate. Therefore, resource analysts view it as a non-market resource because it is not allocated by the price mechanism. It is viewed as a free resource, a public good and factor of production, open to all even though its utilization affects others. Heat, coldness, rainfall, snow and wind are exploitable resources which are renewable.

Hazards are extreme fluctuations in nature. They are acts of both people and nature. These fluctuations become hazards when they disrupt the social and economic well being of people, loss of property and life of people and animals. Understanding human-environmental relationships in extreme fluctuations of climate is of paramount importance to people in their attempt to satisfy their needs. Energy and water are bases of the process of natural hazards. Changes in energy distribution give rise to excess wind and water, snow, hailstorms, heat waves, drought, tornadoes, floods and avalanches. All these have a negative effect on human activities. If rain exceeds the amount required by a crop, the crops will be destroyed. It is, therefore, important for the population to understand the nature, type, frequency, duration, magnitude, speed of climatic hazards and their effects.

1.5.2 Climate and Agriculture

Temperature influences the physical and chemical processes connected with plants and crops including absorption of water, gases and mineral nutrients; diffusion and synthesis; as well as vital processes of growth and production. Climate limits the distribution of plants on earth and largely determines the flora of different regions. It also delimits the areas of agricultural crops. It can cause damage to or destroy plants. It influences the development of diseases, insects and pests. Therefore, temperature affects several aspects of plant germination, vegetative phase, generative phase and flowering response.

In Uganda, temperature does not greatly affect agricultural land use because of minimal variations. The lack of marked variation is attributed to the location within the tropical realm. However, differences in altitude create variations in temperature and give rise to different types of crops grown and animals reared in mountainous areas. Cool temperatures in Kigezi highlands have influenced the growth of oats, barley and pyrethrum and dairy farming typical of temperate lands. High temperatures in Karamoja reduce agricultural effectiveness in this region by limiting it to pastoralism. Understanding rainfall types, effectiveness, distribution, reliability and intensity is of paramount importance to a farmer.

1.6 Types of rainfall

All three types of rainfall are received in Uganda. These include;: relief, convectional and frontal rainfall.

Relief Rainfall

Relief Rainfall is formed when moist air meets a mountain barrier forcing air to rise up the mountain. This triggers conditional instability and air uplift. The rising warm moist air is forced to cool at dry adiabatic lapse rate until the dew point is reached. The air continues and cools at the saturated adiabatic lapse rate. The cooling releases latent heat, which makes the atmosphere unstable and this forces the air to continue rising leading to the formation of cumulonimbus clouds. Rain then falls on the windward side of the mountain. The leeward side lies in the rain shadow and is dry due to the desiccating effect of the dry descending wind (See Figure 1.3). Kisoro, Kabale, Mbale and Rwenzori experience this type of rainfall.

Figure 1.3: Formation of Relief rainfall

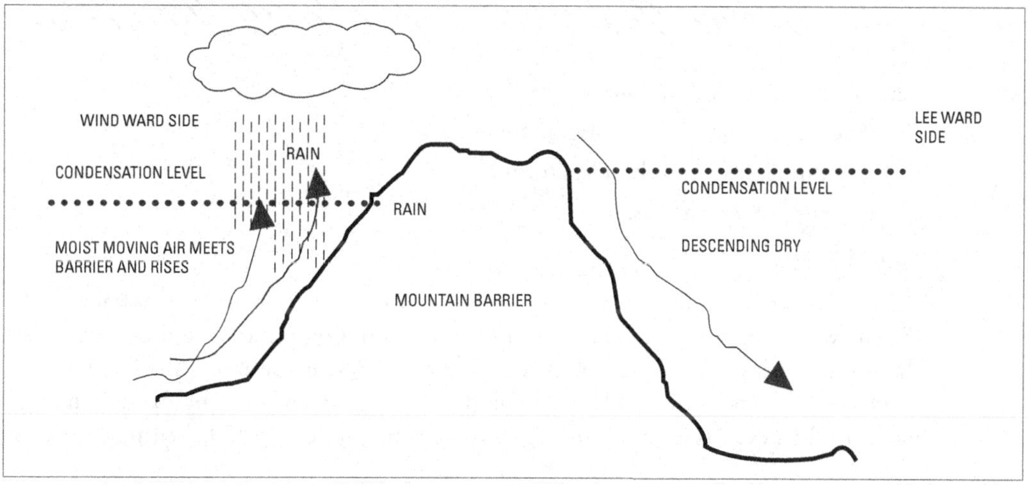

Relief rainfall is characterized by the following:

(i) Rain lasts for a long time;

(ii) Rains in phases. It starts slightly light and gradually increases and is therefore effective in agriculture;

(iii) Rains especially during the night and morning and

(iv) Sometimes accompanied by storms, lightning, hail and thunder.

Convectional rainfall

This type of rainfall occurs when the ground surface is heated leading to upward movement of warm moist air. The air rises and cools to form strato-cumulus clouds. As condensation continues, latent heat is lost and instability occurs. The rising air continues to rise and cools to form cumulonimbus clouds (Figure 1.4). These result in rainfall accompanied by thunder and lightning which occurs throughout the year along the equator.

Figure 1.4: Formation of Conventional rainfall

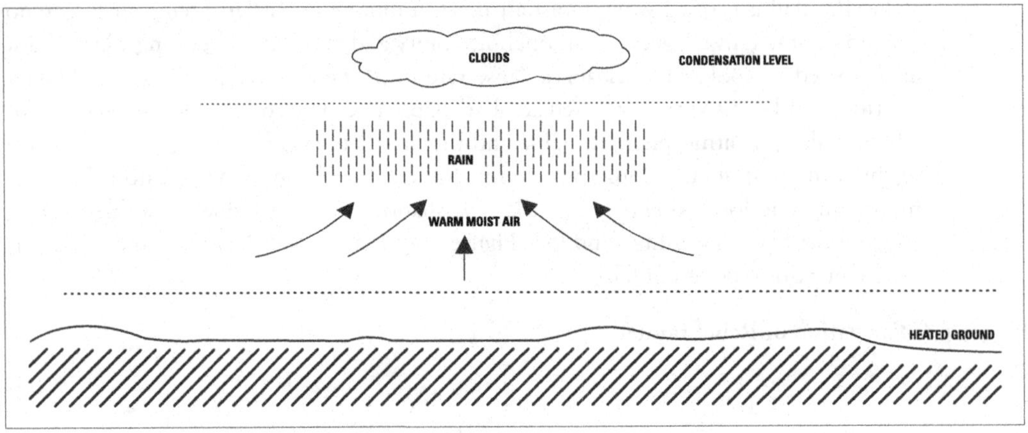

Characteristics of Convectional Rainfall:

(i) Short lived, violent and torrential rains.

(ii) Lightning, heat and thunder storms.

(iii) Occurs late afternoons.

Frontal Rainfall

This occurs at the zones of convergence such as the ITCZ in the equatorial trough. When two air masses of different temperature, humidity and direction of flow meet, the warm moist air rises over the descending cooler less dense moist air. The risen air condenses to form clouds and later rainfall (Figure 1.5). The meeting of NE and SE trade winds in the ITCZ triggers off the upward movement of convectional currents causing cloud formation and rain in the L. Victoria basin.

1.6.1 Effects of rainfall on human activities in Uganda

Rainfall has both positive and negative effects on human activities. The positive effects include;;

(i) Source of water for domestic and industrial use and for animals;

(ii) Rain-based agriculture depends on it e.g. Relief rainfall which is very effective for agriculture;

(iii) Source of water for growth of forests enabling the forestry industry;

(iv) The volume of water in rivers, lakes and swamps is maintained for navigation and fishing;

(v) Source of water for growth of pasture for domestic and wild animals;

(vi) Moderation of climate;

The negative effects include;

(i) Transport is hindered during heavy rains;

(ii) Promotes soil erosion especially in highland areas;

(iii) Triggers and mass wasting eg landslide, soil creep, mudflow in highland areas;

(iv) Heavy rains result into floods, which destroy settlemenst and properies;

(v) Hail storms destroy crops and

(vi) Thunderstorms affect people's lives and propertyies.

Figure 1.5: Formation of Frontal Rainfall

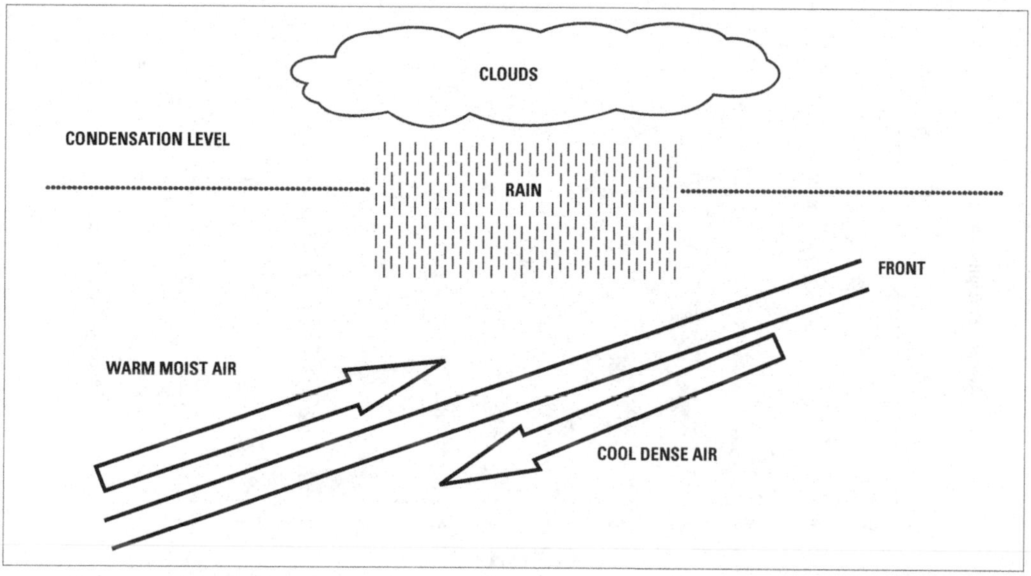

1.6.2 Rainfall distribution

This refers to the way in which rain is distributed in a given place both in time (temporal) and space (spatial). Temporal distribution refers to the way rainfall is distributed in one place over the year. This gives rise to two types of rainfall patterns in Uganda: Bimodal or double maximum and monomodal or unimodal.

Bimodal or double maxima pattern of the equatorial type of rainfall pattern

Rainfall is received throughout the year with two peak periods. Climate stations such as Kampala have this pattern, peaks occurring between March and May and September to November (see Table 1.1 and Figure 1.6).

Table 1.1: Climatic statistics for Kampala, 1999

Months	J	F	M	A	M	J	J	A	S	O	N	D
Rainfall (mm)	29	8	170	293	208	75	133	38	191	202	110	41
Temperature (°C)	29.9	30.	28.9	28.3	27.2	28.0	26.3	27.9	27.9	27.4	28.2	29.0

Source: Department of Meteorology, Kampala

Figure 1.6: Kampala Annual Climagraph 1999

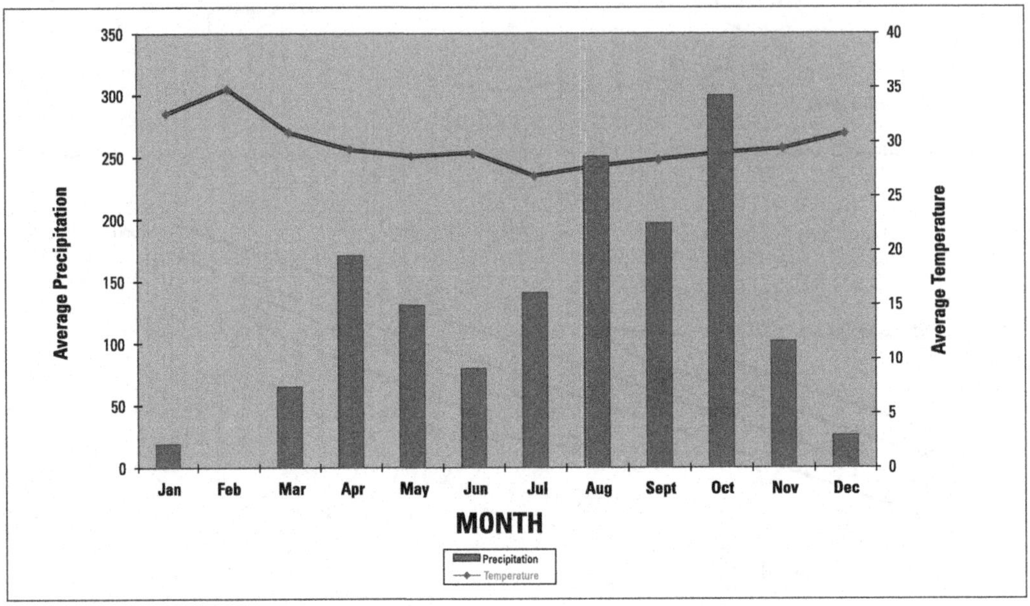

Monomodal or unimodal or the tropical type of rainfall pattern

This is characterized by one wet season and one dry season. The wet season comes when the sun is overhead during the months of December to March in the Southern Hemisphere. In the Northern Hemisphere, one wet season and dry season is experienced. The one rainfall season coincides with the presence of overhead sun in the months of May to August e.g. in Gulu. (See Table 1.2 and Figure 1.7)

Table 1.2: Climatic statistics for Gulu, 1999

Months	J	F	M	A	M	J	J	A	S	O	N	D
Rainfall (mm)	19	0	65	171	131	80	141	251	197	300	102	26
Temperature (°C)	32.6	34.9	30.9	29.3	28.7	29.0	26.9	27.8	28.4	29.0	29.4	30.8

Source: Department of Meteorology, Kampala

The factors responsible for such patterns include;-:

(i) Location either at North or South of the Equator
(ii) Direction of winds and wind bearing rainfall or moisture
(iii) Presence of water bodies, e.g. L. Victoria and the Indian Ocean
(iv) Modifying influence of relief and altitude
(v) Vegetation cover

Figure 1.7: Climagraph for Gulu 1999

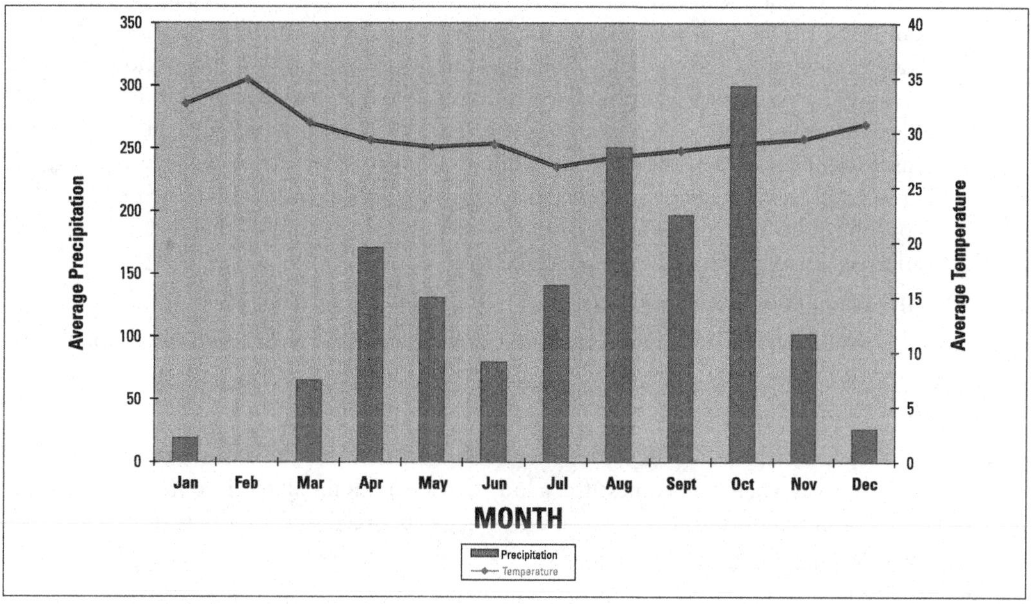

Agricultural practices such as land clearing and ploughing, planting and sowing, weeding and harvesting are based on temporal distribution of rainfall. It assists farmers to identify the type of crop to be grown and when.

Perennial crops tend to grow well in areas with bimodal pattern while the practice is hampered in areas with mono-modal pattern resulting in growing of annual crops. Temporal distribution also relates to plant and animal disease patterns and is therefore important in land use planning.

Spatial distribution of rainfall refers to variations in the distribution of the mean annual rainfall over the surface of the earth in a given period of time. This helps farmers to know the expected amounts of rain received in a particular place and to plan activities accordingly. Climatic stations in the L. Victoria zone receive over 1500mm. Good climatic conditions are one of the reasons for agricultural activities such as tea plantations at Kasaku and sugar plantations at Lugazi and Kakira. These provide raw materials to agro-based industries. In areas with 11 or 12 wet months bananas are grown by 75% of the farmers. Bananas are the main staple food for more than 60% of urban population

(Olupot, M, 2004). Karamoja receives less than 750mm and this has given rise to pastoralism.

1.6.3 Rainfall Effectiveness

This refers to how rainfall of a specific place is useful for agricultural purposes. Two places receiving the same amount of rainfall in a given year may not have the same degree of effectiveness. Near the surface of the earth, there is a zone of aeration in which pore spaces contain both water and air. This zone can be of zero thickness in swamps. Water in this zone is called "soil moisture", which is one of the elements of climate. Soil moisture can be in the form of water vapour, capillary water in the small pores or hygroscopic water adhering to soil particles. Soil moisture plays a big role in the conditioning of biological processes of plants and crops responsible for their growth and development. Plants need sufficient water in all stages of growth in order to sustain optimal growth. Insufficient water can cause wilting while too much water can result in water logging in soil and lack of oxygen to enable plant growth. The amount of water from rain that contributes to soil moisture, not the run offs, influence rainfall effectiveness. Rainfall effectiveness is influenced by several factors.

(i) Amount of rainfall received

Rainfall tends to be more effective in areas that receive large amounts of rainfall and vice versa.

(ii) Seasonal distribution

In areas where rainfall comes after a short dry season, rainfall is more effective than in areas where rain comes after a long dry season. Rainfall received in the L. Victoria basin is more effective as there is no big gap between wet and dry seasons.

(iii) Rainfall intensity

Torrential rain is less effective while gentle showers give the soil a chance of taking in almost all the water and run offs are minimized. Conventional rainfall of South Central Uganda is more effective compared to relief rainfall that drizzles for a long time in mountainous areas of Uganda.

(iv) Degree of infiltration

The greater the degree of infiltration, the more effective the rainfall in a given area. Infiltration is affected by the nature of soil, rock and rate of evaporation.

(v) Evaporation

This is a process whereby water on the earth's surface is changed from liquid state to gaseous state. The higher the rate of evaporation, the less the effectiveness of rainfall. This prevails in parts of the country since the sun is overhead.

(vi) Runoff

This is the amount of rain water that flows off. The greater the run offs, the less effective the rain. Highland areas, e.g. Kabale and Mbale are characterized by too much run offs that flow to the lowlands.

(vii) Retention capacity of the soil

This is the ability of the soil to retain moisture. The retention capacity of sandy soils is low while that of loam soils is high. Sandy soils of Sango bay and lateritic soils of Buganda have low retention capacity and therefore low rainfall effectiveness.

(viii) Capillary action

This is a process by which soils allow water that was previously received by the soil to come back to the surface. This increases the effectiveness of rainfall although it might result in evaporation.

(ix) Rainfall incidence

This refers to the time of the day when it rains. Rain which falls at night is more effective than day light rainfall. This is because during the night there is no evaporation. Morning showers are more effective than afternoon showers.

(x) Technology

Technology used in agricultural areas affects rainfall effectiveness. If a tractor is used it may create a hard pan thereby tampering with the infiltration capacity of the soil.

(xi) Methods of farming

Poor methods of farming decrease the effectiveness of rainfall. The removal of vegetation through burning exposes the soil and consequently increases the rate of evaporation.

1.6.4 Rainfall Reliability

This is a measure of the critical amount of rain a specific crop will tolerate in a period of its growth in a given area. If a crop such as bananas requires 1800mm per year, then rain will be reliable every year if the area receives 1800mm per year or more. Any rain received below 1800mm will lead to crop failure and is unreliable. Excessive rain in a cropping system like cotton and grain can be as damaging as rain failure. El Nino rains in 1997/98 damaged many crops in Mbale, Bundibugyo and Kabale.

Rainfall is reliable where it is the heaviest such as in the Lake Victoria basin, highland and mountainous areas. Rainfall is least reliable where it is light such as in Mbarara, Masaka corridor, Karamoja and the Western Rift Valley. Kakumirizi G.W. (1989) reports that 2/3 of Uganda has a good prospect of receiving more than 750mm in 95 years out of 100 years hence the country has higher reliability.

Knowledge of rainfall reliability is a basis for agricultural policy formulation and extension towards improving crop productivity and designing appropriate land uses. A choice between rainfed and irrigated agriculture is based on rainfall reliability. Unreliability of rainfall in Karamoja has favoured pastoralism.

1.6.5 Rainfall Variability

Rainfall variability refers to the way rainfall of a particular place varies from year to year. It presents the idea of how divergent rainfall is from year to year. If the mean annual rainfall fluctuates with a 50% reduction in a given year, it results into drought which has adverse effects on agriculture. If it fluctuates by doubling in a subsequent year, it results in excessive flooding, silting and mass wasting. In October 1997 and January 1998, El Nino rains resulted in floods that destroyed a variety of crops in Mbale and Kapchorwa (NEMA, 1999).

1.6.6 Rainfall Intensity

This refers to the rate at which rain in a given place pours per hour. High intensity rains supply too much rain in a short time resulting in run offs which enhance soil erosion. Such rains result in a high rate of soil compactness thereby affecting plant root penetration in the soil. Low intensity rains supply little water per hour. The rains take a long time while drizzling. In some places, such as Mbale and Rwenzori mountains, rains sometimes take two days. This increases the chances of water penetration into the soil.

1.7 Climate and Fishing

The renewable nature of fishery resources and the inherent spatial and temporal variation in stock size, are capable of being modified by climate. The link between climate and fishery resources occurs through the transfer of energy from the atmosphere to an open water body. Solar radiation results in energy that is needed by planktons for photosynthesis to take place. Fish feed on planktons. Uganda has 12 hours of sustainable sunshine. Fishermen take advantage of sunshine in fish preservation through sun drying, e.g. at Kasenyi landing site on L. Victoria.

Winds affect fisheries through turbulent mixing and transportation of water and nutrients. The depth in which turbulent mixing occurs depends on wind speed, direction and duration. Wind generated surface currents result in egg and larval transportation away from nursery areas. These currents control productivity through upwelling. Upwelling is a process of water motion whereby cold and deep water moves upwards to the surface as a result of winds displacing warm surface water.

Rainfall received in major fishing grounds and their catchment areas recharges waters in which fish inhabit. The major fishing grounds include; Lakes Victoria, Kyoga, Albert, George, Edward, Wamala, Nakivale and Bunyonyi and numerous rivers and swamps. The fishery resources contribute to gross domestic product, income generation, export earning and nutrition (NEMA, 1996).

On the other hand, solar radiation results in energy that meets water surfaces resulting in evaporation, thereby reducing the quantity of water in a fishing ground. This is particularly true during prolonged drought when water levels of shallow lakes, swamps and rivers shrink. NEMA (1996) reports that L. Wamala is shrinking. High temperature accelerates the decomposition of fish. It is likely that by the time a fish trader transports fish from Gaba landing site to Kampala City on a bicycle it has already decomposed.

1.8 Climate and Disease Control

Some climate and weather conditions help in the propagation of airborne and water borne diseases that affect human beings, plants and animals. Knowledge of excess water and deficit helps in controlling, or mitigating the effects of these diseases.

Excess water provides a suitable environment for the survival of disease vectors such as mosquitoes (malaria), snails (bilharzia), tsetse flies (sleeping sickness) and simulium flies. In Uganda, the occurrence of water borne diseases is positively correlated with wet seasons when there is excess water.

The control of pests and diseases is possible with knowledge of their ecology and physiology. Pests and diseases are not producers but direct or indirect consumers of crops and plants and their direct products. Examples of pests include; birds, small mammals, insects, nematodes and parasites such as fungi, bacteria and viruses. The seasonal activity tends to be associated with the seasonal pattern of growth and development of crops and plants. This pattern is dictated by the seasonal pattern of climatic elements. For example, certain climate conditions may favor dispersal of certain pests, or diseases. The number and size of their host plants which provide a source of chemical energy tends to increase during the growing season. Some pests and diseases tend to attack only specific varieties of crops/plants at a particular time and susceptible period of the cycle of crop growth and development. Potato blight is favoured by periods of high humidity and seasonal susceptibility of the crop. Therefore, to control this disease, the following methods can be adopted:

(i) Seasonal control by early planting or late sowing;

(ii) Spraying the potato crop when temperature and humidity conditions are suitable for spraying of the blight.

Weather conditions for the production and spread of diseases and pests during the dry season influence the amount of viable risk for potential invasion the next year. Nutritional differences and certain metabolic disorders in animals are associated with seasonal variations in the food supply especially vitamins and certain minerals which is dependent on climate conditions. Certain diseases affect livestock of a certain age in a certain physiological condition, e.g. pregnancy or in lactation. The outbreak of an infectious disease may be associated with certain farm operations to coincide with certain stages of growth and development in the animal. For example, tetanus in sheep results from careless shearing and castration whereby the sheep may sustain cuts and abrasions.

Warm months are the periods of greatest activity of many pests although some are in quiescent form in the soil, or in vegetation during warm periods. For example, ticks disappear in the dry season due to sensitivity to bright light. Current weather conditions may influence larval operations because they are susceptible to drought and strong sunlight. Weather variations also influence the intensity and severity with which pests and diseases attack farm livestock and plants. Adverse weather conditions produce changes in nutritional levels of animals and plants thereby affecting the health of the animals and plants.

In summary, knowledge of climate, disease and pest relationship is of paramount importance because it allows us to examine several important factors in the disease ecology relationship. It allows us to measure relevant probabilities of weather patterns/chains and interaction between them and pests and diseases. We can measure the probability in time and intensity of any event or a combination of events. We can examine time and in degree, the components of uncertainty and their interactions, such as disease cycles or pest cycles since these have been highly unpredictable or uncertain.

1.9 Climate and Human Settlement

Human settlement is a place where people live. It ranges from an individual homestead or a hamlet in a country side, to a village, or town or city and then to urban complexes such as metropolitan and cosmopolitan cities (Okello - Oleng, 1989). Human settlement can be rural or urban. People in rural settlements mainly engage in subsistence and processing production activities such as fishing, agriculture, or hunting; and fruit gathering which are dependent on natural factors especially climate. Buildings in rural settlements are predominantly traditional in design; materials used in construction are dictated by the type of climate. Availability of fuel resources and building materials, which depend on climate, are important in choosing rural settlement sites.

Climate influences design and materials used in the construction of houses. For example, houses have stunted roofs in heavy rainfall areas such as Kigezi highlands, while in hot areas such as Karamoja the houses have mud walls and roofs. The hot climate in Karamoja influences the construction of temporary houses called manyattas. While it is true that climate has a hand to play, most housing in Uganda especially in urban areas were built to abide with the building code. Climate also influences settlement patterns. This is because climate influences pollution rates, health and amount of radiation. Thermal considerations, ventilation and wind pressure, day lighting factors, precipitation or dampness are dictated by climate.

1.10 Climate change and variability in Uganda

1.10.1 Climate change

NEMA (2001) defines climate change as a long term change of one or more climatic elements from previously accepted long term mean values. It is the deviation of normal climatic aspects which include; temperature, rainfall, sunshine and others. Uganda is vulnerable to climate change. Asalu A.O. (1995) points out the following evidences for climate change:

1) Geomorphological evidence

   The ongoing glacier recession on Mt. Rwenzori confirms that temperature has risen. The snowline changed from 4500m to 4800m. Climate change as found along L. Victoria lowered the water level and caused raised beaches and raised cliffs.

2) Biological evidence

   Disappearance of animal species and appearance of strange pests and diseases such as skin wrinkling, mental irritability, skin cancer, skin burns and eye cataracts supports this evidence. There is vegetation degradation in which forests degraded to woodlands while woodlands degraded to thorn bush or thicket.

3) Climate data evidence

This is portrayed by reduction in rainfall totals and increase in temperature.

4) Agricultural evidence

This is manifested by low agricultural output due to crop failure, widespread famine, increased establishment of irrigation projects to ensure regular supply of water, valley dam construction in Teso and Karamoja, to overcome harsh climatic conditions, continued flow of humanitarian food assistance and swamp reclamation.

5) Energy budget evidence

This is manifested through global warming and increased heat wave.

6) Limnological evidence

L. Victoria has a history of rapid rise in level possibly due to climate change effect. From 1923-1963 L. Victoria's level was high due to heavy rainfall. Since 1963 the lake level has dropped by one metre. Today's L. Kyoga's level varies between 5-7m in depth due to prolonged droughts; insufficient rainfall and decreased river discharge.

7) Historical evidence

It is believed that Karamoja used to be characterized by wet climate and the Dinka tribe lived there with marine organisms such as crocodiles. And over time, wet conditions dwindled and eventually, the land was occupied by pastoral tribes of Karamajong, Iteso and Jie from Ethiopian highland who overgrazed the area leading to dry conditions.

8) Potamological evidence

Rivers from highlands have been drying slowly, e.g. Malaba, Namatala, Mpologoma, Kerim and Awonja which originate from Mt. Elgon.

1.10.2 Climate variability

NEMA (1996) defines climatic variability as the disruption of normal climatic patterns that results in excessive rainfall totals, or prolonged drought conditions. It is a sharp, short term variations of meteorological events as compared to their long term means. The phenomena are devastating in the short term. There are variations in the yearly mean temperatures and rainfall over Uganda.

Temperature

Average temperatures for 12 stations and their deviations for the years 1994-2000 are shown in Table 1.3. The mean values are based on the data for the years 1994-2000. Deviations are based on these mean values. All the stations indicate positive and negative deviations in temperature during this period. Deviations for all the 12 stations in 1997 are positive. This must have been due to the influence of the 1997 El Nino phenomena that affected Uganda and other parts of the world. Masindi in the western climatic zone had positive deviations between 1995 and 2000, while Gulu in Acholi-Kyoga zone and Mbarara in Ankole southern Uganda zone had positive deviations between 1997 and 2000.

Rainfall

Table 1.4 shows total rainfall, highest and lowest rainfall totals for 12 stations per year from 1994 to 2000. Generally, each station shows wide annual variations during this period. However, the trends of total rainfall amounts for Mbarara and Arua stations tend to decrease. The highest total rainfall was 1115mm in 1994 and the lowest was 532mm in 2000 for Mbarara station. The highest recorded rainfall was 1538mm in 1994 and lowest was 1,052mm in 2000 for Arua. This is particularly due to the effect of overgrazing in Mbarara and deforestation in Arua. Although annual total rainfall amounts for Jinja, Entebbe and Kampala stations, which represent L. Victoria climatic zone show variations in 1994-2004 periods, the totals are relatively higher than those of the stations representing other climatic zones due to the latitudinal effect of the Equator.

Causes of climate variability

Climate largely depends on conditions on the surface of the earth. Any changes on the earth's surface leads to corresponding impact on climate variation in the short term and climate change in the long run. Climatic variation in Uganda is a result of continuous changes of the following factors:

1) Industrialization

Industrialization in Uganda has negative effects on climate. It has contributed to an increased rate of pollution and heat budgets over the earth's surface. The trace elements emitted to the atmosphere through pollution act as a blanket between the atmosphere and earth's surface and this blanket absorbs long wave radiation during the day from the earth's surface thereby raising the temperature of the earth. This phenomenon is called global warming. Industrialization has also contributed to the destruction of vegetation in Uganda. Industrial activities in Kabale (tin works), Tororo and Hima (cement works) have affected vegetation which influences climate. On 28th October 2003, the New Vision, a local newspaper reported that a limestone factory in Moyo has resulted in tree depletion in the area.

Table 1.3: Average, Mean annual temperatures and deviations for 1994-2000

Zone	Station	Average temperature							Deviations							
		1994 2000	1995	1996	1997	1998	1999		Mean	1994	1995	1996	1997	1998	1999	
Acholi-Kyoga Zone	Gulu	21.9	22.4	27.3	30.3	30.3	29.8	30.4	27.5	-5.6	-5.1	-0.2	2.8	2.8	2.3	2.9
	Lira	29.5	28.1	32.9	30.8	28.4	29.6	29.7	29.9	-0.4	-1.8	3.0	0.9	-1.5	-0.3	
										-0.2						
	Soroti	25.6	30.2	27.9	30.5	30.0	30.3	29.4	29.1	-3.5	1.1	-1.2	1.4	0.9	1.2	0.3
Lake Victoria Zone	Entebbe	21.6	26.0	23.5	26.9	27.8	26.3	26.9	25.6	-4.0	0.4	-2.1	1.3	2.2	-0.7	1.3
	Jinja	23.5	28.0	27.9	28.1	25.7	28.1	28.4	27.1	-3.6	0.9	0.8	1.0	-1.4	1.0	1.3
	Kampala	23.5	28.3	28.0	27.9	28.3	26.9	25.0	26.8	-3.3	2.0	1.2	1.1	1.5	0.1	
										-1.8						
Ankole – Southern Uganda	Mbarara	22.3	20.5	19.7	27.0	24.9	27.2	26.9	24.1	-1.8	-3.6	-4.4	2.9	0.8	3.1	2.8
	Tororo	29.2	29.0	29.9	29.9	26.7	24.4	29.4	28.4	0.8	0.6	1.5	1.5	-1.7	-4.0	1.0
Western zone	Arua	23.6	28.9	26.0	28.9	26.1	26.3	26.9	26.7	-3.1	2.2	-0.7	2.2	-0.6	-0.4	0.2
	Masindi	1.1	29.1	28,9	29.2	28.3	29.2	29.2	28,3	-4.2	0.8	0.6	0.9	0.0	0.9	0.9
	Kasese	25.1	26.7	27.2	30.2	25.9	30.7	30.4	28.0	2.9	-1.3	-0.8	2.2	-4.1	2.7	2.4
	Kabale	20.1	22.4	24,2	24.2	24.4	23.9	24.2	28.3	-3.2	-0.9	0.9	0.9	1.1	0.6	0.9

Source: Department of Meteorology, Kampala

Table 1.4: Total annual rainfall, highest and lowest rainfall amounts for 1994-2000

Zone	Station	Annual Rainfall Totals							Highest rainfall	Lowest Rainfall
		1994	1995	1996	1997	1998	1999	2000	(mm)	(mm)
Acholi – Kyoga Zone	Gulu	1828	1308	1670	1296	1159	1483	1371	1828	1296
	Lira	1745	1957	1676	1296	1190	1375	1639	1957	1190
	Soroti	1082	1434	1459	1195	1357	1221	1247	1434	1082
Lake Victoria Zone	Entebbe	1652	1973	1898	2318	1420	2679	1052	2679	1052
	Jinja	1439	1787	1461	1390	1227	1544	1029	1787	1029
	Kampala	1366	1498	1712	1565	1350	1689	1083	1712	1083
Ankole-Southern Uganda	Mbarara	1115	932	535	1070	926	748	532	1115	532
	Tororo	1596	1576	1885	1448	1355	1477	1247	1885	1247
Western Zone	Arua	1538	1430	1283	1298	1148	1403	1052	1538	1052
	Masindi	1320	1697	1309	1596	1106	1221	1567	1697	1106
	Kasese	864	750	1182	1296	811	829	879	1296	750
	Kabale	1079	1195	1218	1110	1167	860	983	1218	860

Source: Department of Meteorology, Kampala

2) Deforestation

Deforestation affects rain formation. It limits evapo-transpiration processes thereby reducing water vapour that goes to the atmosphere hence low cloud and rain formation. In Uganda, the increase in demand for forest products such as fuel wood, charcoal, timber and non timber forest products, combined with the need for more agricultural land, has resulted in deforestation. Whenever land is cleared for agricultural purposes, there is a change in surface character. This change is especially marked when forests are replaced by crop land. Even fruit bearing trees, such as mangoes, have been encroached upon in Tororo, Kumi, Pallisa and Soroti districts for charcoal burning and brick making. The Bamboo forests on the slopes of Mt. Elgon are continuously being destroyed as the Bagisu harvest "Malewa", or bamboo shoots for human consumption.

3) Swamp reclamation

Swamp reclamation affects the re-changing effect of swamps and consequently a dry climate follows. The water vapour that cools to form clouds and then rain will be in limited supply in reclaimed swamp lands. The supply is affected by drainage channels which cause the disappearance of open surface water and the general depletion of swamp vegetation.

4) Poor farming methods

Overgrazing, over-cultivation, overstocking, bush-burning and ploughing up and down hill slopes are contributing to the extension of the Sahel south wards towards Uganda. These farming methods disrupt the natural balance between vegetation cover, pasture and soil. The disruption has negative consequences on climate. For example, overgrazing in Rakai, Kitgum, Kamuli and Ntungamo districts contributes to climatic variations.

5) Drilling of boreholes

This leads to the lowering of the water table, thereby reducing soil moisture needed to sustain vegetation growth. Reduction in soil moisture leads to wilting of plants and consequently immature dying of vegetation and this affects evapo-transpiration processes. This has been experienced in Luwero district.

6) Seasonal changes in ice and snow

The quantity of ice has an impact on global climate. It affects precipitation received. Taylor (2003) reports that the extent of glaciers that cap the Rwenzori Mountains in western Uganda have declined. For example, he points out that the Speke glacier has receded by more than 300m in the last decade. The rate of snow loss is of concern due to its potential impact on water resources, stream flow and ecology of the Rwenzori Mountains and the climate in general. Taylor (2003) points out that there is a very thin layer of snow of less than 20cm and no firm layer, or more dense snow before ice is reached. He states that although there is an overall trend towards reduced snow cover during the 20th Century, there is evidence of a brief expansion of glaciers in the early 1960s that is linked to increased amounts of precipitation when water levels in R. Nile and L. Victoria rose by more than 1m. He points out that recession of alpine glaciers across the tropics suggests that the causes of rapid snow loss are global in nature and therefore little can be done locally to halt this trend in Uganda.

7) Afforestation and Re-afforestation programmes

These influence rainfall through evapo-transpiration which results into a lot of moisture which condenses to form clouds and later rainfall on the slopes of Mt. Elgon forest and Kigezi highland plantation forests.

8) Ozone layer depletion

Ozone is a gas composed of three atoms of oxygen. It surrounds the earth like a delicate veil, protecting the planet and its inhabitants from the direct blaze of the sun. It blankets against ultra violet rays which are harmful to human life. It is a thin layer at about 60km above the ground. Changes in ozone concentration have dramatic effects on climate and life on earth.

Ozone is being depleted because of the introduction of chemicals that increase the rate of conversion of ozone to oxygen. Such chemicals include; chloroflouro carbons (CFCS); halons, methylchloroform, carbon tetrachloride and methylbromide, all of which destroy ozone (NEMA, 1996). The Ozone layer depletion results into utra-violet radiation reaching the earth's surface thereby increasing atmospheric temperature. An increase in atmospheric temperature results into excessive evaporation by water bodies. The water vapour which rises and cools adiabatically results in excessive rainfall. Therefore, ozone layer depletion changes the atmosphere and this in turn results in climate change (Okello-Oleng, 1989).

9) Climate anomalies

Climatic anomalies such as the El Nino and La Nina phenomena contribute to climatic variation. Every year around December, the Peruvian cold current develops into a warm current whose effects are felt through increased rainfall and surface temperature. In November/December 1997, global mean surface temperature was 0.43°C above the 1961-1990 base period average due to the El Nino effect (NEMA, 1996). Rainfall was extremely high resulting into destructive floods and landslides in Mbale, Bundibugyo and Kapchorwa.

1.11 Effects of climatic variations in Uganda

Effects of climatic variations in Uganda include:

1) Vegetation

An increase in temperature in highland areas has caused a gradual disappearance of montane vegetation.

2) Drought and floods

Rainfall amounts below the average leads to drought, while the monthly average above the long term average leads to excessive water supply, resulting in floods, landslides, washing away of roads and bridges, soil erosion, silting of dams and reservoirs (NEMA, 2001). Rainfall duration in Uganda is brief, uncertain and confused. The seasons have changed tremendously.

3) Crop failure

Low rainfall totals lead to crop failure and subsequently food shortages, which in turn lead to famine. Food shortages that are experienced in Soroti, Moroto, Lira, Gulu and Karamoja are the result of low rainfall totals. On July 23, 2004, the government of Uganda warned that the country would face famine due to poor distribution of rain, a prolonged dry spell and that crop production was to drop by 40% (Olupot M, 2004).

4) Changes in farmers' seasonal calendar and a shift to other activities

Delays in rainy seasons or even rain failure have interfered with farmers' seasonal calendar. The planting dates have been affected and yet other practices are based on it. In Ankole, millet used to be planted in mid August when rains began. Now it is difficult for farmers to plan accordingly and stick to the seasonal calendar. As a result, farmers tend to shift from one activity to another. The Basigu banana growers are slowly converting to growing cassava which is drought resistant. The farmers in Apac, Pallisa and Soroti are changing from peasant agriculture to fishing in L. Kyoga. Boat and net-making are associated activities.

5) Changes in levels of lakes, their shrinkage and water use

Water levels of L. Victoria, Kyoga, Opare and Kwania are reported to have fallen since 1963. If the trend continues the impact will be manifested in reduction of hydro electric power, navigation and fishing potentials. A drier or wetter climate affects water use. As evaporation increases due to prevailing drier conditions, many dams, reservoirs and hydro electric power schemes may be rendered useless. Development planning becomes very difficult if there is climate variation.

6) Diseases and pests

Drought results in food shortages and famine. These invariably affect the feeding habits of people. Insufficient feeding results in nutritionally related diseases, such as, kwashiorkor, marasmas, stunted growth, under weight and anemia. NEMA (2001) reports that malnutrition ranks among the 10 killers in Uganda. Malnutrition also lowers people's immunity, thus exposing them to other diseases e.g. diarrhoea and respiratory infections.

Excess rainfall results in outbreaks of waterborne diseases such as cholera and malaria. Outbreaks of cholera were recorded in Karamoja (1973-1980), Kasese (1978-1984), Kampala (1985), Kasese (1991), Kampala and other parts of the country (1997/98) due to the effects of El Nino rains (NEMA, 2001). Malaria is a leading cause of death in Uganda being responsible for 17.9% of all deaths in health units (NEMA, 2001).

Changes of weather conditions contribute to the occurrence of new human diseases such as skin cancer, drying, wrinkling and itching of human skin. The occurrence of animal diseases such as foot and mouth in Mbarara and plant diseases such as cassava mosaic in Mbale and millet disease in Teso are partly due to climate variations. The prevalence of pests such as army worms, green hoppers and potato-pests in Soroti is also due to climate variations.

7) Wide spread changes in the natural ecosystems

With grasslands decreasing, arid areas expanding and forested areas shrinking there is decrease of vegetation species hence loss of habitat for wildlife. NEMA (2001) reports that between 1960 and 1998 Uganda lost 71% and 76% of its antelope species and other larger animal population respectively. The worst hit animals are the black and white rhinos.

8) Conflict

Climate variations can be associated with scarcity of resources. Conflict arises when it comes to sharing and utilization of scarce resources. The competition for pasture and water in drought-stricken pastoral areas can generate conflict, including cattle rustling.

9) Decline in crop and animal production

Climate variations can be a result of ultra violet rays reaching the earth. Ultra violet radiation affects plant growth, reduces crop yield as it slows down photosynthesis and delays germination in plants. Climate variations lead to variations in rangeland plants which may have negative consequences on the grazing of livestock and hence on animal production.

10) Acceleration of Desertification

Okello-Oleng (1989) defines desertification as the process by which desert like conditions encroach, appear and spread to formally non-desert areas as a result of the destruction of the equilibrium of fragile ecosystems through excessive human activities. Drought (little and/or irregular rainfall) accelerate desertification. In Uganda, severe and widespread occurrence of desertification was formally recognized for the first time in 1977 at the United Nations conference on Desertification (NEMA, 2001).

11) Poverty

Droughts and the resultant famine situations compound poverty among the peasants, incapacitating the communities from participating in economic development.

1.12 Proposed future directions in mitigating the effects of climate variations

Practical actions are urgently needed to minimize the effects of climate variations in Uganda.

1) Planting more trees

The quantity of carbon dioxide already emitted to the atmosphere can be mopped by planting more trees through afforestation and re-afforestation programmes. More trees imply more water vapour which cools and forms rain. Ibuge County in Apac district and Ruhama County in Mbarara district, practise afforestation while re-afforestation takes place in Mbale and Arua districts. Other areas in the county should similarly take action.

2) Agro-forestry

Farmers should be encouraged to adopt this farming system which is a deliberate integration of wood perennials, or trees and shrub, with crops and livestock on the same piece of land. This is currently practised in Kabale district with a demonstration center at Kacwekano but can be extended to other areas. Perennial trees and shrubs on a farm have a significant impact on climate.

3) Development of alternative sources of energy

There has been massive use of biomass energy in industries, institutions and homes hence the destruction of vegetation which controls climate. Alternative sources of energy, such as hydro electric power, wind, solar and geothermal energy need to be developed fully so as to minimize the cutting of trees. Uganda has an estimated 2700mw power potential along R. Nile and other 22 small hydro sites in different parts of the country which can be exploited to meet the power demand of 4mw per month (NEMA 2001). Areas around L. Victoria, central and N.E. parts of the country have the highest potential of wind energy (NEMA 2001). Uganda has a high level of solar insolation and sunshine all the year round. Incident radiation is estimated to be 5-6 kw.h/m^2/day (NEMA, 1998). Potential areas for geothermal power include; Katwe, Buranga and Kibiro in western Uganda. All of these can be exploited without impacting the environment. The Government of Uganda should encourage local and foreign private investment in the energy sector.

4) Environmental Education promotion

In Uganda environmental education is part of the formal education curriculum. Nonetheless, it should be strengthened through mass media, televisions, radios, workshops, seminars and wild life clubs. Environmental education is essential to all planners, engineers, scientists and the general public who are involved in various management activities. Mt. Elgon Conservation and Development project promoted environmental education among the Bagisu on the slopes of Mt. Elgon.

5) Cooperation between Uganda and her neighbours

Uganda and her neighbours should emphasize ecologically sound land use practices and this requires cooperation. This is because people's actions in one country affect people in other countries either positively or negatively.

6) Capacity building

There is a need for building institutional and financial infrastructure, as well as human resources capacity in order to deal with issues of climate change and variations. More meteorological stations should be established in various parts of the country for further climatic analysis, forecasting and predictions of weather and climatic conditions.

7) Adjustment of activities

Ugandans need to accept that climate change and variations are real and therefore necessitates a need to adapt to these conditions by adjusting their activities. In agriculture adjustment activities can aim at increasing "water use efficiency" in crops. The following two methods can be used to increase water utilization in crops:

a) Plant species adaptations

Farmers need to select plant species adapted to the total amount and distribution of water. Early maturing crop species can escape drought or the dry part of a season. Through research drought resistant species can be developed.

b) Planting dates

The main reason for choosing optimum dates for planting is to ensure good germination, optimum growth and root penetration for desired yield. Constant research is needed for the determination of optimum planting dates for different varieties of crop species.

8) Environment Impact Assessment (EIA)

Before undertaking any development project in Uganda, an EIA should be carried out in order to determine side effects and devise remedial measures. National Management Authority (NEMA) is mandated with protecting the environment.

9) Carrying capacity

The exploitation of range lands on the basis of carrying capacity of the natural pastures is being implemented. This will result in reduction of livestock numbers to avoid overstocking.

10) Legislation

Enact laws against bush burning of vegetation especially in pastoral areas.

11) Farming methods

Use of proper farming methods such as crop rotation, inter-cropping and mulching so as to control soil erosion.

Key Terms

Climagraph	Condensation	Conventional Rainfall
Continental Climate	Equatorial Climate	Evaporation
Frontal rainfall	Lake breeze	Land breeze
Montane climate	Relief rainfall	Rainfall reliability
Rainfall Variability		

Questions

1. What is the difference between weather and climate?
2. Why do temperatures vary from place to place?
3. How is the atmosphere heated? What is lapse rate?
4. What are different types of precipitation?
5. What factors influence climatic conditions in Uganda?
6. What are the different climatic types in Uganda?
7. How does the population influence weather and climate of Uganda?
8. What is the relationship between climate and disease spread in Uganda?
9. How has the climate in Uganda changed? What evidence supports this position?

References and further readings

Asalu A.O. 1996. *Meteorology and Climatology Including Vegetation and Soils.* Unpublished pamphlet, Kampala, Uganda.

Beckinsale R.P. 1955. *Land, air and ocean.* Gerald Duck Worth and Co. Ltd. London, p.370.

Griffiths J.E. 1972. Climate. In Morgan W.T.W. (eds) *East Africa: Its peoples and Resources.* Oxford University Press, Nairobi p.107-118.

Kakumirizi G.W. 1989. *Physical and Human Geography of Uganda.* Unpublished pamphlet, Kampala, Uganda p.312.

Mawejje A.B. 2004. *Uganda Human and Economic Geography.* Unpublished pamphlet, Kampala, Uganda.

National Environment Management Authority. 1996. *State of the Environment Report for Uganda, 1996,* Kampala, Uganda p.271.

National Environment Management Authority. 1999. *State of the Environment Report for Uganda 1998.* Kampala, Uganda p.236.

National Environment Management Authority. 2001).*State of the Environment Report for Uganda 2000/2001.* Kampala, Uganda p.153.

Okello-Oleng C. 1989. *World Problems and Development: Incorporating Environmental Issues.* Unpublished pamphlet, Kampala, Uganda p.407.

Olupot M. 2004 'Museveni to address workshop.' *The New Vision* 18th June, 2 t M. 2004 'Ugandfaces food shortage.' *The New vision* July 23, 2004 Vol. 175 p.19.

Robinson P.J. and Sellers A.H. 1999 *Contemporary Climatology.* Prentice Hall, Pearson Education Ltd, England.

Taylor R. 2003. "Rwenzori Glaciers Melting." *The New Vision* Tuesday 28th October, 2003, p.19.

The New Vision. 2003. *The New Vision.* Tuesday 28th October, 2003. *The New Vision* Printing and Publishing Corporation, Kampala, Uganda p.19.

Wamala E. 1995. *A Comprehensive Geography of Uganda.* Unpublished pamphlet, Kampala, Uganda.

Chapter 2

GEOMORPHOLOGY OF UGANDA

YAZIDHI BAMUTAZE

2.1 Introduction

There are two sources of energy that help to build landforms and mold landscapes. These sources are internal (endogenic processes) and external (exogenic processes). These two processes are evident in Uganda and in concert with human (anthropogenic) activities; they have been molding the land surface and therefore account for the present morphology of Uganda. Uganda has a complex and unique land surface that has gone through cycles. This chapter examines Uganda's landforms and the forces that have shaped the different landforms.

Uganda is located at the heart of the great African plateau and is a country of varying relief, extensive swamps, forests and large water bodies. The land surface and the structural changes of Uganda have largely been molded by the processes of sub-aerial erosion and river action (exogenic processes). There is a tiny area on the summit of Mt. Rwenzori which is still covered by ice and the top of Mt. Elgon also depicts features of glacial erosion and morainic deposition. In the Lake Victoria and Lake Albert (Mobutu) regions, the immediate coastal shores show lacustrine sand deposits. Around Lake Victoria, a secession of raised beaches occur at approximately 3.5m, 15m and 20m and on part of the shores of Lake Albert (Mobutu), bars and cuspate fire lands have been created by the prevailing winds from the west.

For the rest of Uganda, weathering, erosion and river action have created a succession of level surfaces. There are some difficulties in determining the relative heights of the surfaces in Uganda, as well as their relationship to those recognized on the African plateau. Equally difficult is determining the date of formation of each surface.

Earlier attempts described the surface topography of Uganda as peneplain, implying multiple cycles of erosion, basing on the Davisian cycle of erosion. This conceptualization is largely the product of Wayland's observations and is in line with the nature of geomorphological analysis of the 1930's (Wayland, 1930). However, the characterization of surfaces in Uganda as peneplains is rather problematic and the correlation of level patches at different heights depended very much upon speculation. Other discussions have also considered the surfaces as being a product of pediplanation and parallel retreat erosion as invoked by King (Groves, 1934).

Recent writings on the surfaces of Uganda avoid using the term peneplain and instead use the term comparative heights. For example, upland or lowland are terms used to label the various surfaces. Although flat surfaces exist in Uganda, they are not sufficient to warrant the term peneplain. It is, however, clear that Uganda is to a large extent a product of multiple cycles of weathering, erosion and river action and it presents various geomorphological level surfaces. It may well have been a plain which is now represented by summits, or was buried beneath volcanoes or otherwise modified by other forces or concealed and does not necessarily imply flatness.

2.2 Landforms

2.2.1 Residuals on Upland Surface

These occur in the tiny areas of some of the hills of Bushenyi District. They represent residual features of an early surface. The surface could possibly be of late Mesozoic age which stands above the main upland surface. Basing on the Davisian concepts, Wayland categorized this surface as Peneplain I, although its extent and coverage in relation to adjacent surfaces is very limited.

2.2.2 Remnants of Upland Surface

This landform is characterized by the major hilly features and flat-topped hills found in Buganda region and some parts of Ankole region (especially Mbarara district). This surface is what Wayland categorized as Peneplain II (Buganda surface). It is important to note that there must have been sufficient variation in relief, even before this area was disturbed by uplifting and downwarping processes. This is why the present elevation differs considerably from place to place. It must have been on this surface, initially sloping from east to west that the former dominant east-west drainage pattern developed prior to its later distortion by uplifting and downwarping. The period for the formation of this upland surface is given tentatively as Upper Cretaceous. In the course of time, many of the uplands have become lateriticised, but the distribution of duricrust and near laterite conditions is very complex.

2.2.3 Remnants of Lowland Surface

At low altitudes, an extensive surface occurs, indicating a rejuvenation of the erosion action and creation of an incision into the upper surface sometimes of 130m to 170m. To the north of the country (especially in Acholi region), the Lake Kyoga basin and to the east of the country (especially in Busoga region), this surface widens out into an extensive plain. This surface is what Wayland characterized as Peneplain III. This lowland surface is not well developed and it is of weaker geological formations as is the case on the gneissic complex on which inselbergs and tors stand out to break the monotony of the landscape. The floors of areas of Ankole region (especially Mbarara district) form part of this lowland surface. This surface does not occur at a uniform height since it has also been affected by the crystal warping. A tentative date of Early Oligocene (early Tertiary) is given for the initiation of this surface, which reached maturity of the Miocene. This has allowed time for very deep weathering of the soil and advanced lateritization in some places.

2.2.4 Surfaces of Rift Edge and Achwa

The lowland surface seems to have been formed by rejuvenation following the rift formation to form the low-lying lands in the vicinity of the Achwa shear zone. This region is different from the 'lowland surface' with laterite. Similar incision seems to have occurred elsewhere on the edge of the rift. This area is still in a state of active formation.

2.2.5 Bevels in Eastern Upwarp

After the commencement of crystal warping and the associated volcanic activity and faulting of early Tertiary times, the up-tilted land of the area of eastern Karamoja became subject to powerful erosion and major bevels (cliff like landscapes) were formed.

2.2.6 Zones of Inselbergs and Tors

Landscapes characterized by inselbergs and tors occur in four areas:-

i) The mountains of Karamoja - this high ground represents remnants of an older landscape, even older than the Buganda surface, though it is not laterised.

ii) The southeastern tip of Busoga and Samia-Bugwe regions - a lowered remnant of the upland surface.

iii) The areas of massive granites which have proved to be very resistant to erosion in the Kitara upland.

iv) Small and scattered inselbergs around the perimeter of Lake Kyoga.

2.2.7 Sediments of the Western Rift Valley

The rift valley zone contains sedimentary deposits to great depth. The formation of these must have begun during the Miocene times when former rivers that used to flow to the Congo basin were terminated in rift valleys and rift valley lakes.

2.2.8 Alluvial Infills and Outwash Fans

Rivers flowing into the downwarped lakes of Central Uganda (Lake Victoria, Lake Kyoga, Lake Wamala, etc.) produce a wide range of alluvial infills in the form of superficial riverine depositions of recent age. There are areas of sedimentation around the bases of the main mountains, though the application of the word fan to them is quite appropriate.

Uganda's geomorphology is also characterized by other land surfaces whose characteristic features have nothing to do with various successive levels of weathering, erosion and river action as indicated at the beginning of this section i.e. Glacial erosion and morainic deposit surfaces; Lacustrine sand deposits and raised beaches; the rift valley system and block uplift surfaces and the Volcanic belts and associated volcanic and volcanic features.

Murchison Waterfalls

Source : Uganda Travels, 2000

Ruwenzori Mountain in Uganda

Source: Uganda Travels, 2000.

Map 2.1: The Geomorphology of Uganda

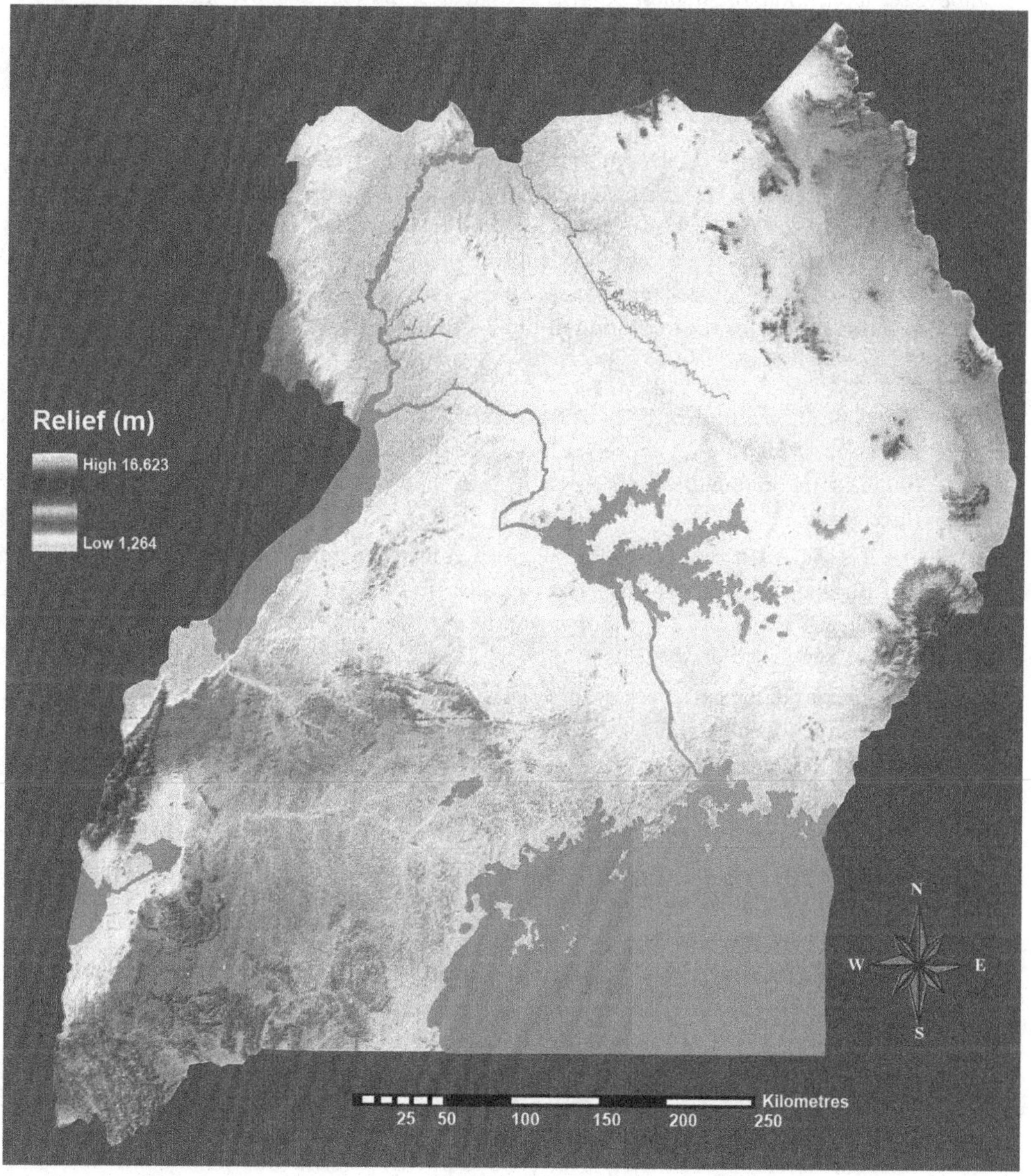

Source: Langlands, 1971

2.3 Relief and Physiographic Regions of Uganda

Although Uganda is largely a plateau, there are discernable altitudinal variations. The flat landscape of the country is found in the central part around Lake Kyoga, while the hilly landscape is in the southwest. In terms of relief and altitude, Uganda largely lies between 900 metres and

1500 metres above sea level. The lowest point (Lake Albert region) drops as low as 620 metres above sea level, while the highest point is as high as 5,029 metres above sea level (Magherita peak on Mt. Rwenzori). Physiographically, Uganda can be divided into four regions namely; Lowlands, Plateau, Highlands and Mountains.

2.3.1 Lowlands

The lowlands range between 600 metres and 900 metres above sea level and comprise approximately 9% of the country. All areas in Uganda which can be categorized physiographically as lowlands occur within the Western Rift Valley. They also cover the Lake Edward flats, the Semliki lowlands, the Albert flats, the Albert-Nile as well as the Aswa lowlands. The lowlands owe their existence to the tectonic movements which are also responsible for the formation of the Rift valley.

2.3.2 Plateau

The plateau accounts for 85% of the country. In terms of altitude, it lies between 900 metres and 1500 metres above sea level. The plateau is highest in the south and lowest in the centre (mainly the Lake Kyoga region). The Ugandan plateau can be divided into three zones:

i) The Lake Victoria plateau zone which consists of the flat topped hills of southern Buganda and Busoga, the dissected plateau of central Buganda especially around Lake Wamala and the Katonga and Sango Bay flats. These flats are also associated with impeded drainage.

ii) The interior plateau zone which is a zone of relatively uniform topography, except for a few inselbergs. The region consists of the Lake Kyoga southern interior plateau and the inselberg zone, the Bisisna-Okolitwom swamp lands, the northern interior plateau and the Karamoja plateau.

iii) The western plateau which is less differentiated because of the highland nature of the topography. It include;s the rolling ridges and valleys of Katonga and the Rwizi basin. The Rwizi basin also depicts various arenas, i.e. low lying terrains encircled by downland landscapes.

The Ugandan plateau was, however, greatly affected by tectonic movements along the eastern edge of the rift valley resulting in river reversal and the formation of Lakes Kyoga and Victoria.

2.3.3 The Highlands (Upland)

The altitude of the highlands varies between 1500 metres and 2000 metres above sea level and covers approximately 5% of the country. There are four major zones that form Uganda's highland regions. They include; (1) the south western highlands which are composed of the Kigezi highlands (may go as high as 2000 metres and 2500 metres above sea level), the Rwampara ridges and down lands (may go as high as 1,900 metres above sea level) and the northern Bushenyi hills (Buhweju); (2) The Western uplands which include; the central Bunyoro hills and the Kitara uplands; (3) The West Nile Uplands; (4) The interior hills located in the border lands between the plateau and the highlands. They mostly include; the Koki-Isingiro hills and the Ssingo hills which both rise up to 1,500 metres above sea level.

2.3. 4 Mountains

Areas classified as mountainous in Uganda are generally 2,000 metres above sea level and cover 2% of the country. 2,000 metres above sea level is more or less the upper limit for agriculture and human settlement. One of the most conspicuous phenomenons about mountains in Uganda is that they are almost all found on the border. The mountains of Uganda are either up-tilted horst blocks, volcanoes or up-tilted inselbergs. They are Mt. Rwenzori, second highest in Africa; Mt. Elgon consisting mostly of tertiary volcanic ash and is the highest volcanic mountain in Uganda; the Mufumbiro ranges in extreme south western Uganda, also a tertiary volcanic belt; the volcanic mountains of Kotido and Moroto districts as well as the isolated inselbergs of Madi, Acholi and Karamoja. The tectonic movements (endogenic processes) had a direct influence on the formation of Mt. Rwenzori and indirectly on the volcanic mountains which were formed in the lines of weaknesses along the rifts in the crust.

The relief and physiographic regions of Uganda have been influenced by various factors such as the structural and tectonic history of the country, erosion and rejuvenation aspects, variations in rock types and rock age. For example, the tectonic movements which took place in Western Uganda resulted in the following:

i) The up warping of landforms close to the present day rift valley area, creating depressions in Central Uganda. The upwarp also reversed river flow so that rivers which had previously flown westward reversed and filled the depressions in central Uganda culminating in the formation of Lakes Kyoga and Victoria.

ii) The formation of the western arm of the East African Rift Valley also housing Lakes Albert, Edward and George.

iii) The formation of Mt. Rwenzori which is a block mountain.

iv) The indirect formation of volcanic mountains in South Western Uganda including Mt. Muhavura (4,127 metres), Mt. Gahinga (3,447 metres) and Mt. Sabino (3,647 metres). In this case magma poured out taking advantage of the weaknesses created by tectonism in the earth crust.

Relief and physiography in general are very important as far as the Geography of Uganda is concerned because:

- They are an important factor underlying the nature of the physical (soils) and biological (flora and fauna) environment.
- They have both a direct and indirect effect on climate especially rainfall and temperature of various areas in Uganda.
- They have a direct bearing on speed and flow of rivers (drainage system) and therefore indirectly control the erosion rates and sedimentation.
- They have a direct bearing on diseases and other epidemiological aspects. For example, the normal altitudinal limit for transmission of malaria (mosquitoes) and tsetse flies is approximately 1,500 metres above sea level. Highlands do not offer favourable

grounds for bilharzias. It is also speculated that there is a relationship between altitude and human health and human fertility.

- They have an indirect influence on the economic activities in the country especially when you relate relief to the agricultural sector.

Lake Bunyonyi and Terraced Slopes

Source: Uganda Tourism Board

Mt. Ruwenzori

Source: Bakama BakamaNume, 2005

A view of the Ruwenzori from Queen Elizabeth National Park

The park extends from the edge of the plateau to the valley. Lake George and Lake Edward are lakes found within the park boundaries. See Figure 2.2 below.

Figure 2.2: Topographic cross-Sections across the East African rifts

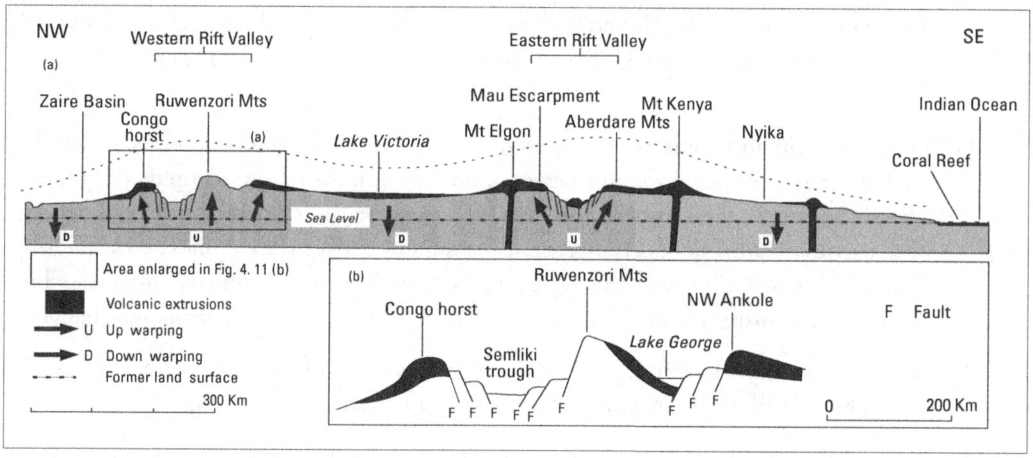

Source: Waters and Odero (1986)

Figure 2.3: Rift Valley and Block Mountain Formation

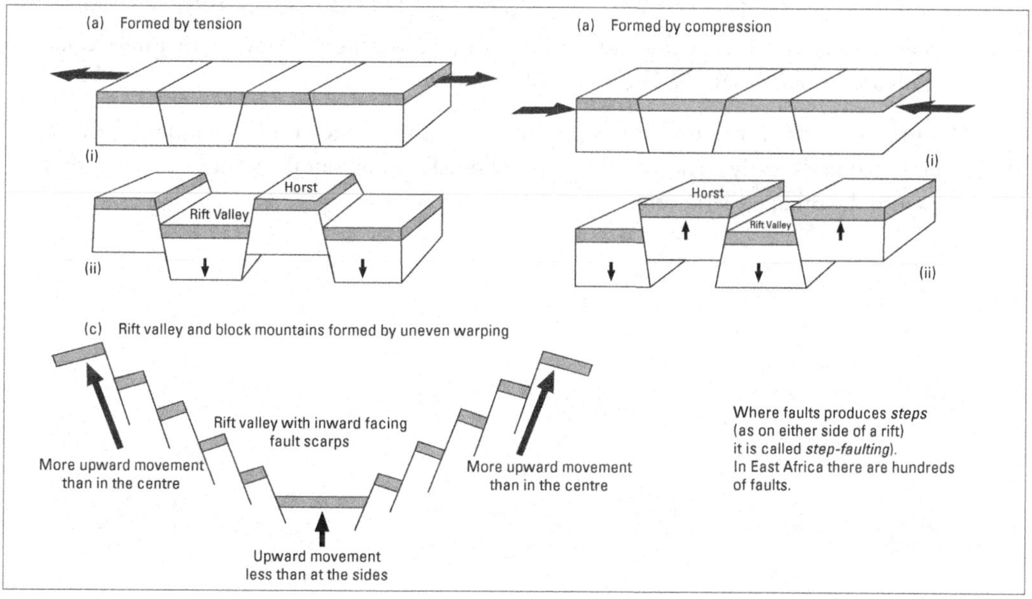

Source: Waters and Odero (1986)

2.4. The Structure of Uganda

The structure of Uganda is basically very simple, especially if one ignores pre-Cambrian folding which would seem to have little geomorphological significance (Beadle, 1981). Apart from the west and partly central Uganda, folding had minimal impact on the landscape of Uganda mainly due to the hard basement complex rocks. Uganda is largely a product of the extensive uplift of the continent of Africa as a whole and was later affected by rifting and warping as part of earth movements that affected eastern Africa. The dominant geomorphological features of Uganda's structure are:

1) The western rift and the associated uplift of Mt. Rwenzori; the axes of uplift parallel to this rift and the zone of down warp in-between, the Mufumbiro volcanoes, uplifted shoulders (West Nile, Kigezi, West Ankole, Toro, etc.). The western rift is marked by a line of faulting with a series of short faults arranged in echelon and between these, short cross faulting occurs. The susceptibility of this region to earthquakes and volcanicity (manifested by numerous hot springs) is an indicator that Uganda is still undergoing structural changes.

2) The uplifted areas of eastern Uganda which are associated with the Eastern Kenya (see cross section fig.2.2) rift valley including the uplands of Eastern Karamoja region.

3) The tertiary and Pleistocene volcanic formations in Southwestern Uganda and Eastern Uganda volcanic belt.

4) The down warped region of Central Uganda including Lake Victoria and Lake Kyoga basins. This region arose from the impeded drainage especially the pounding of major rivers.

5) The Aswa shear zone, a belt of intensive fracturing/faulting as a sign of structural weakness. This area is older than the Great East African Rift System.

The process of rift formation in Uganda was very prolonged. Faulting is thought to have commenced in the early tertiary times and continued throughout the tertiary, early quaternary times into the middle Pleistocene.

Map 2.2: Drainage in Uganda

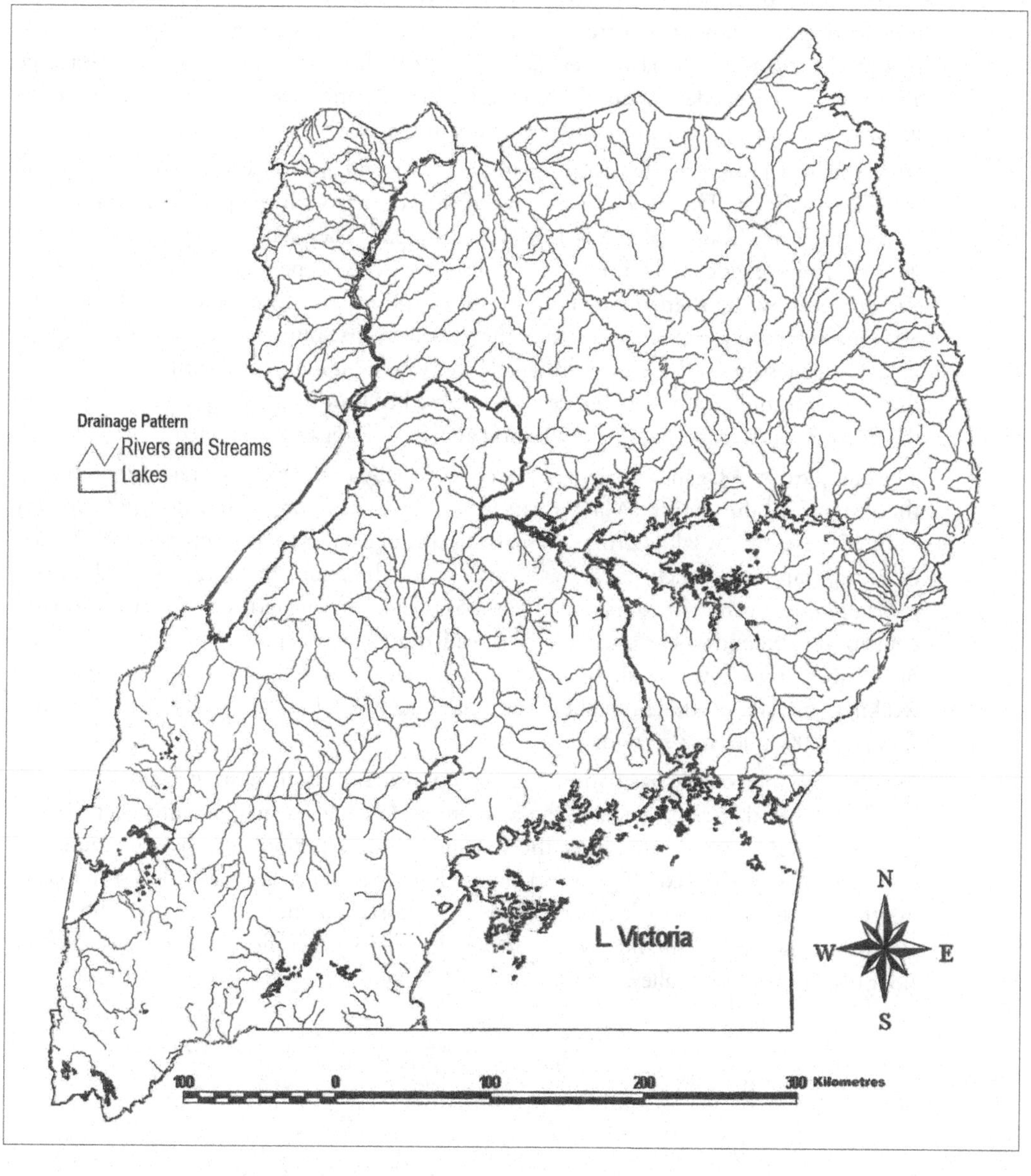

Source: Uganda Geological Department

2.4.1 The Influence of Structure upon the Drainage Pattern of Uganda

The structure of Uganda is very important to geographers especially geomorphologists because over time it has had a great bearing on the country's drainage through creation of new catchment areas and a complex drainage system characterized by reversal of drainage and associated impeded drainage, river rejuvenation and river capture. The impeded drainage accounts for the numerous wetlands in Uganda.

With the initial uplift of the African plateau, the Ugandan surface was gradually sloping from East to West. The major rivers crossed the country with the parallel rivers pattern forming the headwaters of the Congo River system. They had their origin over what is now Kenya. Evidence to this effect is shown by the deepening in the floor of Lake Victoria and by the shape of its islands and coastline. The river valleys of the Kagera, Katonga and Kafu are much wider than the present rivers could have cut and the shape of Lake Kyoga testifies to it being a drowned river valley. Subsequently with the creation of the uplifted shoulder, these rivers became reversed to flow back, along their courses into the down warped axis where Lake Victoria and Lake Kyoga were formed.

The complexity of Uganda's drainage system especially in the west is largely attributed to the structure of the country. Many rivers have tributaries joining them at angles contrary to their present flow, while others are passing through valleys with imperceptible breaks in the swamp from which water flows out in each direction. In Eastern Uganda, rivers generally flow in the old courses of the initial rivers. In Northern Uganda, the old river courses are evident where once a widespread drainage system formed the headwaters of whatever river lowered along the course of the Kyoga. In addition, the structural weakness marked by the Aswa sheer zone has been exploited by the Aswa River which flows for 200 miles in a straight course.

Finally, volcanic activity has had its bearing upon drainage. In Eastern Uganda, there is some evidence that rivers such as the Sironko and Manafwa may be antecedent rivers, retaining their courses throughout the volcano building period. It is also probable that these rivers may have contributed to the wide embayment into the volcanic mass. In Southwest Uganda, the volcanoes effectively interrupted the river system. Thus Lake Bunyonyi and Lake Mutanda were partially formed by lava flow damming across the floor of a former river valley.

Map 2.3: Map of Uganda Relief

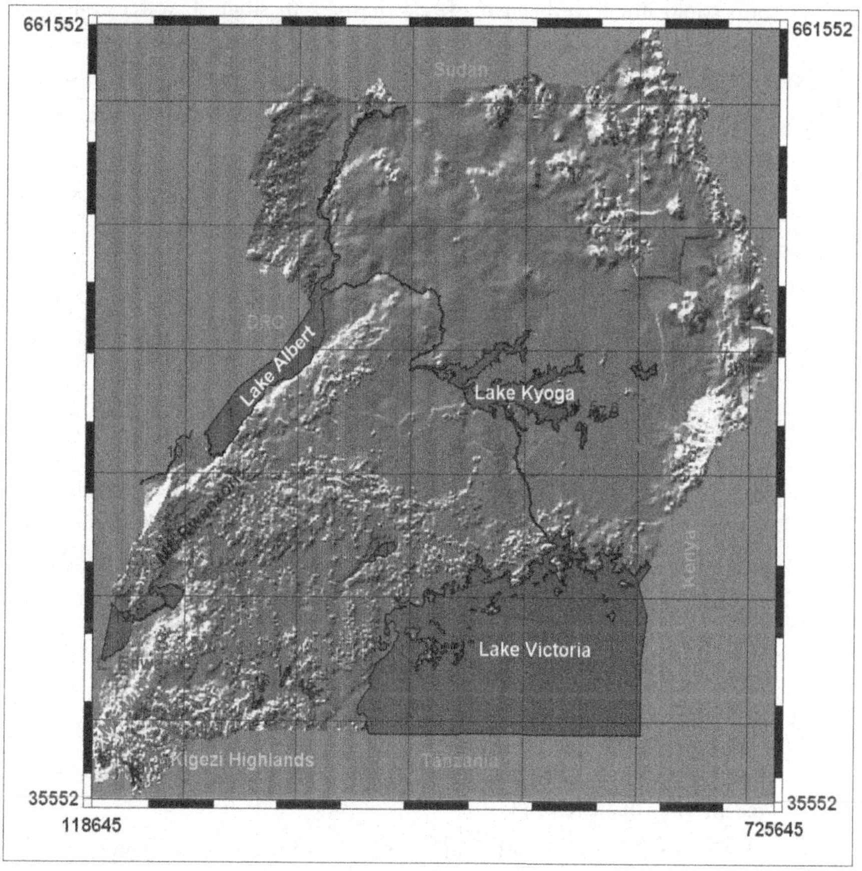

Source: Uganda Geological Department

2.4.2 The Drainage pattern of Uganda and its evolution

There are two theories that attempt to explain the drainage system in Uganda. Uganda in the pre-Pleistocene time appears to have been drained by a series of sub-parallel rivers which had their headwaters originated near the present day East of Mt. Elgon. The rivers flowed westwards to join river Congo and onwards into the Atlantic ocean (Temple). The second theory argues that the sub-parallel rivers joined river Lualaka and then eventually flowed to the Red Sea (Beadle, 1960; 1981). The enormous evidence in support of river reversal in Uganda include;s:

i) On the Western side of the line of upwarp, steeply graded streams (Rivers Mpanga and Ntuusi) descend abruptly to the floor of the rift valley. This abruptness manifests disturbances in former river courses.

ii) Rivers Kafu and Katonga start their eastward journey from swampy watersheds. These rivers are quite long and become progressively less steep as they cross the plateau. These rivers pass through wide valleys and they are sluggish by the time they drain into Lakes Kyoga and Victoria. The wide valleys choked with swamp vegetation could not have been made by the present sluggish water courses. They must have been cut by rivers that flowed at a greater pace from the east.

iii) There are rivers flowing in different directions, but sharing the same watershed and river valleys that seem to be formerly one. This points to the fact that these rivers must at one time have been the same, as indicated below.

   a) River Katonga shares its course with rivers Rutungu, Nyabisheke and Oruyubu. However, Katonga flows eastwards to join Lake Victoria while the rest flow westward to join Lake George.

   b) Whereas River Rwizi seems to share the same river course with River Bwira, River Rwizi flows eastwards towards Lake Mburo, while River Bwira flows westwards to Lake Edward.

   c) River Ishasha flows westwards to Lake Edward while River Kikitumba flows eastwards to Lake Victoria.

   d) River Mpanga seems to share the same river course with River Kafu, yet the former flows westwards to Lake Albert while the latter flows eastwards to Lake Kyoga.

iv) River Kafu and Victoria Nile flow in different directions despite being in the same region. This pattern of river flow is indicative of some disturbances in river flow, probably for River Kafu.

v) The filling up of Lake Victoria was a slow process. It is likely that the filled lake was flowing southwest to Tanzania via Mwanza. Later, land seems to have tilted and water started flowing northwards through Rippon Falls to Lake Kyoga and subsequently to Sudan. The Nile water course also shows signs that the river course is a recent one (Beadle, 1981).

   a) There are many rapids and waterfalls within a short distance, such as the Owen Falls, Bujjagali Falls, north Agago and south Agago Falls, Murchison Falls, etc. If the water course was old, it is likely that waterfalls and rapids would have been reduced in size and number.

   b) The narrow cleft (fissure), of 20 feet at Murchison Falls through which River Nile passes indicates that the river course is young. If it were old, the cleft would have widened.

vi) Beadle has observed that River Semliki which joins Lake Edward to Lake Albert is a young river which is yet to make impressions on the valley through which it flows. Thus River Semliki seems to be one the rivers which have found newer outlets when the old river course was changed.

vii) The preponderance of Lakes Nakivali, Mburo, Kachira and Kijanobarora in Mbarara with extensive swamps indicates a river flow that was interrupted.

viii) Until recently, about 23 species of cichlid fish have been recorded from Lakes Edward and George. Many of these are closely related to species in Lake Victoria and five were identified to be identical in both lakes. Beadle concluded that there was a very recent connection by which these species moved from Lake Victoria to Lake Edward. According to this view, the newly formed Lake Victoria probably continued for some time to drain, through the Katonga valley westwards into the Edward basin. The connection between Edward and Victoria seems to have gone on as far as the late Pleistocene.

Key Terms

African Plateau	Alluvial infill	Bevels
Peneplains	Glacial erosion	Morainic erosion
Outwash	Capture of overflow	Gorge
Fault line	Drainage basin	Wind gaps
Flats	Inselbergs	Tectonic movements
pre-Cambrian		

Conclusions

The present morphology of Uganda's landscape is a product of both the endogenic and exogenic processes, which have shaped the surface over millions of years. Uganda's landscape has gonc through transformations most notably tectonism which reversed the drainage system in the west and the volcanism which produced outstanding features in Eastern and Southwestern Uganda. Uganda's landscape also mirrors multiple cycles of weathering, erosion, river action and increasingly anthropogenic activities. Most of the historical gigantic cataclysmic processes are dormant or very slow, although earthquakes in western Uganda and landslides in Eastern Uganda are frequent phenomenon dynamically creating new features. Noteworthy is that the diverse geomorphology of Uganda offers a great resource potential and supports many livelihoods in different settings as much as it poses development challenges in many regions.

Questions

1. What are exogenic and endogenic processes?
2. What is the dominant set of processes that have shaped the landforms of Uganda?
3. Review the trends in the evolution of the current morphology of Uganda.
4. Using accurate illustrations give a detailed explanation of Uganda's structure and geomorphology.
5. Examine the impact of the pre-pleistocene process on the morphology of Uganda.
6. To what extent do morpho climatic processes explain the landscape of Uganda?
7. The geomorphology of Uganda cannot be uniquely understood without putting it in the context of the African morphology. Discuss.

8. Examine the view that 'variations in the levels of regional development in Uganda are a function of the regional geomorphology'.

9. Examine the importance of Uganda's morphology to her economic development.

References and further readings

Beadle, L.C. 1981. *Lakes Edward and George*: in *Inland Water of Tropical Africa*, ed. Beadle, Longman, London, p.231-244.

Beadle, L.C. and Lind, E.M. 1960. *The Uganda Journal*, 24.

Byamugisha, B. 1994. *Geography of Uganda*, Revised Edition, Makerere University Printery.

GOU. 1973. *Geochemical Atlas of Uganda.* Published by the Geological Survey and Mines Department, Entebbe, First Edition.

GOU. 1967. *Atlas of Uganda.* Published by the Department of Lands and Surveys, Uganda, Second Edition.

Langlands, B.W. 1974 *Uganda in Maps.* Preliminary edition, Parts two and three

Langlands, B.W. 1971. *The Gospels According to Langlands.* Department of Geography, Makerere University.

Osmaston, H., *eds.* 1996. *The Rwenzori Mountains National Park, Uganda.* Fountain Publishers.

Stheeman, H.A. 1932. *The Geology of Southwestern Uganda.* Temple

Waters, G. and J. Odero. 1986. *The Geography of Kenya and the East African Region.* Macmillan Publishers Ltd. First Edition.

Waylan, E. J. 1930. *Rift Valley, Lake Victoria: International Geological Congress*, XV, 323-352.

Yeoman, G. 1989. *Africa's Mountains of the Moon.* Universe Books, London.

Chapter 3

SOILS AND SOIL DEGRADATION IN UGANDA

Bob Nakileza

3.1 Introduction

This chapter provides an analysis of the distribution of major soil types and soil degradation including the types, causes, trends and effects in Uganda. Response strategies to the problem of soil degradation are also examined.

Soils are dynamic ecological systems providing plants with support, nutrients, water and air and home for a large proportion of micro- and macro-organisms that play a crucial role in recycling of materials. They are storage sites for water, carbon and plant nutrients. Soil forms as a result of decay of living matter, weathering of solid rocks and sediments deposition. Understanding soils and managing them well are important for human welfare, particularly in consideration of the increasing population.

Soils constitute an important resource supporting agriculture, which contributes 60% of Gross Domestic Product (GDP) and employs over 80% of the population in Uganda. Soils play other important ecological and hydrological roles. Thus, their sound management is very crucial for sustainable development of Uganda. Knowledge and understanding of the soils including their distribution and constraints is vital for promoting and achieving sustainable development.

Five major interactive factors namely, (1) climate, (2) parent material, (3) topography, (4) biological activity and (5) time influence soil forming processes and emerging soil properties. Management of the land is the other factor. Understanding these factors helps in identifying soil distribution occurring in any place.

1. Climate is a major factor controlling the type and rate of soil formation including vegetation distribution. The key climatic parameters are temperature and rainfall, which influence water, or moisture availability as well as biological activities. The rate of weathering of rocks and minerals and biological activities increases with temperature. Moisture plays an important role in the soil formation. For instance, it participates in numerous chemical reactions, helps in translocation of substances, thus the differentiation of horizons. Areas experiencing hot conditions with little moisture are less leached compared to wet and cool areas such as the humid highland areas. In general, therefore climate plays an important role in affecting the rate of leaching and development of horizons.

2. Parent material is the material from which the soil forms, i.e. the mineral composition of the soil. The parent material may be the underlying rock, or may be transported by wind or water. The "residual" soils form from the underlying rocks have similar chemistry as the original rocks. The parent material influences soil properties such as texture, colour and nutrients.

3. Topography is the shape of the land surface, its slope and position on the landscape, greatly influences the kinds of soils formed. Slope and aspect affect the moisture and temperature of the soil. Steep slopes facing the sun are warmer. Steep soils may be eroded and lose their topsoil as they form. Thus, such soils may be thinner than nearly level soils that receive deposits from upslope areas. Such soils are deeper and darker coloured and may form in the valley areas or depressions.

4. Plants, animals and organic matter of the soil influence biological activities. However, the biotic and climatic factors are interrelated; organisms in the soil respond to climatic conditions, particularly temperature and moisture. Organisms in the soil help in mixing the soil (bioturbation), in adding organic matter and nutrients, in creating pores and in the physical breakdown of the soils. The organic matter from both the vegetation and dead animals is added to the soil. This improves soil aggregation, cation exchange capacity and water holding capacity. Human activities such as agriculture and forestry are also important in influencing soil properties. On the other hand, other human activities like pollution and compaction and those that accelerate soil erosion are, however, detrimental.

5. Time is the period it takes for soils to form. Soil formation is a continuous process and generally takes several thousand years for significant changes to take place. The soils are considered to be relatively young soils with slight alteration of parent material and weak soil horizon development. Though the age of soil may be considered to be the length of time in years since the land surface became relatively stable, thus enabling the soil development to proceed, it is how much development the soil has undergone that matters most. The length of time required for soil formation depends on the intensity of the other active soil forming factors of climate and organisms and how topography and parent material modify their effect. Young soils (e.g. lithosols) have minimal soil development and few horizons while old soils (Ferralsols in central Uganda) have well developed horizons.

Soil forming factors that hasten the rate of soil development are;

- permeable and unconsolidated parent material;
- warm and humid climate;
- forest vegetation;
- summit or back slope landscape position that is well drained.

Conditions that are likely to impede soil development include;

- impermeable, hard, consolidated, parent material;
- cold and dry climate;
- temperate vegetation;
- steeply sloping back slopes or shoulders.

Soils do not have uniform depth but rather are differentiated into layers or horizons (Figure 3:1). The soil horizons are formed as a result of additions, losses, translocation and transformation. The upper horizons A and B constitute the solum. Horizons A, B, C and R are known as mater horizons.

Figure 3.1: Soil profile

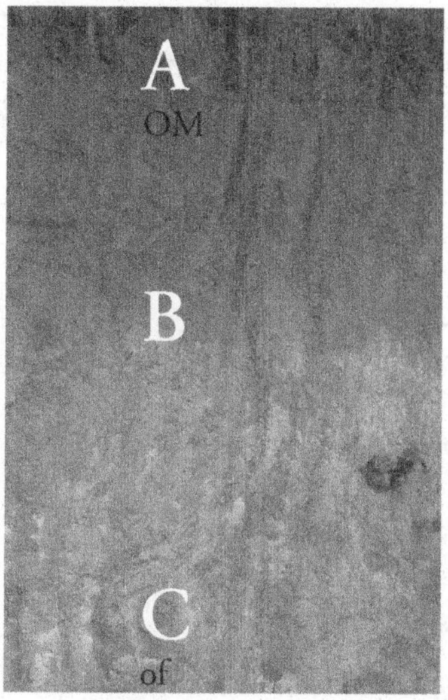

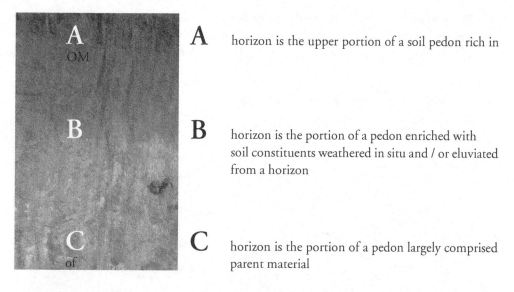

A horizon is the upper portion of a soil pedon rich in

B horizon is the portion of a pedon enriched with soil constituents weathered in situ and / or eluviated from a horizon

C horizon is the portion of a pedon largely comprised parent material

Some soils have the E horizon, from which large amounts of materials (clay and iron) have been eluviated.

Soils constitute an important resource supporting agriculture, which contributes 60% of GDP and employs over 80% of the population in Uganda. Soils play other important ecological and hydrological roles, e.g. water purification. Knowledge and understanding of the soils including their distribution and constraints is vital for promoting and achieving sustainable development.

3.2 Distribution of major soil types

Soils of Uganda were first studied and classified in the 1950s and 1960s, based mainly on the morphological characteristics, geology, phytogeography and minimal field observations by the soil survey unit (GOU, 1967). A countrywide reconnaissance survey was conducted from 1958 to 1960. It provided important information on soil series and mapping units. The mapping units were grouped according to the geological and landscape or geomorphological features (erosion surface and their dissected remnants). The original mapping units were then regrouped according to the classification scheme devised for the 1:5,000,000 soils map for Africa, by the D' Hoore cited in GOU (1967). Chenery (1960) as head of the Uganda Soil Survey reviewed 12 systems of soil classification and found the D'Hoore and Sys' system to be the most appropriate for Ugandan soils. He mapped the soils of Uganda based on D'Hoore classification at a scale of 1:1,500,000 using combined techniques including aerial photography. Details of these earlier efforts regarding soil studies and classification are available in the form of six memoirs for the different regions of the country (Radwanski, 1960). The major soil categories generated early were revised recently according to the FAO system.

Soil names and to some extent the boundaries, have changed overtime in view of further understanding of the soils and changes to different classification systems. The distribution of major types of soil groups based on the FAO-Unesco system is shown in Figures 3.1, 3.2 and Table 3.1.

Map 3.1: Soil types based on the FAO-Unesco system

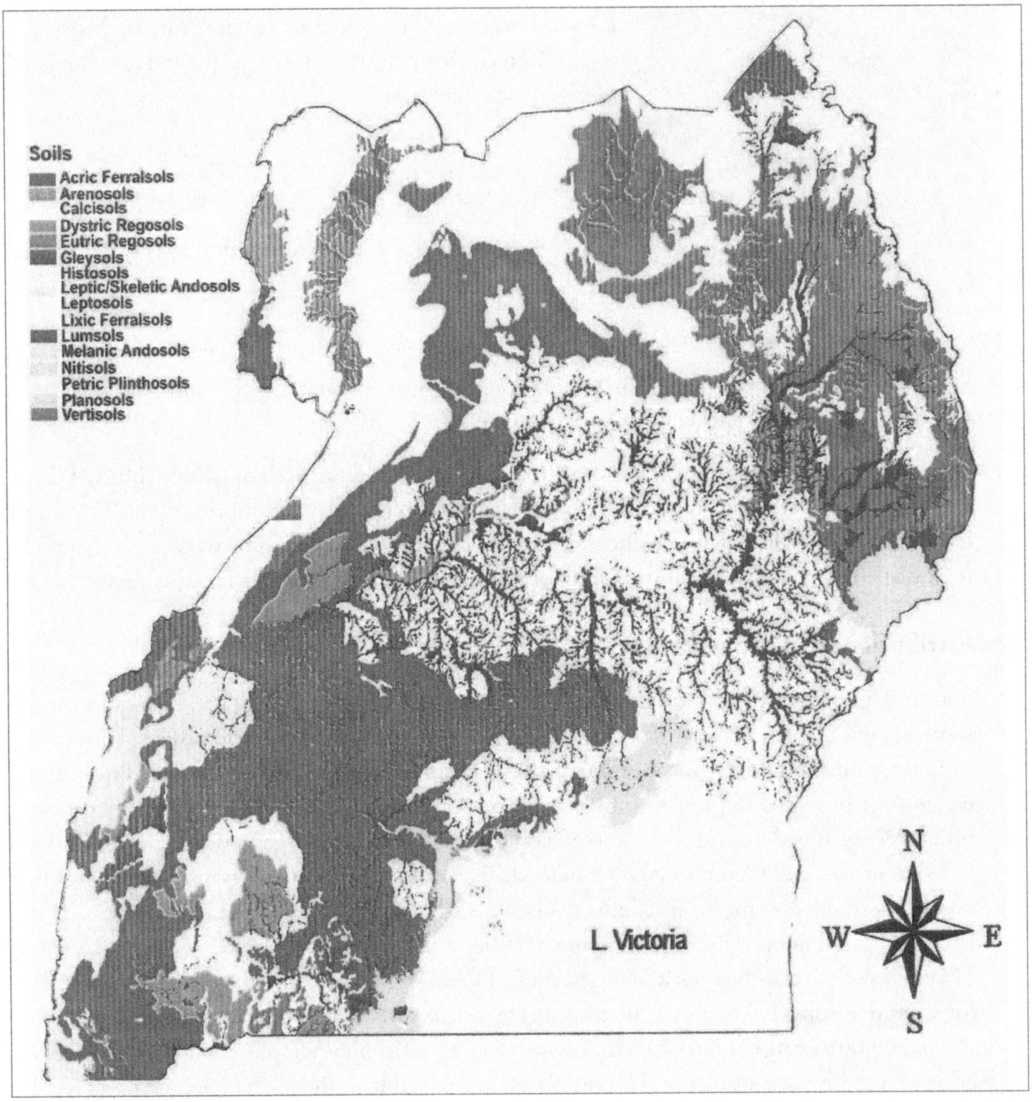

Source: Uganda Geological Department

The area distribution of the soil orders in Uganda is shown in Table 3.1 and Figure 3.2. Table 3.2 shows the distribution by the district and regions in Uganda.

Table 3.1: Soil types and aerial distribution in Uganda

Soil order (FAO system)	D'Hoore system	*Area in km2	% area
Ferralsols	Ferrallitic	67,572	28.02
Nitisols	Ferrisols	3,902	1.62
Plinthosols	Ferruginous tropical soils	35,205	14.61
Vertisols	Vertisols	19,067	7.91
Andosols	Eutrophic soils of tropical regions	5,523	2.29
Gleysols	Hydromorphic soils	24,019	9.97
Podzols	Podsolic soils		
Histosols	Hydromorphic soils	580	0.24
Solonetz	Halomorphic soils		
Leptosols	Weakly developed soils	22,086	9.16
Luvisols		5,275	2.19
Planosols		1,735	0.72
Regosols		14,157	5.45

**Water occupies 36,804 km² (15.27%)*

Source: Isabirye (2000).

Figure 3.2: Percentage distribution of soils by orders

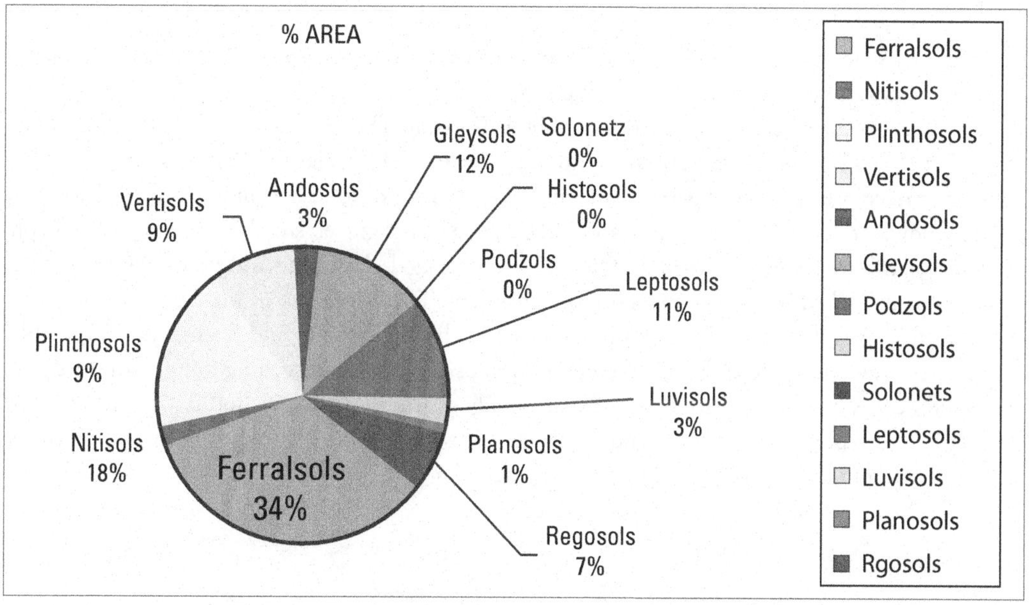

Source: Uganda Geological Department

Table 3.2: Soil Types by district/region in Uganda

Soil type	District/region	Comment
Ferralsols	Lira, Apac	Widespread in the central and northern parts of the country
Nitisols	Kamuli, Iganga, Jinja, Mukono, Kabalore, Rukungiri, Mbale and Manafwa	
Vertisols	Kotido, Moroto & Nakapiripirit, Arua, Nebbi, a few places in Kasese, Hoima and Kabalore	
Andosols	Kisoro, Mbale, Manafwa, Sironko, Moyo and Kasese	
Plithosols	Tororo, Gulu	They are found scattered in other districts in the country
Histols	Mbale, Sironko, Manawa, Kapchora, Bukwa, Kisoro	Histols are confined at high altitude areas or mountains
Leptosols	Moyo, Kitgum, Rukungiri, Kotido, Moroto	Occur mostly in the rift valley
Podzols	Bushenyi (western part)	
Gleysols	Rakai, Masaka, Mpigi	Confined mostly to depressions or along the valleys

Source: Uganda Tourist Board

A brief acc ount of the ten different soil types of Uganda is presented in the following section. Emphasis is placed on the distribution and the main soil characteristics.

Ferralsols

Ferralsols are the most predominant soil category covering more than 28% of the surface area in Uganda. They are widespread in the central, south-western, northern and north-western Uganda. These soils are deep with a ferralic B-horizon. They occur at a depth of 25-200 cm from the soil surface. They are made up of moderate to high clay content including a strong water retention capacity at permanent wilting point. In addition, a ferralic B horizon must have a sandy loam or finer particle size and <90% (by weight) gravel, stones or petroplinthic (iron and manganese) concretions and a CEC of 16 cmolc kg-1 clay or less and effective CEC of <12 cmolc kg-1 clay; and a thickness of at least 30 cm. These soils are in the advanced stages of weathering and are associated with low cation exchange capacity and negligible weatherable mineral reserve. The criteria of differentiating soils as ferralsols according to D'Hoore and Sys' method was intended to group soils that are highly weathered with low productivity. Thus, their agricultural potential use is dependent on the bases held in the organic and clay complexes as well as high rainfall and proper soil management.

Nitisols

These soils are found mainly in the districts of Mukono, Jinja, Mbale, on the lower slopes of Mt. Rwenzori in Kabalore and Rukungiri and to a small extent in Iganga and Tororo. Unlike the Ferralsols the Nitisols are deep, red clay (Kaolinite), less weathered and porous with a stable sub-angular blocky, or granular structure. Thus, Nitisols are amongst the most fertile soils with favourable agronomic properties that support a wide range of crops.

Vertisols

Vertisols are dark coloured clayey soils occurring in the low-lying seasonally wet areas in much of the north-eastern Uganda, in the districts of Moroto, Kotido, Nakapiripirit and along the Albert Nile in Arua and Nebbi, as well as in small patches in Kabalore, Kasese and Hoima in the southwest. These soils have 30% or more clay, usually dominated by montmorillonite, which causes shrinking and cracking during the dry season and swelling during the wet season. Vertisols have high productivity as long as the inherent constraints are tackled; they are very hard and difficult to work in the dry season and excessively moist and sticky during the wet season.

Plinthosols

These soils are found scattered in different parts of Uganda, particularly in Tororo and Gulu districts. Generally, Plinthosols occur in low-lying areas experiencing fluctuating water levels, which causemottle formation. The soils have a high content of Aluminium and or Iron. Basically, they are unsuitable for arable farming due to inherently low nutrients and water logging conditions.

Gleysols

Gleysols are largely found in depressions along river valleys and wetlands on the western shores of Lake Victoria in Masaka and around Lake Kyoga in Pallisa, Nakasongola and Luwero districts. These soils are developed from various unconsolidated materials under the aquic moisture regime. They are usually drained for growing paddy rice and other crops, or for grazing. Permanent exposure of these Gleysols rich in sulphur content leads to acidification.

Andosols

Andosols mainly occur in the districts of Mbale, Kisoro, Kasese and Moyo. They are derived from alluvial deposits and volcanic ash, have a generally rich organic matter A1 Horizon and are sometimes saturated with bases to more than 50% of its exchange capacity (Briggs et al, 1998). Generally, these soils have a rich weatherable mineral reserve and therefore are fertile and productive. Due to the high natural fertility, they are intensively cultivated and they support a variety of agricultural crops. Andosols, however, have a high content of allophane, which causes low bulk density and gives a smeary consistence. Some of these soils have low fertility due to high fixation capacity for phosphorous. Those occurring on steep slopes, on mountains, or hills are likely to be prone to soil loss by erosion.

Histosols

Histosols occur predominantly under cool and cold temperatures in poorly drained areas such as swamps and marshlands on the mountains of Elgon, Rwenzori and Muhavura. They have an h-horizon of at least 40 cm and organic matter accumulation due to wet conditions. The soils are extensively utilised for grazing, forestry and horticulture.

Leptosols

These soils are mainly distributed in the Rift valleys in Kabalore and in other areas such as Moyo, Kitgum, Kotido and Moroto districts. They are young, or skeletal shallow soils developed on hard rocks or calcerous material on steep slopes and elevated terrain. Low temperatures and soil erosion influence their development. Generally, Leptosols are unsuitable for growing crops but have a greater potential for extensive grazing and tree crop production.

Solonetz

Solonetz soils are mainly confined in the Semiliki valley, in the western Rift Valley, in Bundibugyo and Kabalore districts. These soils have developed from unconsolidated and fine materials in the dry land area. They are alkaline soils with a high percentage of exchangeable sodium, a property which limits agricultural use.

Podzols

Podzols soils are very limited in their occurrence in Uganda. They are found in Lubare in western Bushenyi. They are associated with quartzite rocks. Podzols are characterised by a surface peat soil horizon underlain by a highly leached sub-surface in which Iron and Aluminium have been translocated, thus producing an ash bleached horizon. These soils have limited agricultural use due to their poor nutrient status and presence of a thin iron, or fragipan. However, these limitations can be overcome through use of appropriate fertilizers and deep ploughing. Where the pressure for land cultivation is low, the land can be left under forest, or grazing.

3.3 Soil Degradation

Soil degradation is defined as the loss in potential productivity due to the processes of erosion by water or wind (Lal, 1997). Soil degradation is one of the major threats to environmental quality. It reduces agricultural production. Reduced agricultural production leads, among other things, to food insecurity and low incomes, thereby affecting people's livelihood. Issues of soil degradation, its causes and impacts in Uganda are analysed in this section.

3.3.1 Types, Distribution and Trends of Soil Degradation

The major categories of soil degradation recognised in Uganda are water erosion, nutrient depletion and soil compaction.

Water erosion

Soil erosion by water occurs in most parts of the country from virtually every land use in both high- and low- lands. Soil erosion by water is not a new problem in Uganda, but the magnitude has increased in recent times. The increase is due to deforestation, over grazing and poor agricultural practices. Recognition of soil erosion in Uganda, dates back to the 14th and 15th centuries (Hamilton, 1985). Soil erosion and degradation in the highlands of Uganda was officially recognised as a problem in the 1920 (Martin, 1926). For example, concern over soil erosion and declining soil fertility in south-western highlands was first expressed by the British colonial government in the 1920's (Farley, 1996).

Earlier studies during the colonial governance revealed that erosion by water was a real problem and there was uneven distribution of soil erosion, particularly in humid areas of central Uganda, in the mountainous/highland parts of south-western and in the relatively dry grazing region of Karamoja in eastern Uganda (Wayland, 1938). Tothill carried out a review of small agricultural areas later to ascertain the status of soil deterioration in 1938. Gully and sheet erosion were reported to have seriously affected the land in some northern districts in the 1950s (Chenery, 1960). Experimental set ups in the 1960s at Namulonge and Serere research stations showed relatively higher rates of soil erosion in the annual monocropping systems of cotton and maize (Temple, 1972). However, the actual picture on the farmers' lands remained unclear since no experiments were established at that time anywhere in the country.

In the recent years, starting from 1990, water erosion in Uganda has been assessed more quantitatively based on runoff plots. Table 3.3 depicts soil erosion measurements on runoff plots in different parts of Uganda. Generally the findings reveal high rates of soil erosion occurrence in the humid parts of Lake Victoria basin (e.g. Magunda, et al., 1999; Lufafa, 1999) and in the mountainous and/or highland areas (e.g. Nakileza, 1992; Bagoora, 1995; Tukahirwa, 1997) on ferralsols. It is widely accepted that the degradation of cultivated land is increasing due to high population pressure particularly on the steeply sloping highland areas, which receive high rainfall and are inadequately conserved (Briggs et al., 1998). However, a more comprehensive description of soil degradation cannot be given in view of limited and short term uncoordinated research. Further studies are required to monitor the problem particularly at experimental scale in different areas to provide input in modelling or identifying priority areas for useful conservation planning.

Table 3.3: Soil erosion measurements in different parts of Uganda

Source	Date	Method/location	Erosion rate (t/ha)/remarks
Rose	1958	Simulation, Central Uganda	Erodibility for soils in central Uganda
Temple 1972		Micro-plots, Namulonge	High erosion in cotton
UNEP	1987	National, Erosion hazard Spatial erosion hazard using GIS, Uganda	
Nakileza	1992	Plot, Elgon slopes	6.4 in maize sole crop on 21° slope; 3.03 on 8° slope
Magunda	1992	Micro plot, erodibility	Kabanyolo clay more erodible unstable & than Kachwekano
Kakuru	1993	Plot, Kabale	56% & 68% reduction in soil loss by calliandra & nappier grass
Tenywa 1993		Simulation, Elgon, Kabanyolo	Low erodibility on Elgon soils
Tukahirwa	1995	Plot & WEPP model, Kabale,	1.4 on 10° slope in sorghum
Majaliwa	1997	plot, Kabanyolo	43-75 on 22° slope
Bagoora 1997		Plot, Kabale,	10 -14; 14 -129; 23 -107 on lower, middle & upper slope respectively
Lufafa	1999	Micro-catchment, GIS	93 annual crops, 52 mixed
Magunda	1999	Plots, L. Victoria basin	49.8m^3/ha/yr runoff in banana, 1089.6 m^3/ha/yr in pastures; 27-126t/ha/yr soil loss

Source: Uganda Tourist Board

Soil degradation varies greatly in space and time. To obtain realistic spatial estimates, prediction methods have been applied, although each method has its own limitations. The Universal Soil Loss Equation (USLE) was used to estimate the annual erosion rates for Uganda (UNEP, 1987). The USLE model showed areas of high, moderate and low erosion (see Figure 3.2) and the data is probably good for general planning purposes. It does not, however, provide details of soil degradation by erosion at a finer scale. Some studies (e.g. Lufafa, et. al., 1999) have attempted to map soil erosion at large scale using plot measurements. However, more studies at a plot scale are needed to validate erosion prediction in other parts of the country.

The USLE does not give estimates of erosion by storm. This requires a model such as Water Erosion Prediction Project (WEPP), which is based on hydrological concepts. Dorsey (1990), using the available maps and data, compiled a first generation of WEPP slope, soil and climate input files to represent Ugandan conditions. He recommended that these files and an appropriate management input file, should be tested and compared with field measurements. Tukahirwa (1995) picked up this work later and applied the WEPP to test the predictive capacity of the hill slope version of this model.

He obtained results that were comparable to those obtained empirically when some of the parameters (e.g. Infiltration capacity) were modified for the site. However, there are significant limitations in the application of the WEPP model, particularly the detailed climatic data that are often lacking, or not measured, at the weather stations.

3.3.2 Chemical and physical deterioration

Chemical deterioration refers to the irreversible change in the soil properties due to loss of nutrients, increase in sodium content and change in soil pH. Physical deterioration is the adverse change caused by deterioration of soil due to compaction and loss of organic matter. A number of studies carried out in Uganda indicate a downward spiral in the nutrients across numerous cropping systems, particularly where there are low inputs and inadequate soil management practices. Experimental work in South Uganda on nine experimental sites investigating the effects of four year continuous cropping on yields (targeted topsoil pH, total N, exchangeable Manganese and organic carbon) showed significant changes in these soil properties (Stephens, 1969). Studies by Zake and Nkwine, (1992) have reported continued nutrient decline confirming observations by Sanchez and Leakey (1996) in many parts of Africa. Negative nutrient budgets have been reported in soils of Uganda since they are continuously mined of nutrients (Wortman, 1999). High rates of nutrient loss ranging from 60-100 Kg/ha/yr are reported in the densely populated drylands in Africa, including Uganda (Henao and Baanante, 1999). This is particularly witnessed in areas where there is shallow, highly weathered soils subjected to intensive cultivation, with low levels of fertiliser application. It is important to note that despite this awareness about soil fertility, soil management is still low in Uganda (Tenywa, 1999).

3.3.3 Causes of soil degradation

The causes of soil degradation are numerous and diverse. However, the main causes include; the biophysical, socio-economic and politically related factors. These factors contribute to high variability and dynamism in the soil degradation processes that occur in the country.

The socio-economic factors include; land tenure, poverty, population pressure and land use practices. Poor agricultural practices, such as cultivation on steep slopes without adequate soil water conservation technologies often lead to tremendous loss of soil. Zake (1999) argues that due to inept resource management prevalent in the country, desertification in dry areas, soil erosion and deforestation on hillsides have continued to increase. He further states that loss of topsoil fertility in many areas is exacerbated by poverty, rapid population growth, land fragmentation and inadequate progress in increasing crop yields.

Reduced plant cover particularly due to intensive grazing and trampling in the dryland areas of Karamoja, Mbarara and Nakasongola exposes the soil to erosive storms that lead to soil erosion (NEMA,1997; Nakileza, 2004) (see Figure 3.4).

Severe sheet erosion on the exposed red ferralsols in the dryland area of Nakasongola district

Source: Bob Nakileza, 2004

Soil compaction is not a major problem in Uganda except in areas where there is frequent ploughing by tractors such as in the districts of Kumi, Lira and Soroti and trampling on overstocked, or overgrazed grazing lands particularly in the areas of Karamoja, Nakasongola and Mbarara. In some parts of the districts of Mukono and Masindi, where there is use of heavy agricultural machinery, problems of soil compaction have also emerged.

Agrochemical pollution is not a major problem in Uganda due to low use of agrochemicals (fertilisers and pesticides). There are, however, some on-site and off-site pollution of water. The districts most likely to be affected are Masaka and Bushenyi (Tukahirwa, 1992). There are no major salinity problems in Uganda except for Solonetz soils in Semliki valley to the south of Lake Albert, which are naturally saline.

3.3.4 Impacts of soil degradation

Soil degradation has a range of environmental, socio-economic and political consequences at national and local levels and with immediate to long term implications. Soil erosion accounts for about 80% of the total cost of environmental degradation in Uganda and is conservatively estimated at 4% -12% of the GNP (Slade and Weitz, 1991). The cost of soil erosion has been estimated between US $ 132-

396 million per annum. This is a high cost indeed, impacting on the local and national economy of the country. However, it is unfortunate that there are no other studies undertaken yet to build on these findings, particularly in view of the rehabilitation efforts initiated by NEMA and land management in the MAAIF.

Like many tropical countries, Uganda's agriculture largely depends on topsoil where most nutrients are concentrated; hence loss of this topsoil is synonymous with soil productivity loss (Tenywa, 1999). Soil erosion undermines the agricultural productivity mainly through structural destruction, loss of sediments and nutrients. Nutrient imbalances have been reported in soils of Uganda due to mining and erosion (Wortman, 1999). Reduction in natural fertility of soils and consequently on crop yields is observed widely in different parts of the districts of Tororo, Mbale and Masaka. The production level of farms in Uganda is barely up to about 30% of the levels obtained at research stations (RELMA, 1999). Reduced agricultural crop yields and pastures affect people's incomes and investment in land and/or soil management.

Soil erosion from agricultural lands constitutes an important source of sediment pollution of water sources in Uganda. This is probably more serious in highlands/mountain areas and the drylands. Studies in Northeastern Uganda (Kotido, Moroto) by NEMA (1997) and Nakileza (2004) in Nakasongola indicate high siltation of the valley dams/water reservoirs; and studies by Tenywa et al (2004) on Mt. Elgon point to the serious threat posed by sedimentation of rivers. Silting of the Maziba Hydropower dam in Kabale is further evidence of the danger caused by environmental degradation of the upstream-cultivated and poorly conserved catchments.

3.4 Strategies for addressing soil degradation in Uganda

There is neither a specific strategy nor a single institution dealing with the soil degradation problem in the country. Issues of soil degradation addressed are incorporated in the activities of various sectoral institutions. NEMA is charged with the responsibilities of coordinating environmental management including soil resources in the country.

Historically, the high productivity of the soils of Uganda resulting from favourable rainfall and as evidenced by the lush vegetation was misunderstood for inherent natural fertility, which led to some degree of soil neglect (Zake, 1999). Little attention was paid to soil fertility management in the country to an extent that no comprehensive plan was adopted to address soil degradation. Soils in Uganda were traditionally cultivated until exhaustion, then the farmer shifted to another area to allow for natural regeneration, or restoration of soil fertility; this is what can be described as a natural fallow.

Application of both organic and inorganic fertilisers as sources of fertilisers is noted to be inadequate to cope with the rate of soil degradation (Zake, 1999). The level of inorganic fertiliser use has remained very low and inadequate. In colonial days, throughout much of Africa, the enforcement of soil fertility and land management practices incorporated a strong element of coercion in a top down approach leading to resentment of the policies and the subsequent neglect of the practices after independence. In Uganda, most areas have neglected soil conservation methods. A few places such as the Kigezi highlands have taken steps to control soil degradation.

Increasing efforts are being made to involve various stakeholders in participatory design and planning of soil conservation practices. A few cases to illustrate this include; the Vi-Agroforestry project in Masaka; IUCN-MECDP on Mt. Elgon; AFRENA-ICRAF in Kabale, etc. However, it should be realised that resources to tackle the soil resource degradation problem are still scarce; for instance farmers have little resources to invest in integrated nutrient management (Wortman, 1999).

The Plan for Modernisation of Agriculture (PMA) is a broad framework/policy addressing agricultural productivity for improved livelihood and environmental management (MAAIF, 2002). Soils are recognized as an important factor that should be properly managed to ensure high crop productivity. Therefore, issues of soil management are targeted in the PMA though in general.

The most recent efforts dealing with soil degradation are well outlined in the draft Soils Policy (NEMA, 2004). Due to wide variations in the physical and socio-economic conditions in the country and the severity and extent of soil degradation, there is need to formulate appropriate measures for each area. However, greater achievement will be made if appropriate research is undertaken on soil degradation related problems and results integrated in conservation planning.

3.5 Conclusion

Uganda is endowed with diverse soil resources that support different vegetation types and production of various crops. These soils, however, are threatened by degradation due to poor maintenance practices, particularly, in the mountainous/highland areas and drylands where there is high human population pressure. Degradation of the soils by erosion and nutrient loss causes a lot of concern because it undermines the crop production capacity, threatens food security and people's livelihood. There are scattered attempts by NGOs and government departments, to restore the degraded areas, but this is rather inadequate. Further efforts by the government, farmers and other stakeholders are required to ensure adequate soil conservation for sustainable development.

Key Terms

Mapping unit	Soil horizon
Soil profile	Soil erosion
Soil classification	Soil degradation
Nutrient balance	Integrated Nutrient Management
Sustainable development	Soil texture

Questions

1. What is soil texture? How does it affect the movement and storage of water in the soil? How are soil-formation processes affected by water availability?
2. How do soil types affect soil erosion?
3. What are the impacts of soil degradation in Uganda?
4. What other strategies should Uganda adopt in SWC?
5. How has the Ugandan population diminished the productivity of the soil?
6. What methods are used to prevent soil degradation?
7. What is soil texture? How does it affect the movement and storage of water in the soil? How are soil-formation processes affected by water availability?
8. How do soil types affect soil erosion?
9. What are the impacts of soil degradation in Uganda?

10. What other strategies should Uganda adopt in SWC?
11. How has the Ugandan population diminished the productivity of the soil?
12. What methods are used to prevent soil degradation?

References and further readings

Briggs, S.R., etal1998. *A review of past and present agricultural and environmental research in the highland areas of Uganda, with a view to developing natural resources research priorities for the Mt. Elgon hillside; Mbale and Kapchorwa areas.* National Agricultural Research Organisation and the Department for International Development, Renewable natural Resources Research System; Hillside Production System.

Chenery, C.S. 1960. *An introduction to the soils of Uganda Protectorate.* Memoirs of the research division, Department of Agriculture, Uganda, Series I. Number 1. Soils.

Farley, C.S. 1996. *Smallholder knowledge, soil resource management and land use change in the highlands of south-western Uganda.* Unpublished PhD. Thesis.

GOU. 1967. *Atlas of Uganda.* Lands and Survey Department. Government Printer, Entebbe.

Henao, J. and Baanante, C. 1999. *Nutrient depletion in the Agricultural soils of Africa.* 2002 Brief no. 62.

Lufafa, A. 1999. *Validation of pedotransfer functions for soil erosion and management in a banana-based microcatchment of Lake Victoria Basin.* Unpublished MSc. Dissertation, Makerere University.

Magunda, M.K., etal1999. Soil loss and runoff from agricultural landuse systems in the Sango Bay icrocatchment of Lake Victoria. Preliminary report.

Nakileza, B. 1992. *The influence of cropping systems and management practices on soil erosion on Mt. Elgon.*

MSc is, Makerere University. 2004. *Soil degradation and socio-ecological impact in the drylands of central Uganda.* PhD thesis. University of Cape Town, South Africa. In press.

NEMA, 2001. *State of Environment Report for Uganda,* 2000/2001. Republic of Uganda.

NEMA, 2004. *Soils Policy.* Republic of Uganda.

RELMA, 1999. *Soil fertility and productivity status in East Africa region: Proceedings of the Soil Water Conservation Society of Uganda.* 13-16, 5th –6th May, 1999.

SWCSU technical report No. 2. Rose, C.W. 1958. *Effects of rainfall and soil factors on soil detachment and the rate of water penetration into the soil and the theory of moisture movements caused by temperature gradients in soils.* PhD thesis. University of London.

Rwandasnki, A. 1960. Memoirs of the research division series 1. Soils No. 2 pp 112-11. *The soils and landuse of Buganda.* Uganda Protectorate. Department of Agriculture. Reconnaissance survey.

Slade, G. and Weitz, K. 1991. *Uganda: Environmental options.* Duke University. North Carolina, United States of America.

Stephens, D.1970. Soil fertility. In: Jameson, J.D. (ed.) *Agriculture in Uganda.* Oxford University Press. London, UK.

Tenywa, M.M. 1993. *Overland flow and soil erosion processes on spatially variable soils.* Unpublished PhD dissertation. The Ohio State University.

Tenywa et al 2004. *Soil erosion and runoff under different land use types on Mt. Elgon slopes in Uganda.* Consultancy report submitted to NEMA.

Tukahirwa, J.B.M. 1995. *Measurement, prediction and assessment of socio-economic factors of soil erosion in south-western Uganda.* Unpublished PhD thesis. Makerere University.

Wayland, J. 1938. *Interim report on soil erosion determination and water supplies in Uganda and the method of combating one and conserving the other.* Government of Uganda. Entebbe.

Wortmann, C.S. 1999. *Nutrient budgets: Understanding the problems, causes and trends* of soil resource degradation: *Proceedings of the Soil Water Conservation Society of Uganda.* 47- 55, 5th –6th May 1999. SWCSU technical report No. 2.

Zake, J.Y.K.1999. Towards building a participatory soil fertility management initiative for Uganda. *Proceedings of the Soil Water Conservation Society of Uganda.* 1-12. 5th –6th May 1999. SWCSU technical report No. 2.

Chapter 4

FORESTRY SECTOR IN UGANDA'S NATIONAL DEVELOPMENT

MUKADASI BUYINZA AND JOCKEY B. NYAKAANA

4.1 Introduction

Forests and woodlands constitute one of the important elements of the Ugandan landscape. They are at the same time vital to people's livelihoods, particularly in poor rural areas and conversely vital to development. This is because forests provide a wide range of products and ecological services which a large section of the population depend on for basic subsistence and agricultural production. Through its strong forward, backward and horizontal linkages, commercial forestry can also contribute significantly to employment and economic growth, not to mention further improvement upon development. This chapter deals with forestry in Uganda. The discussion focuses on availability of forest resources, exploitation and deforestation.

Figure 4.1: Land cover in Uganda

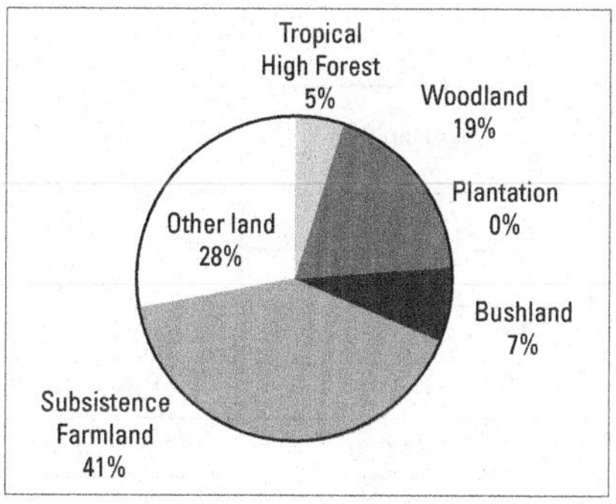

Uganda has 4.9 million hectares of forests covering about 24% of the land area (see Figure 4.1). The large portion of this forest area (81%) is woodland, 19% is tropical high forest and less than 1% is forest plantations. The tropical high forest represents only 5% of Uganda's land area, but holds 35% of the country's total biomass resource and produces a net growth of 15 tonnes of wood on each hectare, every year (see Figure 4.2). The plantation resource is currently very small (0.2%) but also quite productive (16 tonnes/ha/year), with great potential for expansion in area and yields.

Figure 4.2: Biomass in Uganda

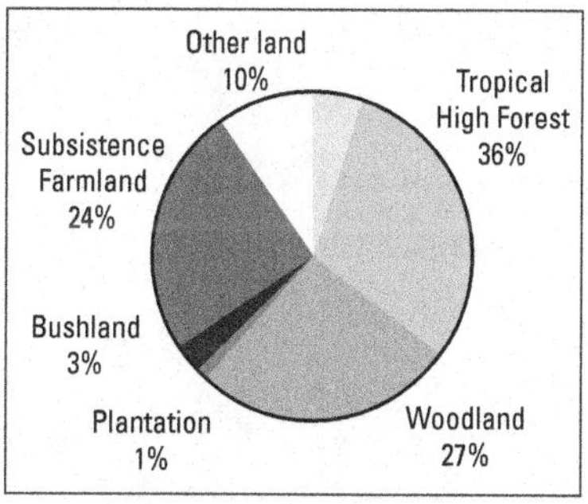

In addition to the 4.9 million hectares of natural forests and woodlands, there are also substantial forest resources on-farm. Over 40% of the land is under subsistence agriculture. This holds 24% of national biomass in the form of scattered trees, forest patches and agroforestry crops include;d within farming systems. There is thus almost as much forest biomass on-farm as in the country's natural woodlands.

4.2 Land and Tree Tenure

The majority of the forest area (70%) is on private land, consisting mainly of woodlands and scattered trees. The remainder is held in trust by the government for the citizens of Uganda, almost equally distributed between Uganda Wildlife Authority (UWA) and Forestry Department or its successor National Forestry Agency (NFA) as shown in Table 4.1. NFA manages about 417,000 ha of tropical high forests and montane forests and 720,000 ha of savanna woodlands (15% of forest cover), while UWA manages about 321,000 ha of forested land (15%). The two organizations manage a protected forest area system that accounts for 6.7% of total land area of Uganda. Over 50% of the forests are, however, located outside the protected area network on private and customary land. The districts manage the Local Forest Reserves which cover a smaller area of about 5,000 hectares (MWLE, 2001). Other protected areas are managed jointly by NFA and UWA and are called joint management areas. These include; South and North Maramagambo, Morungole, Zulia and Namatale Forest Reserves. Most of the deforestation or forest degradation takes place on the unprotected areas that is 70% of forest and woodland cover in Uganda.

Table 4.1: Area of forest and woodland under different categories of ownership and management

	Government land (ha)		Private land & Customary land	Total* (ha)
	Forest Reserves	National Park and Wildlife Reserves		
Tropical High Forest	320,354	2-67,000	351,000	924,000
Woodlands	411,578	462,000	3,102,000	3,975,000
Plantations	20,041	2,000	11,000	33,000
Total Forest	737,000	731,000	3,464,000	4,932,000
Other cover types	414,000	1,167,000	13,901,000	15,482,000
Total land	1,151,000	1,898,000	17,365,000	20,414,000

* Total land area excludes area covered by water

Source: MWLE (2002)

Map 4.1: Central Forest Reserves

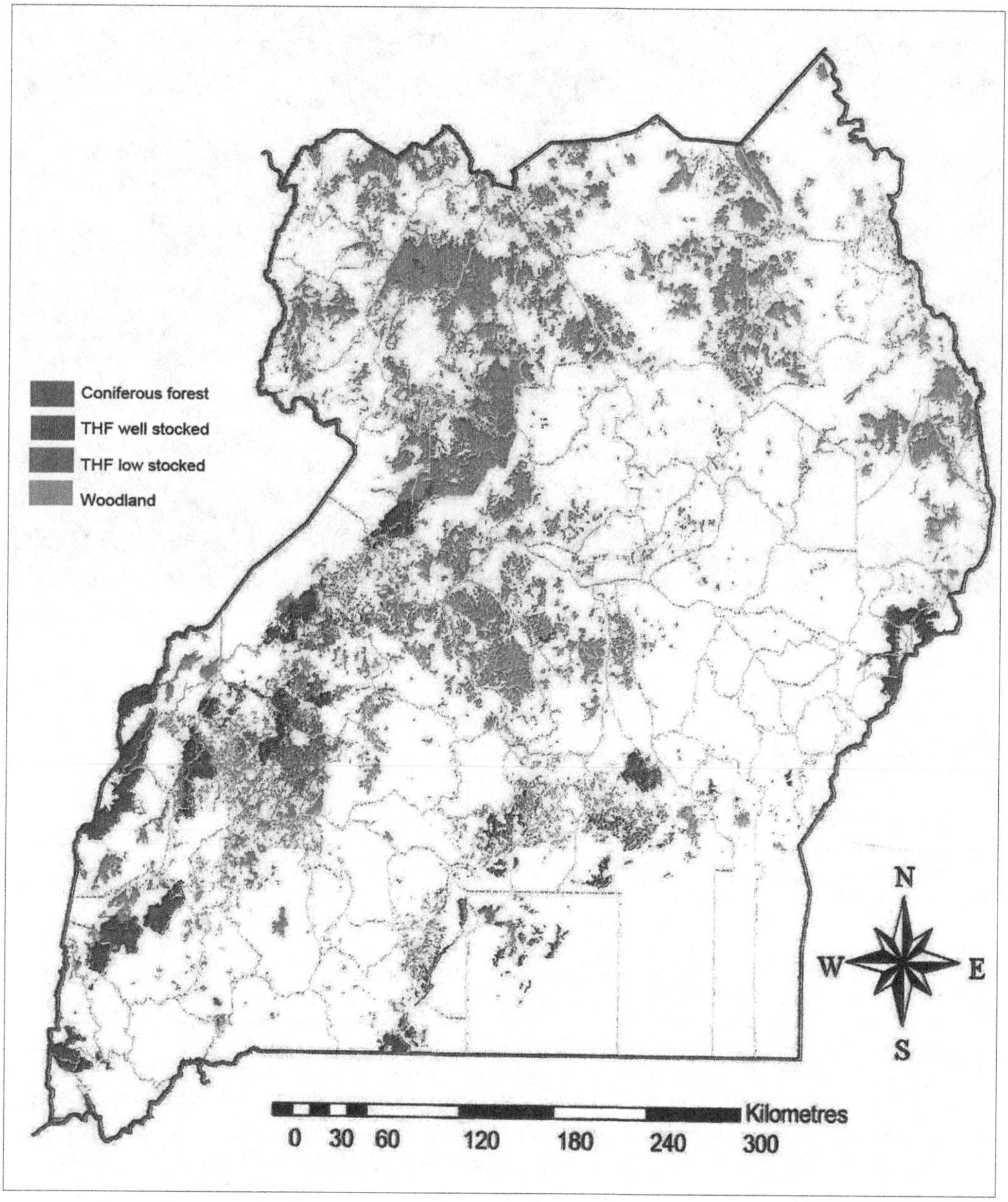

Source: Forest Department (2002a)

The distribution of forests varies greatly by region. Northern Uganda is dominated by woodlands for example, while most of the tropical high forest ecosystems are found in western Uganda (Table 4:1 and 4.2).A

Figure 4.3: Natural Tropical Rain Forest

The potential for nature-based economic activities therefore appears to be higher in Uganda's moist savanna areas (IFPRI, 2003).

Source: Forest Department

Table 4.2: Forest distribution by region (ha)

Strata	Central	Eastern	Northern	Western	Total
Hard woods	4,370	4,856	2,628	6,827	18,682
Conifers Plantations	2,746	2,140	3,238	8,259	16,384
THF (normal)	136,874	29,987	1,458	481,830	650,150
THF (depleted)	134,177	48,868	5	91,007	274,058
Woodlands	715,449	224,685	2,194,463	839,505	3,974,102
Total	993,616	310,536	2,201,792	1,427,428	4,933,376

Source: MWLE (2002)

In the absence of compensatory planting of trees, current trends show that Uganda is increasingly losing its forest cover through deforestation. Estimates of annual deforestation rates vary from 550 km^2 per year (FAO, 2000) to 700 km^2-2,000 km^2 /year (FD, 2000; MoFPED, 1994).

Deforestation is not only caused by increased demand for agricultural land. Poor land failures have also contributed to the shrinking of the forest cover. For example, peri-urban forest plantations around towns have been degazetted in favour of urban development with no provision for green belts. Some peri-urban forest plantations have been cleared to deny rebels hiding places. Other factors contributing to deforestation are: lack of manageable alternative energy sources which leads to over-harvesting of trees for fuelwood; high population growth and large families; absence of effective systems for enforcement of forest and other environmental laws; and increased demand for construction materials (MoFPED, 2003).

4.3 Importance of Forestry in Uganda

4.3.1 Forest revenue

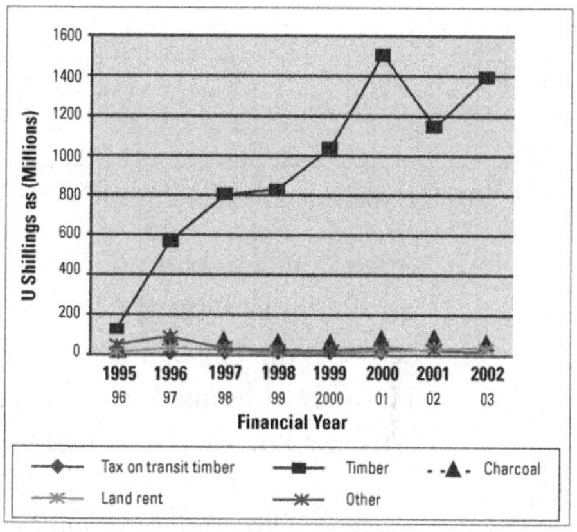

Forests and trees are richly endowed with assets from which revenue is collected and pooled in the treasury of both local and central governments to fund development programmes like provision of services (health, education, roads). Most of the public revenue from the forest sector is collected by the Forestry Department through the sale of forest products, licences and taxes levied on timber, charcoal and other forest products and from rent charged on forest reserve land for permitted activities like forest plantation establishment. Figure 4.4 shows the revenue collected from 1995/96 to 2002/03 financial years. On average, the Forestry Department has been collecting about Ushs. 1 billion annually.

According to Uganda Bureau of Statistics (2004), the contribution of forestry to GDP is still low (about 2%). A number of studies have demonstrated that forestry has the potential of generating more revenue than is currently earned, which would boost the contribution of forestry to GDP. It has been estimated that when the contribution from the informal sector is include;d, a 6% estimate will be achieved (Falkenberg & Sepp, 1999).

4.3.2 Fuelwood

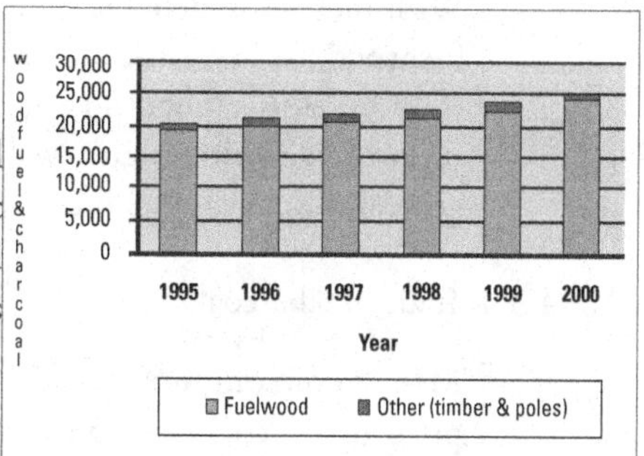

The majority of Ugandans depend on fuelwood (fuelwood and charcoal) for their domestic heating and lighting. It is well documented that over 90% of the population of Ugandans use wood for fuel (UBOS, 2002).

Total wood production (monetary and non-monetary) has registered a steady increase over the period 1995-2000. Fuel wood used by households constituted a big percentage (about 95%) of the total wood production in

Uganda (Figure 4.5). This position is likely to be maintained for the next ten years, given that there is lack of cheaper alternative sources of energy to the rural areas. The energy sector is characterized by a heavy dependence on biomass resources (UBOS, 2002) and studies have demonstrated that fuelwood (fuelwood and charcoal) accounts for more than 90% of energy used in Uganda (Table 4.3).

Table 4.3: Contribution of different energy sources

Fuel type	Percentage contribution (%)
Fuelwood	88
Charcoal	05
Petroleum	06
Electricity	01
Total	100

Source: Yakobo Moyini & E. Muramira, 2001

Apart from domestic use, fuelwood is used in industrial processes, such as lime production, brick and tile industry, fish smoking, tea and tobacco curing and sugar refining. These activities directly contribute to raising the incomes of the poor and government revenue. However, although fuelwood is a key input in these sectors, the contribution is not clearly attributed to forestry in national accounting systems.

4.3.3 Fuelwood for rural households

The rural communities particularly depend on the use of fuelwood for their domestic cooking and heating. It is estimated that rural households consume about 97% of the household energy requirements in Uganda (both rural & urban) and about 3% of the rural population have access to other forms of energy (electricity and petroleum products).

Over 90% of rural households still use fuelwood for cooking and this percentage has almost been constant since 1992 (UBOS, 2001). The combined percentage for fuelwood and charcoal usage in 1999/2000 was 98%. However, there has been a small shift from fuelwood use to charcoal in urban areas (UBOS, 2001). It is also predicted that fuelwood will remain the main source of energy for the next 10 years.

With increasing deforestation, the average distance traveled by households to collect fuelwood has increased dramatically between 1992 and 2000 from 0.06 km to 0.73 km (UBOS, 2001). There are significant regional variations with the northern and eastern regions registering higher walking distances than the western and central regions. In general, women and children are the ones responsible for collecting fuelwood.

4.3.4 Trade in Charcoal

Many poor communities derive their incomes from charcoal burning and trade. The prices of charcoal vary from district to district (Table 4.4). It also depends on the distance to the main market centre (Kampala).

Table 4.4: Charcoal Prices per district (October 03)

District	Wholesale price (Ushs.)	Retail price (Ushs.)
Arua	7,000	8,000
Jinja	8,000	10,000
Kabale	6000	7,000
Kalangala	2,000	3,000
Kampala	10,000	15,000
Kyenjojo	4,000	5,000
Luwero	6,000	8,000
Masindi	6,000	7,000
Mpigi	4,000	5,000
Mukono	5,000	10,000
Nakasongola	6,000	7,000
Tororo	10,000	14,000

Source: Forest Department

Out of 270,000 tonnes of charcoal consumed in all urban centres in Uganda, Kampala demand alone accounted for 205,825 tonnes (76%). The prices of charcoal in Kampala vary depending on quality as shown by the prices for charcoal in Entebbe and Kampala markets (Table 4.5).

Table 4.5: Charcoal prices in Kampala markets

Market Location	Charcoal Source	Wholesale Shs/bag	Retail Ushs./bag
Kampala			
Busega	Mpigi (THF – Light)	10,000/=	12,000/=
Nateete	Buruuli/Kakooge	11,000/=	13,000/=
Bwaise	Masindi	12,500/=	15,000/=
Kawempe	Buruuli	10,000/=	8,5000/=
Kawempe			10,000/=
Owino	Buruuli		10,000/=
Owino	Kyagwe/Mukono	7,000/=	8,000/=
Entebbe			
Kitooro	Buruuli	10,000/=	13,000/=
Kitooro	Buwaya Islands	8,500/=	10,000/=
Kitooro	Mpigi	7,000/=	8,000/=
Nakiwogo	Mpigi	6,000/=	7,500/=

Source: Falkenberg and Sepp, 1999 (2000)

4.3.5 Timber

There has generally been an increase in construction of private homes and buildings in Uganda from about 50% in 1992, to almost 60% in 1999/2000, with the biggest increase in the rural areas (UBOS, 2002), see also Table 4.6. Although statistics for the timber consumed directly by the construction industry are lacking, there is a direct implication on the demand for timber to meet the requirements for house construction. Through better housing facilities and improved health environment, forestry inevitably contributes to the quality of life of the people.

Table 4.6: Dwelling by locality (Percentage)

	1992/93			1999/2000		
	Rural	Urban	Total	Rural	Urban	Total
Independent House	55	30	52	63	34	58
Tenement (Muzigo)	-	-	-	5	56	13
Huts	39	6	34	31	5	27
Other	6	64	14	1	5	2
Total	100	100	100	100	100	100

Source: Forest Department

In addition, the increased demand for school facilities under the Universal Primary Education (UPE), has generally increased the demand for timber for classroom construction and school furniture. Overall, the total enrolment at primary level rose from about 3.6 million in 1994/95 to nearly 7 million during the period 1999/2000. There is no clear documentation on the consumption of timber under Universal Primary Education. However, there is evidence on the contribution of forestry to UPE since its inception in 1997, in form of timber used in the construction of the large number of classrooms and volumes of furniture supplied to schools per district.

4.3.6 Trade in timber

Timber provides business opportunities to a range of people, including pit sawyers, saw millers, timber traders and transporters. Saw milling and pit sawing activities are some of the biggest and most profitable economic activities around most of the forests. The market prices for timber and timber products differ within the country depending on the proximity between the supply and demand areas and the spending power of the consumers. The prices are reported as retail or wholesale. The prices of sawn wood have increased in the eastern part of the country where the saw log resource is not abundant.

The average sawn wood value in 1991 for Kampala was Ushs. 85,000/m3 whereas in 1999 it was around Ushs. 120,000/m3 for softwood and Ushs. 140,000/m3 for hardwood (Jacovelli and Caevalho, 1999). This is a nominal increase of only around 4-5% on average per year. The current prices range between Ushs. 120,000/m3 – Ushs. 160,000/m^3.

4.3.7 Non-wood forest products

Rattan, bamboo and medicinal plants are currently increasingly looked up to by communities as commodities for raising income. The tropical high forest yields bush meat, medicine, rattan, bamboo, fruits, vegetables and bark cloth. Woodlands are a source of thatching grass, medicine, nuts, oil, gum arabic, fruits, mushrooms, honey and weaving materials

(Box 1). Access for harvesting on private land is usually negotiated with the owner. In protected areas, harvesting is generally allowed for domestic use under negotiated agreements, as is the case for Bwindi National Park.

Box 1: NWFPs - a few examples

Gum arabic, harvested from *Acacia* trees, is an important product in the north-east. Gum arabic was important in food-processing, paper binding and printing, textile manufacturing, cosmetics, pharmaceuticals and lithography. In the 1970's the paper binding and printing industry was the greatest consumer, with the Government Printer taking 1-2 tonnes annually. A number of government departments, institutions and private industries were becoming interested in locally produced gum arabic to reduce import costs. However, this was abandoned following the political upheaval of the 1970's and 80's. The Government Printer has resorted to other alternatives such as cassava and many other users have ceased to exist, or have stopped using gum arabic.

The domestic requirement for gum arabic in the early 1990s was 5-8t per year, but the farm gate price in 1991 was only Ushs. 300 per kg. It is generally collected by herdsmen at a typical rate of 0.5-1.5kg per day. An average yield of 85-120kg per ha is estimated for Karamoja.

Medicinal plants - Over 100 different types of medicines are collected from natural forests. Trade in medicines is not normally recorded, but in Nyimbwa Sub-county, Luwero district it was estimated that in 1999 a monthly revenue of around Ushs. 300,000 was generated by medicine collectors in the area.

Shea butter for oil and medicinal use is an important multiple-use product for people in northern Uganda. Surveys indicate that the shea tree has over 100 different uses, the trees produce an average of 7-10kg of dry nuts per annum and there are an average of 7 trees per ha on farmland in the north-east of Lira District. The major product is shea butter oil. In 2000 this was valued at Ushs. 2,000 to 5,500 per litre.

Moringa and Neem are the other important multiple product trees and are becoming popular for their medicinal value against common ailments such as malaria, skin diseases and AIDS-related opportunistic diseases. A mature *moringa* or *neem* tree produces 30-100kg of seed per fruiting season and fruits 1-2 times a year. These are sold at Ushs. 60,000 per kg by the National Tree Seed Centre.

Bushmeat is significant, but largely unrecorded. In Kafu, Masindi District communities survive on bushmeat trade.

Rattan is of considerable socio-economic importance and there is concern regarding over-harvesting.

In Ngogwe Sub-county, Mukono, the percentage of people involved in rattan trading has grown significantly, from 7% in the 1950s to around 30% in the 1990s.

Rattan harvesting is already shifting away from the depleted lakeshore region and it is now mostly gathered from the western districts of Kibale, Hoima and Masindi. However, most rattan products end up in Kampala or Mukono. It is estimated that there are close to 120 rattan-based enterprises in Mukono, 100 in Hoima/ Masindi and 25 in Mpigi.

Bamboo - In a 1999 study by FORRI, it was estimated that the monthly net income of bamboo collectors in Mbale and Kabale, from bamboo alone, was around Ushs. 48,000 and 38,000 respectively.

4.3.8 Watershed management and soil conservation

Natural forests and to some extent trees on farm, provide important environmental services of soils protection and water conservation. The value of this can easily be appreciated when one looks at steep, hilly and mountainous areas like the districts of Mbale/Kapchorwa, Kabale, Mbarara and Kasese/Bundibujo. Today many watershed areas including riverbanks are suffering from deforestation and degradation. Through overgrazing, agricultural expansion, bush fires, commercialisation of charcoal and fuelwood and urbanization, loss of trees and vegetation cover has exposed these environmentally fragile ecosystems to soil erosion and floods. The areas within the mountains of Elgon and Rwenzori have in particular experienced landslides, which have been a common feature especially in recent times.

The importance of tree cover on the vulnerable landscapes is hardly perceived by the people, until they experience and suffer the consequences of such degradation and disasters. These range from reduced water flows in streams and rivers, to displacements and loss of lives and property where landslides have occurred. It also triggers off impediments to other productive sectors that contribute to people's livelihoods and ability to raise income.

4.3.9 Soil protection and soil fertility

Forests and trees are characteristically able to reduce water runoff, reduce the erosion of topsoil and increase water infiltration. In so doing, they increase soil water conservation and nutrient retention. This in turn improves soil fertility and crop yields and contributes to people's ability to raise income, food security and improvement of the quality of life. Recent research by ICRAF/FORRI (Siriri *et el*, 2000) demonstrates the substantial impact of tree management in farming systems. In the hilly areas of Kigezi, seven out of ten farms with contour hedgerows have on average 14cm more topsoil than those without hedgerows after 3-6 years of growth. This represents 79 tonnes of soil conserved for every 100 metres of hedgerow, or Us.700,000 worth of available nutrients (at market price) per one million trees/shrubs planted (Nema, 2002).

4.3.10 Forests improve local-climates

Forests influence micro-climates and possibly local rainfall pattern. This has a direct impact on the agricultural productivity, which depends on climatic regimes. Some farmers clearly recognize the linkage between forests and their impact on climate. Adverse climatic conditions like drought affect agricultural productivity and hence food security and income generation for households. Forests and trees also provide agricultural support and environmental services in ways that are taken for granted. These are especially important to the poor as they cannot afford alternatives such as piped water or fertilizers. Forests and trees protect and improve soils, which substantially increases crop yields. Trees can add up to 150kg of nitrogen per hectare, increasing maize yields up to five times while also producing 25 tonnes of fuelwood per hectare - enough fuelwood for seven families for a year.

4.3.11 Biodiversity support systems

Forests contain a rich biodiversity of national and international importance. Uganda is one of the most species-rich countries in the world for its size, with strong tourism potential and considerable economic value for medicines and agricultural crops. Based on the bio-diversity resources, the Forestry Department initiated eco-tourism projects in Budongo, Mabira, Mpanga and Kasyoha-Kitomi forest reserves. The communities have received benefits from the tourist revenues and in the form of employment (Box 2).

This has had a great impact on the social and economic development programmes initiated by the local people and supported by the Ecotourism Project. It has demonstrated the importance of biodiversity, conservation and protection of forested areas at a local level.

Box 2: Contribution of eco-tourism to people's livelihoods

Budongo Forest Ecotourism Project was initiated by the Forestry Department in 1993 with the zoning of part of the Budongo Forest Reserve for recreation, as a foundation for sustainable community development through Ecotourism.

The 8,020 visitors (as from 1995 to December 2002) that came to the two sites of Budongo (Kaniyo Pabidi and Busingiro) have contributed close to Ushs. 140 million (equivalent to US$ 70,800) mainly from forest walks and accommodation fees. Of these, about Ushs. 28 million (US$ 14,000) representing 20% of the revenue collected has been re-invested in local community development programmes particularly in school construction and construction of a maternity unit. Using participatory methods of problem identification and analysis, the communities around the project area identified education, water and health as their major problems. The project contributes cement, iron sheets, metal bars and construction timber, while the communities contribute bricks, sand stones and unskilled labour during construction work.

The decision to allow the people to decide on the project has brought benefits to the whole community. The project has scored major points in confidence and relation building among the local people. The local communities have become part and parcel of the project development and management. They have taken upon themselves to report to forestry department staff any cases of illegal activities within the forest.

Similarly, the local communities around Mabira Forest Reserve ecotourism site have similarly derived a number of benefits, which include;:

- Sales of fruits and vegetables to visitors from Najjembe market. Tourists are known to spend at least Ushs. 5,000/= at the market compared with ordinary passers-by who spend an average of Ushs. 1,000/=. It was noted that about 30% of the tourists make purchases at the market.

- Part of the tourist revenue has been reinvested in social services selected and co-financed by the local communities. So far, six schools have been supported and the total grant totalled to about Ushs. 17,000,000/= (which is approximately 20% of revenue generated from ecotourism between 1996 and2002).
- Employment opportunities have been created for the local people as guides, caretakers, Askaris, trail cutter/maintenance crew and accommodation rehabilitation workers.
- By working together with FD, the MFTC and most of the local communities CBOs working in and around the FR formed an umbrella organisation called Mabira Forest Community Development Organisation (MAFICO) and the NGO is promoting conservation of forests for supporting community livelihoods.

4.3.12 Employment

Forestry consists of a range of activities that provide employment opportunities to the people of Uganda, especially to the poor communities. Such opportunities include; extractive activities, (sawmilling, pit sawing and harvesting of various products from forests and trees), processing (carpentry, joinery) and construction. The forest-based activities in the formal sector employ approximately 100,000 people, while the informal sector employs about 750,000 people (Falkenberg and Sepp, 1999). Tables 4.7(a) and 4.7(b) below give the distribution across the main activities.

Table 4.7(a): Employment in the forest sector - Formal sector

Employment arae	Person Year/annum
Fuelwood and charcoal production	89,150
Plantation establishment and management	1,400
Forest industry (incl harvesting)	3,200
Non-timber forest products	n.a.
Institutions	2,600
TOTAL	96,350

Source: Yakobo Moyini & E. Muramira, 2001

Table 4.7 (b): Employment in the forest sector - Informal sector

Employment arae	Person Year/annum
Fuelwood collection by households	710,000
Fuelwood production for commerce and industry	36,000
Pole production	1,000
Total Informal	747,000

Source: Yakobo Moyini & E. Muramira, 2001

Basing on the above, field observations tend to re-emphasize the potential of the forest sub-sector in providing employment opportunities for poverty alleviation, especially for the rural poor.

4.4 The policy framework for Collaborative Forest Management in Uganda

African governments are increasingly realizing that centralized planning and management of forest resources is failing to conserve the resources and failing to meet the basic needs of other interested parties, particularly those of local people living next to the resource (Mupada, 2002). In Uganda, for the last 99 years, 57% of the prime forest resource has been under the management of the government.

The history of forest management in Uganda has gone through four distinct periods:

- The colonial period (1898-1961) when the government established a network of central forest reserves and local forest reserves that were well managed using rigorous scientific methods and elaborate management plans.
- The post-independence era (1960-71) when the government centralized forest resources but was able to maintain the forest estate in reasonably good condition through a process of command and control.
- The military dictatorship era (1971-1985) when there was no effective forest management by the state due to the political and economic instability. The forest estate was severely degraded.
- The decentralization period (1987 to present), when the government embarked on a process of devolution of power to districts and local councils including the management of natural resources.

The policy environment in Uganda is ripe for collaborative forest management. The Constitution of Uganda (1995) recognizes for the first time the significance of the environment sectors. The National Environment Management statute (1996) which established the National Environment Management Authority (NEMA) emphasizes the importance of involving and empowering local councils and local communities in environmental management. In addition, the Wildlife Bill (1996) which formed the Uganda Wildlife Authority has a stated policy and legislation that recognizes the need to collaborate with and consult a wide variety of possible stakeholders, including local authorities and communities. The Forest Department recognized the need to involve communities in the management of forest resources as early as 1988. With the help of funds from the European Union (EU) and NGOs such as CARE, IUCN, the department initiated pilot projects in Semliki, Rwenzori, Kibale, Bwindi and Mt. Elgon. Bwindi Impenetrable Forest and Rwenzori were later transferred to the management of National Parks in 1991 while Mount Elgon, Kibale and Semliki were transferred in 1993.

4.4.1 Mount Elgon National Park

Mount Elgon National Park (114,000 ha) was gazetted as a Forest Reserve in the 1930s. The area is basically a montane afro-forest and moorland ecosystem. It ranges between 2.000m and 4,200m above sea level. The higher altitudinal range is dominated by heath and moorland vegetation. Between 2,400m and 3,000m lie extensive bamboo forests. Mount Elgon has a high conservation value for the global and local communities (FD,

1996). It is an important watershed supplying water to thousands of people in Uganda and Kenya (Howard, 1991). The area has several endemic species of flora characteristic of the East African montane ecosystems and provides many products such as bamboo shoots, medicinal herbs, fuelwood and timber, to the local community. Globally, Mount Elgon is important because of its unique species, especially in the alpine and ericaceous zone and its role as a carbon sink. Mount Elgon National Park has gone through the turbulent times of protected area management in the country and severely suffered from agricultural encroachment in the 1970s and 80s (Box 3)

Mt. Elgon National Park is one of a few areas in Uganda where collaborative management has been implemented. Others are Bwindi, Rwenzori, Kibale and Semliki National Parks and Buto-Buvuma, Budongo and Mabira Forest Reserves (Mupanda, 2002; Cunnington, 1996). Collaborative management activities in Mt. Elgon National Park have so far concentrated on exploring the opportunities for making local people active managers of the forest belt of the Mt. Elgon ecosystem, to promote sustainable development activities for substitutes to the current unsustainable forest uses. The present collaborative management at Mt. Elgon National Park is based on the following principles (Onyango, 1996):

- Addressing the major threats of destruction to the Mount Elgon ecosystem.
- Provision of long-term solutions for park conflicts that arise from harsh law enforcement.
- Provision of access to agreed upon forest resources for local communities bordering the park.
- Changing the role of park staff from strict law enforcement to one that facilitates and monitors activities in the forest zone.

The communities depend on natural resources. They also have the interest and capacity to conserve them as long as their rights, duties and responsibilities are defined and consensus reached by all concerned parties. The other rationale behind the introduction of collaborative management is the realization that protective policies have not worked. The two organizations (UWA and FD) are aiming at uplifting the socio-economic status of communities living next to protected areas and to give them 'sustainable' access or alternatives to natural resources so that they can satisfy their basic needs (Mupanda, 2002).

Bwindi Forest

Source: Uganda Tourism Board

Box 3 A chronological overview of the change from Forest Reserve to Mt. Elgon National Park	
1983	The government proposed to convert some forest reserves to national parks. This was followed by a series of intra-governmental consultations. The FD started a process of cleaning up forest areas, clarifying boundaries and evicting encroachers.
1986	Strengthening of public institutions by the government; the FD made a policy decision to create forest parks to be managed under the Man and Biosphere principles. The FD invested and prepared six forest reserves and also suggested a re-name to forest parks.
1987	The government secured a multi-donor loan to sponsor the rehabilitation of the forestry sector under World Bank auspices. Its mandates were; Rehabilitation of FD and coniferous industrial plantations, Natural Forest Management and Conservation, Peri-Urban Plantations, Training, Farm Forestry. FD still continued the eviction of forest encroachers while implementing the rehabilitation projects.
1988	A decision to implement a government proposal of turning forest reserves into national parks was taken. It was agreed that for Mt. Elgon and Rwenzori, the areas above the tree line would remain forest reserves. Bwindi was to remain a forest reserve as FD was regaining its capacity to effectively manage this particular reserve. The Ministry of Environment Protection and IUCN started Mt. Elgon Conservation Development Project through NORAD funding. FD recognized the need for collaborative management and started pilot projects in Semliki, Bwindi, Mt. Elgon, Kibale and Rwenzori. This was done with funding help from the European Union (EU) and international NGOs such as the World Conservation Union (IUCN) and Carry American Relief Everywhere (CARE).
1989	UNP insisted on a cabinet endorsement of the transfer of the forest reserves to UNP.

1990	FD completed the exercise of removing encroachers and started enrichment planting.
1991	UNP board of trustees, cabinet and parliament debated and endorsed the decision of converting the concerned forest reserves into national parks. FD carried out biodiversity inventories in 66 Tropical High Forests (THF) with an aim to determine what percentage should be set aside as strict nature reserves. The inventory also led to a Nature Conservation Master Plan (NCMP). The President instructed his prime minister to direct parliament to turn the remaining parks of Mt. Elgon and Rwenzori and in addition Bwindi into national parks. The National Environmental Action Plan (NEAP) process was launched to integrate environmental concerns into the overall social and economic development processes.
1992	The establishment of the National Agricultural Research Organisation (NARO) implied transfer of all forestry research activities from FD. Budongo Forest Reserve was earmarked for the change, but remained as a forest reserve after substantial internal struggle.
1993	Mt Elgon, Semliki and Kibale were converted to national parks.
1994	The National Environmental Action Plan resolved to have an integrated national policy framework and legislation for sustainable exploitation and management of natural resources.
1995	There was a constitutional amendment to provide for sustainable environmental management (National Environmental Statute (NES) enacted to coordinate environmental issues. This statute established National Environmental Management Authority (NEMA) to coordinate, monitor and supervise all activities in the field of environment. NEMA was also to provide the framework for integrating environmental issues to the overall national social economic development plan.
1996	The Uganda Wildlife Statute (1996) established Uganda Wildlife Agency (UWA) as a new organization with a merge of the Game Department (GD) and the Uganda National Parks (UNP). One started pilot schemes of CFM at Ulukusi and Mutushet, with assistance from the Mount Elgon Community Development Programme (MECDP).

4.4.2 Budongo Forest Reserve

Budongo forest was one of the first forest reserves to be gazetted in 1932, with an area of 825,000 ha, of which 487,000 ha is actual forest and the rest is grasslands. It is basically low lying forest within a range of 700m-1250m above sea level. Budongo is the largest mahogany forest in Uganda and has about 60% of the country's species, making it a very important conservation and production forest (Howard 1991). It is believed to be home to one of the largest chimpanzee populations in the country and has been the focus of many researchers, scholars and tourists. It is a very interesting forest with one of the best management records in the world, dating back to the 1910s.

Budongo forest did not suffer from agricultural encroachment as did other reserves, such as Mabira. Nevertheless, there are a number of threats such as illegal pit sawing, extraction of particular species and habitat destruction by saw millers and pit sawyers which may lead to the extinction of species such as the chimpanzee. Consequently, a number of strategies have been initiated to increase the conservation status of Budongo. One such effort was started by the Budongo Forest Project through its studies of the chimpanzees of Budongo Forest. The Budongo Forest Ecotourism Development Project (BFEP), which started in 1992, built on the initial work of the Budongo Forest Project. The main objective of BFEP was to promote ecotourism as a means of raising revenue from non-consumptive uses and encourage community participation in conservation through creating awareness, education, sharing responsibilities and returns.

The project has received most of its funding from the European Union (EU) and has involved the community in the neighbouring six parishes from the beginning. Many meetings were held with the neighbouring communities to increase their awareness of the values of conservation and to solicit their participation and collaboration in the conservation of Budongo Forest, especially through the ecotourism project. It was realized through the meetings that the neighbouring communities were interested in the conservation of the forest. As a way of understanding the relationship between the community and the forest, a Participatory Rural Appraisal (PRA) was carried out in 1993. Through the PRA study, it was discovered that the neighbouring communities obtained many resources from the forest which include; water, fuelwood, building poles, timber, medicinal herbs, mushrooms, vegetables, honey, fruits and wild animals such as duiker, bush pigs and occasionally monkeys. It was easy to build on indigenous management practices because, even without any explicit institutional guidelines for collaborative management, the Forests Act (Chapter 246, 1964) allows the local community to collect resources in `reasonable' quantities for their domestic use. Each of the six parishes where the project is operating is represented by two people on the committee that makes decisions regarding the development of collaboration with the Forest Department. The communities, one at each of the two sites of the project (Busingiro and Kaniyo-Pabidi) meet once a month to decide on management issues.

The ecotourism committee, in addition to the two representatives from each of the parishes, has two members of staff. It has several responsibilities which include;

1. linking the committee with the community in order to promote a better image of the committee;
2. monitoring revenue collection;
3. encouraging community involvement in income generating activities; and
4. appointing workers employed in ecotourism development.

On the other hand, the Forest Department is responsible for making sure the project continues and that the process is completed. The Forest Department is expected to pay permanent staff seconded to the project and to publicize the project nationally and internationally. The project has gained a lot of support through local community participation and conservation education programmes that now involve school children. The community has also benefited through employment and revenue sharing. The project has initiated income generating activities such as bee-keeping and growing to create alternative sources of income for local people. Local people have started reporting cases of illegal harvesting of forest resources.

No long-term agreements have been signed with the community, partly because of the past management practice of the Forest Department. The command and control method of management alienated the neighbouring community complained that their survival depended on more than just the non-timber products of the forest. Therefore, in 1994, the Forest Department asked the members of the community to form a pit sawyers' association to negotiate with the Forest Department. The Department agreed to give a concession to the pit sawyers on condition that they would operate only in specified felling areas, undertake stock-mapping using Forest Department personnel, finance the enrichment planting of harvested areas and pay timber royalties for any wood removed. On the other hand, the Forest Department was expected to continue its supervisory role in all operations and collect revenue for the government. It has been a process of consensus building. It is probable that even if agreements had been signed in the initial stages they would not hold because of the high level of suspicion between the Forest Department and the neighbouring community.

4.4.3 Mabira Forest Reserve

Mabira Forest Reserve has an area of 30,600 ha (Forest Department 1996). It is a low lying forest between 1070 and 1340m above sea level with a wide range of species of fauna and flora. It is estimated that Mabira has 312 species of trees, 287 of birds, 23 of small mammals, 218 of butterflies and 97 of moths (Howard, 1996). The forest has, therefore, been ranked as a good conservation area to protect the various assemblages of species. Mabira forest has a very interesting history of management dating back to its time of being gazetted when villages were enclosed within the forest as enclaves. At that time, people co-existed with the forest and had access to the forest for basic needs such as water, fuelwood, vegetables, honey and medicine for domestic use (Forest Act, 2002: Chapter 246, 1964).

Mabira Forest, just like Mount Elgon, was seriously degraded in the 1970s and 1980s. By 1988 about 25% of the forest had been cut down for agriculture and over 3,000 families had settled in the forest (FD, 1988). The work force reduction in the Forest Department authority over the years, coupled with corruption and political interference led to the fast disappearance of the forest. When the Government of Uganda made a commitment to the conservation of its forest resources, the biggest task was for the Forest Department to regain control of the resources and to remove all the encroachers from the forest. A number of donors came in to help the country to achieve its goal of forest and biodiversity conservation, among which the EU has played a dominant role.

Mabira Forest

Source: Uganda Tourism Board

The process of regaining control of the forest reserves involved removal of encroachers. This exacerbated the conflict between the Forest Department and local communities. It was against this background that the Forest Department decided to promote the conservation of Mabira through a number of strategies, which include;d zoning into Strict Nature Reserves, Buffer and Production zones and the development of ecotourism to raise revenue from non-consumptive uses; to involve the community in the management of the reserve and revenue sharing. Unlike in Budongo, the initial stages of ecotourism development did not directly involve the community. However, the Forest Department still believed that this was a good entry point for collaboration with the community. A study was therefore undertaken by the Forestry Research Institute (FORI 1995) to lay a firm ground for a collaborative forest management process. The following interest groups were identified: fuelwood collectors, timber harvesters, collectors of building poles, sand collectors, edible wild plant gatherers, livestock grazers, wild game hunters, herbalists (traditional healers), handicraft makers, wood carvers, boat makers, bee keepers and wild honey collectors, grass/fodder collectors and alcohol brewers.

> Generally, the various groups are not organized around definite institutional frameworks. The local people were in general agreement that they had some rights to the forest to collect resources such as water, medicinal plants, grass and fuelwood although they claimed to have no control over the forest and expressed a desire to manage the forest with the Forest Department in a collaborative forest management process. The peoples' expectations

include;d promotion of rural development, sharing of revenue, obtaining access to credit facilities and loans and assistance in marketing goods such as handicrafts. However, it was an unanimous opinion that there was need for mobilization and sensitization to make all stakeholders appreciate the need for conservation of the forests and its biodiversity and the role of the forest in socio-economic development. The process of mobilization, sensitization and collaboration started in 1996 through the ecotourism project, with the principles already stated by the Forest Department. The next stage should be to bring the parishes neighbouring the reserve and the Forest Department through the whole process of consensus building and signing memoranda for collaboration.

4.5 Conclusion

Forests and trees are a natural resource that provides numerous goods and services, which are important to the livelihoods of the majority of the people of Uganda. For a long time, Ugandans have harnessed fuelwood, timber and poles, or used their derivatives for their energy needs, domestic comfort, security, or development. Of particular importance are the non-wood benefits from the forests and the environmental values that contribute significantly to the people's livelihoods, especially women and yet are not reflected in the national accounting systems.

Many people depend on forestry for all or part of their livelihoods. It is the poorest, however, who often depend most critically on forest resources for their well-being and survival in the absence of other livelihood assets and opportunities. Forest products are some of the most important free goods and safety nets produced in nature providing shelter and food security that are critical to poor subsistence households, especially the children and the elderly.

The quantity and quality of the forestry resource base is important for the continuous supply of goods and services to the people of Uganda. However, it is noted that there is an increasing decline of the resource base through a number of social, economic and political factors. The major factors include; forest clearance for agriculture, over-harvesting and degradation of forests, encroachment of government reserves and the degazetting of forest reserves for alternative purposes, such as, industries and urban development.

The contribution of forestry to poverty eradication is poorly understood among policy and decision-makers in government. The lack of recognition or the poor perception of forestry shows clearly in a lack of a concerted national policy to promote investments in the forest sector. There has been very little recognition of the economic importance of the forest sub-sector both as a source of rural incomes, a major source of energy, not to mention the environmental benefits. Many of the forestry-related services, including environmental services, are of public good in nature and their contribution to poor people's incomes and livelihoods is currently undervalued.

There is a need to institute effective forestry management mechanisms to reduce resource degradation and to increase forest products and services for the benefit of the people of

Uganda. For over 100 years, the Forestry Department (FD) has been responsible for the management of all forest resources in Uganda, including the forest reserves, extension services and regulation of forestry activities on private and customary lands.

Agreements should reflect the compromises reached, define the rights, responsibilities and benefits that accrue to each party and be a tool for constant dialogue and compromise. Good agreements must, of necessity, be built on trust and cater for the relevant needs of each party. All the partners need to be aware that constraints are likely to affect the smooth implementation of any programme and that they should therefore participate in defining the criteria for monitoring and evaluation of the successes and failures; to work out remedial measures together.

Key Terms

Forest: A forest is an area of land containing a vegetation association that is predominantly composed of trees, which may be naturally occurring, or planted. To be considered a forest, an area should have a tree cover of at least 20% and more and the area must not be less than 0.5 ha. Forests can be classified according to their regeneration systems, the most common distinction is between natural forests, commonly termed as original, or virgin forests. The second type is the plantation forests which are artificial, for example social forests, village woodlots and farm forests.

Forestry: Includes all activities related to forests, tree growing, forest produce, forest conservation, forest management and forest utilization.

Forest management: Forest management encompasses all those decisions needed to operate a forest on a continuing basis. In the broadest sense, forest management integrates the biological, social and economic factors that affect management decisions about a forest.

Reserve trees: The Forest Act (1964) gives powers to the concerned Minister to prescribe the species of trees, or group of trees on private land to be protected. This is for purposes of scenic beauty, conserving a distinctive specimen of any tree species, preventing soil erosion, conserving biological or species diversity and for conservation, protection and development of natural resources.

Land tenure: Refers to a bundle of rights which may be held by different people at different times. There are four categories of rights which make up the bundle that comprises of land and tree tenure: the right to own, or inherit, the right to plant, the right to use and the right to dispose.

Private forests: A natural forest, or a plantation forest, or an area dedicated to forestry, registered under section 21 or 22 of the Forest Act. 70% of total forest and woodlands in Uganda make up private and customary land.

Collaborative forest management approaches: The Central government and local communities jointly share roles and responsibilities in the management of the forest resources. When local people are involved in the management of the forest, without pressure from the government, the result is the mutual benefits for both the local communities and the forest department.

Poverty: According to the Ministry of Finance, Planning and Economic Development (MFPED) report (2000), poverty is characterized by lack of income and consumption, physical insecurity, poor health, low levels of education, heavy burden of work or unemployment and isolation (both socio-economically and geographically).

Questions

1. Why has the private sector been reluctant to participate in forestry management?
2. Why has the forestry sector in Uganda failed to achieve its primary goal of improving the general welfare of the rural people by providing basic needs amidst the conflicting objectives of land and forestry resources conservation?
3. Describe the measures that can be taken to increase access of the poor local communities to forestry resources in the protected areas (forest reserves, national parks and wildlife reserves).
4. In what ways has the land and tree tenure affected forestry development programmes in Uganda?
5. Explain the institutional challenges facing the forestry sector in Uganda.
6. What are the benefits and challenges of the collaborative forest management approach in Uganda?
7. Why is there a need to develop institutional collaboration in the management of the forestry resources?
8. What are the causes and effects of deforestation in Uganda?

References and further readings

Cunningham, A.B. 1992. *People, Park and Plant Use.* Research and recommendations for multiple-use zones and development alternatives around Bwindi-Impenetrable National Park. Report prepared for CARE - International, Kampala, Uganda.

Egelling W.J. 1947. *Journal Ecology* Observations on the ecology of Budongo rain forest, Uganda. Kampala, Uganda.

FAO. 1992. *Forests, Trees and Food.* Forestry Policy and Planning Unit, Forest Department, Rome, Italy.

FD (Forest Department). 1996b. *Budongo Forest Reserve Biodiversity Report.* Forest Department, Kampala, Uganda.

FD (Forest Department). 1988. *The Status of Mabira Forest Reserve.* Forest Department, Kampala, Uganda.

FD (Forest Department). 1996a. *Mt. Elgon National Park Biodiversity Report.* Forest Department, Kampala, Uganda.

FD (Forest Department). 2002. *National Biomass Study Technical Report 2002 – Draft.* Ministry of Water, Lands and Environment. Kampala, Uganda.

FD (Forest Department). 1951. *A history of the Uganda Forest Department, 1898-1929.* Entebbe: Uganda Forest Department.

Falkenberg, C.M., S. Sepp. 1999, *Economic Evaluation of the forest sector in Uganda*; A study carried out as part of the Forest Sector Review.

Gombya-Ssembajjwe, W. S., etal J. 1999. Property Rights: Access and Forest Resources in Uganda. In: *Access to Land, Rural Poverty and Public Action.* A Study Prepared for the Institute of the World Institute for Development Economics Research of the United Nations, Janvry, A., Gordilo, G., Platteau, J. and Sadoulet, E. (editors), University (UNU/WIDER) Oxford University Press.

GoU (Government of Uganda). 1998. *The Land Act.* Kampala, Uganda.

GoU (Government of Uganda). 1964. *The Forests Act.* Laws of Uganda, Chapter 246. Republic of Uganda, Entebbe, Uganda.

Jacovelli, P., J. Caevalho. 1999. *The private forest sector in Uganda – opportunities for greater involvement.* A study carried out as part of the Forest Sector Review, Ministry of Water, Lands and Environment.

Hamilton, A.C 1984. *Deforestation in Uganda.* Oxford University Press Nairobi.

Hoefsloot, H. 1996. *Collaborative Management on Mount Elgon*; an account of first experiences (for publication). The IUCN Tropical Forest Conservation Programme.

Howard, P.C. & Davenport, T.R.B. (eds). 1996. *Forest Biodiversity Reports.* Kampala: Uganda Forest Department.

Howard, P.C. 1991. *Nature Conservation in Uganda's Tropical Forest Reserves.* IUCN, Gland, Switzerland.

IFPRI (International Food Policy Research Institute) 2003. *Strategic Criteria for Rural Investments in Productivity (SCRP). Phase II Completion Report.* Main Report submitted to USAID

Kamugisha, J.R. 1993. *Management of Natural Resources and Environment in Uganda. Policy and Legislation Landmarks, 1890-1990.* Regional Soil Conservation Unit/SIDA, Nairobi.

MoFPEF (Ministry of Finance, Planning and Economic Development). 1997. *Poverty Eradication Action Plan (PEAP). A National Challenge for Uganda. Kampala, Uganda.*

MoFPED (Ministry of Finance Planning and Economic Development) *Uganda Poverty Status Report, 2003 (Achievements and Pointers for the PEAP Revision),* Kampala, Uganda.

Moyini Y, E. Muramira. 2002. *The Cost of Environmental Degradation and Loss to Uganda's Economy with Particular Reference to Poverty Eradication. Policy Brief* No.3 IUCN.

MUIENR (Makerere University Institute of Environment and Natural Resources). 2000. *National Biodiversity Data Bank Report* 2000. Makerere University. Kampala, Uganda.

Mupanda, E. 2002. *Towards Collaborative Forest Management in the Conservation of Uganda's Rain Forests.* In url: www.earthwatch.org/europe7limbe7collabformgMt.htm#heading153, 06.02.03

MWLE (Ministry of Water, Lands and Environment). 2002. *The Uganda Forestry Policy 2001.* Kampala, Uganda.

NEMA (National Environment Management Authority). 2001. The State of the Environment Report Uganda, 2002.

Onyango, G. 1996. *Collaboration Between Public Agencies and Adjacent Communities in Managing and Conserving Natural Resources.* A Paper Prepared for the UNEP/World Bank African Forest Policy Forum, Nairobi-Kenya. Unpublished

Scott, P. 1998. *From Conflict to Collaboration: People and Forest at Mount Elgon, Uganda.* IUCN East Africa Regional Office, Nairobi Kenya.

Scott, P.J. 1994. *Bamboo utilisation on the slopes of Mount Elgon.* Consultancy Report for Uganda National Park/The Mount Elgon Conservation and Development Project.

Siriri,D., Raussen, T. and Poncelet, P. 2000. *The development potential of Agroforestry technologies in Southwestern Uganda.* Agroforestry Trends 2000 (ISBN)

Tumuhimise J, J Kuteesakwe. 2003. *Sustainable Charcoal Production and Licencing System in Masindi District.* Ministry of Energy and Mineral Development

UBOS (Uganda Bureau of Statistics). 2002. *Provisional Population Census Results.* Entebbe, Uganda.

Chapter 5

WATER AND WETLAND RESOURCES IN UGANDA

BAKAMA BAKAMANUME AND HANNINGTON SENGENDO

5.1 Introduction

This chapter examines water and wetland resources in Uganda. The chapter is divided into two sections. The first section deals with water resources and the second examines wetland resources.

SECTION 1: WATER IN UGANDA

5.2 Water Resources

Water resources include; water itself in liquid form, fish and water generated power (hydro electric power). Water resources are abundant in most parts of Uganda. This is partly due to the favourable climatic conditions (see discussion in Chapter 1). There are, however, some areas in Uganda where water scarcity does occur. Ugandans get their water from several sources. Table 5.1, shows that the number of sources with clean water supply have increased since 1999. Unlike in developed countries where agriculture is the leading water user, in Uganda, the public is the leading user followed by industry. Agriculture in Uganda depends on rainfall rather than irrigation.

Sources of Water

Ground water (boreholes and shallow wells) is the most common source of water in Uganda. Surface water (springs) is the second. Piped water (taps) is the other important source. Of these sources, surface water resources have been well documented. There is very little research done on ground water table in Uganda.

Table 5.1: Number of Sources of Clean Water and Percentage Rural Population Served

Source	1999	2000	2001	2002
Spring	17,282	17,842	19,029	20,224
Borehole	15,374	16,520	17,915	17,846
Shallow Well	2,774	3,417	4,734	5,998
Gravitation Flow	82	125	134	138
Taps	2,655	3,866	4,058	4,233
% Rural Population Covered	46.6	47.5	50.3	51.2

Source: Uganda Bureau of Statistics 2004

5.2.1 Surface Water Resources

Twenty percent of Uganda's surface area is covered by water bodies – lakes and rivers. Uganda is in the middle of what is often called the Great Lakes region of Africa. The Great Lakes are a series of lakes in and around the Great Rift Valley. Three of the five lakes provide natural boundaries for Uganda. The three lakes are Victoria, Albert and Edward. Besides these large lakes, Uganda has 160 small lakes.

Albert Nyanza, is 2,064 mi^2 (5,346 km^2) wide, located on the Congo (Kinshasa)-Uganda border. The lake is 100 mi (160 km) long and 19 mi (30 km) wide, with a maximum depth of 168 ft (51 m). It is located in the Western branch of the Great Rift Valley, 2,030 ft (619 m) above sea level. River Semliki and the Victoria Nile drain into Lake Albert and it is drained by the Albert Nile. The Albert Nile becomes the Bahr-el-Jebel when it enters Sudan. During the rule of President Mobutu Sese Seko, the official name of the lake in Zaïre (now Congo) was Lake Mobutu Sese Seko.

Edward Nyanza is 830 mi^2 (2,150 km^2) and is located in the Great Rift Valley. It is south of Lake Albert in the Rift Valley. The lake forms part of the boundary between Congo and Uganda. It lies at an altitude of 3,000 ft (910 m). It is 50 mi (80 km) long and has a maximum width of 30 mi (48 km). Lake Edward is connected to the Nile system by the Semliki River, which drains the lake in the north and flows into Lake Albert. Many fish and hippopotamuses abound on Lake Edward's southern shores.

Victoria Nyanza is the largest lake of Africa and the world's second largest freshwater lake. It is 26,830 mi^2 (69,490 km^2). Lake Victoria is 255 mi/410 km long and 155 mi/250 km wide. It occupies a shallow depression (250 ft/75 m deep) on the Equatorial Plateau (altitude 3,725 ft/1,135 m) between two branches of the Great Rift Valley. It has an irregular shoreline and many small islands. Numerous streams, including the Kagera River, drain into the lake.

Lake Victoria is the main headwater reservoir of the Nile River. The Victoria Nile River drains the lake. At Owen Falls Dam on the Victoria Nile the lake's waters are used to generate hydroelectricity. The lake basin is densely populated and intensely cultivated and the lake is an important fishery, but fish stocks and diversity have declined since the 1980s as a result of over fishing and the introduction of the Nile perch. Ships regularly call at lakeside towns, including Port Bell (Luzira), Entebbe, Mwanza, Bukoba, Jinja and Kisumu.

Lake Kyoga (Kioga) is 100 miles (160 km) long. It is formed by the Victoria Nile and located in central Uganda. It occupies part of the same depression as Lake Victoria, to which it was once joined. The shallow lake has large areas of papyrus swamp. Lake Kyoga used to provide transportation for a large cotton-growing region.

Uganda is part of the Nile basin. The Nile is the longest river in the world. It is 4,160 mi (6,695 km) long from its remotest headstream, the Luvironza River in Burundi, to its delta on the Mediterranean Sea in Egypt. The Nile flows northward and its basin areas is 1,100,000 mi^2 (2,850,000 km^2), about one tenth of Africa, including parts of Egypt and Sudan.

The Nile is an extension of the Luvironza River in Burundi (Langlands, 1975). The Luvironza flows into the Ruvuvu River which, in turn, is a tributary of the Kagera River, one of the principal headstreams feeding into Lake Victoria. It drains from the northern end of Lake Victoria at Jinja and flows generally north and west, over Ripon Falls and Owen Falls (both now submerged), through shallow Lake Kyoga and then over Kabalega (formerly Murchison) Falls to Lake Albert. Hydroelectric plants are located at Owen and Kabalega Falls. The river is navigable from Lake Albert to Kabalega Falls.

The first section of the river between Lake Albert and Lake Victoria is known as the Victoria Nile. It leaves Lake Albert as the Albert Nile and flows north to Nimule, where it enters Sudan and becomes the BahrelJebel. From Nimule to Rejaf is a zone of rapids.

Map 5.1: Population per borehole

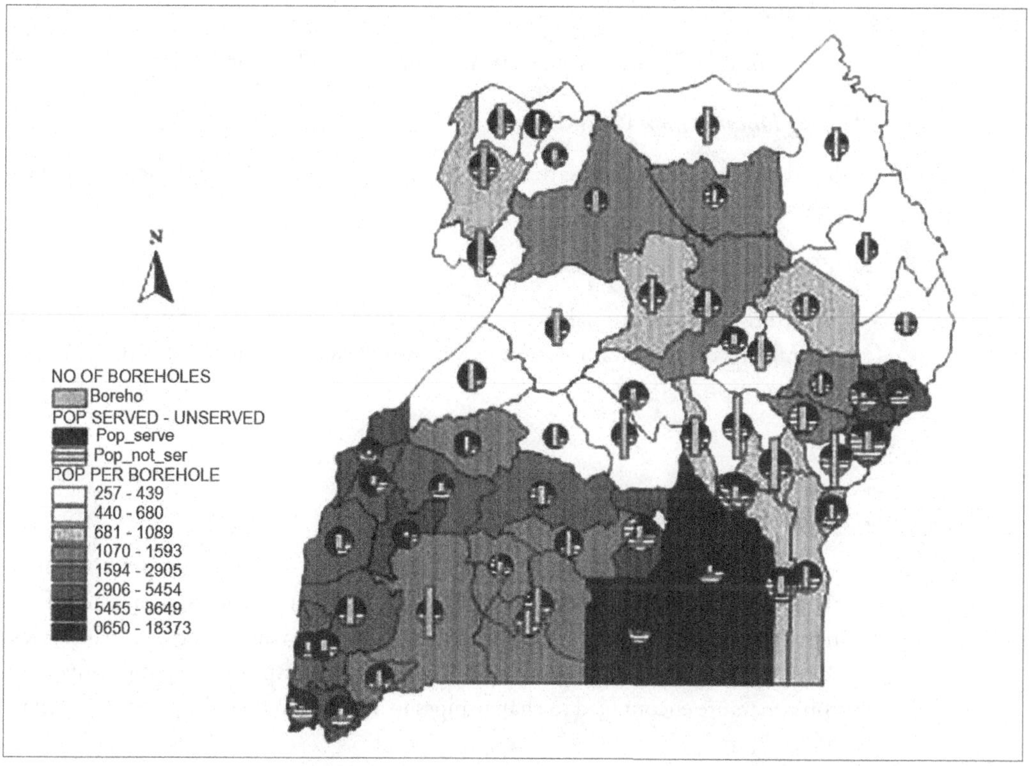

Source: Uganda Bureau of Statistics 2004

Borehole Sources

Boreholes are the basic underground water source. This source of water is common in places where there is no piped water. Therefore, borehole access is partly a function of economic well being of an area. Urbanized areas have access to piped water. The areas without access to pipe water sources require borehole facilities. The map above shows the distribution of boreholes by district, population per borehole, population served, as well as population not served. The central region, Karamoja, the Northwest and part of the eastern region are better served.

5.2.2 Water Resource Problems

1) *Droughts*

There have been periods in which precipitation is much lower and evaporation higher than normal. In water-shortage areas of Uganda, women and children travel/walk long distances to fetch water. Droughts often affect agricultural production. For example, it is not uncommon for cattle keepers to sell cheaply most of their animals in dry periods.

2) *Excessive Amount of Water*

Downpours in Uganda often cause flooding especially in urban areas.

3) *Long Distance from Water Sources*

Many villagers travel more than one mile to get water. Map 5.2 shows causes of contamination.

4) *Contaminated Water*

According to the National Water and Sewarage Corporation (NWSC), there are several causes of contamination of water between the source and the point of consumption.

The causes are

a) *Old water pipes to and within premises*

It is apparent that many water pipes leading to and within many premises from the NWSC main lines have been in place for more than 30 years. As a result of the metal aging, there may be leakage on the pipes. Such leakage could cause suck-back resulting in sucking in of waste water into clean water pipes. It is imperative that consumers are encouraged to change pipes to and within their premises every ten to fifteen years.

b) *Poor plumbing within premises*

Laying of pipes within premises, if poorly done, can cause a mix up of clean water and waste water. Such a mix up is a source of direct contamination of clean water. Consumers should be advised to employ professionals to do plumbing work.

c) Unclean roof tanks

In its annual report, the National Water Sewerage Corporation indicated that many households do not routinely or never clean their roof tanks, while some have even been found uncovered. The fact that there is no cover makes the tanks a water source for birds, lizards, etc. Clean water flows into the tank but once inside the tank contamination occurs. Consumers should be advised to clean and disinfect their roof tanks at least once a year and should ensure that the tanks remain covered at all times.

d) Water from non-pipe sources

Besides the NWSC water, there is water from other sources in and around Kampala, e.g, channels, springs and swamps. These sources are suspect as far as quality is concerned. When water containers used to carry or hold such water are later used to carry, or hold water from NWSC, there is a possibility of contamination. Where NWSC water is available consumers are advised to use it exclusively.

e) Additional fixtures to stand pipe taps

Communally used stand pipes are one metre high and as a measure to cub wastage of water many owners add a piece of horse pipe to form an extension from the tap. This may cause clean water coming from the tap to be contaminated as it flows through

the hose pipe which may have been handled by many unclean hands.

5.2.3 Hydropolitics of the Nile

In 1929, The British signed an agreement which gave exclusive rights of the Nile Basin waters to Egypt. In 1959, Egypt and Sudan signed a second agreement dividing the waters of the Nile Basin between the two countries. These agreements excluded the other eight countries of the Nile Basin (See map of the Nile Basin).

The Nile River is the longest international waterway in the world (6,695 kilometres). Its drainage basin covers ten countries (Rwanda, Burundi, Congo, Tanzania, Kenya, Uganda, Ethiopia, Eritrea, Sudan and Egypt). The Nile water supports all agriculture in the densely populated part of Egypt and provides water for 20% of Sudan's agricultural area. The Nile river is also used for navigation and generating electricity.

Egypt depends on the Nile waters. It has dominated the water usage through an agreement signed with Britain. The first agreement was signed in 1929 between Egypt and Sudan (represented by Great Britain). The agreement allocated forty eight billion cubic metres of water to Egypt and four billion to Sudan. From the early 1930s, Sudan gradually adopted irrigational agriculture and the demand for water increased. As a result, on the eve of its independence and following the Egyptian revolution in 1952, the administration in Sudan started demanding that the agreement with Egypt be renegotiated. After a period of bilateral tension, negotiations resumed and a new agreement was signed in 1959. Based on new calculations that put the annual runoff at eighty two billion cubic metres, Egypt has an allocation of 55.5 billion cubic metres, 18.5 billion being allotted to Sudan (the rest being lost to evaporation and swamp areas). The new agreement also include;d some provisions regarding the replenishment of the reservoir of the planned Aswan Dam.

The implication of the 1959 Agreement is that the eight countries cannot develop their water resources without the consent of Egypt. It is no wonder that those eight countries find this agreement unacceptable. In the last few years, there has been some cooperation. The Nile River Basin Initiative was established to deal with the issue of the Nile water. It is an international body with a mission to find common and agreeable management of the waters of the Nile Basin.

5.2.4 Managing Water Resources

Uganda has limited resources to develop the water storage and distribution systems required to increase supply. It also lacks the resources to harness the power potential from its rivers.

5.3 Fishery Resources

Water resources are important as a source of food. Fishery provides an important economic activity to a substantial section of the population. Most of the fisherfolk communities live on land that belongs to other people. Map5.3 below shows the fish catch by water body. Over 80% of the fish catch comes from Lake Victoria and Lake Kyoga. Lakes Albert, Edward and George are also important sources of fish.

The current fish resources base is made up of capture fisheries (artisana) and acquaculture (Ministry of Agriculture, 2004). There are several fish species in Uganda lakes, but the most important are Nile perch and Tilapia (see pie chart). The major commercial species are Nile Perch (Late niloticus), Nile Tilapia (Oreochromis nilotica), Mukene (Rastreneobola agenta), Bagrus docmac and Protopterus. Fish exports to the European Union (EU) countries, is now one of the leading nontraditional source of foreign exchange.

Figure 5.1: Different Fish Species in Uganda

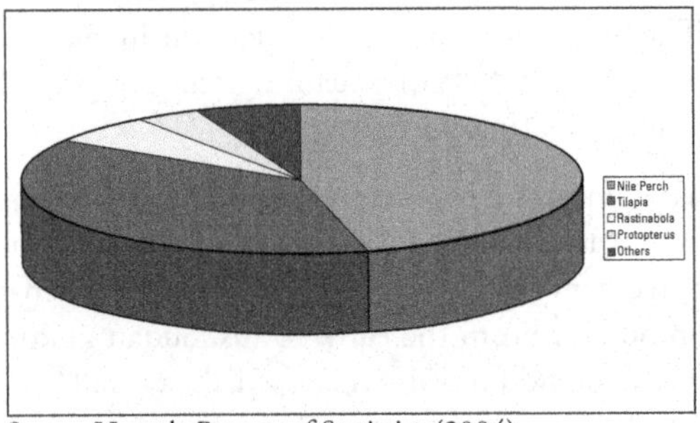

Source: Uganda Bureau of Statistics (2004).

The country is estimated to have a capacity to produce approximately 300,000 metric tonnes of fish per year. However, the maximum catch was 245,000 tonnes in 1990. The catch levels in the recent years have been rather low. Besides fisher men, this sector employs individuals in the fish processing, fish net making, boat making, fish equipment trade and fisheries research.

Fishermen on Lake Victoria near the source of the Nile

Source: Uganda Tourism Board

Map 5.3: Map Uganda Fish Catch

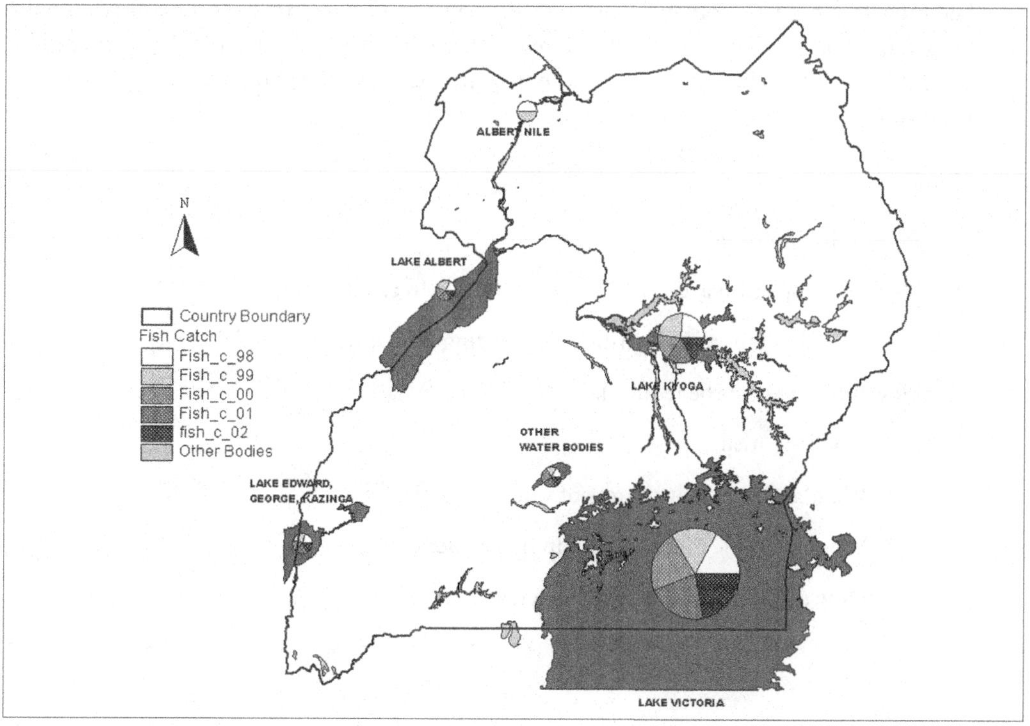

Source Uganda Bureau of Statistics 2004

The map shows fish catch from 1998 to 2002. The fishery industry is based on inland catch from lakes and rivers which cover 20% of the total area (Kaelin, 2002). There has been very minimal increase in fish catch over the years. But exports have steadily increased (see graph). Fish population in most lakes have declined (Ministry of Agriculture, Animal Industry and Fisheries, 2004), partly because of over fishing (East African Fishery, 2003).

Fisheries play an important economic role, contributing to food security, employment, income and foreign exchange earnings. The average annual per capita consumption of fish is approximately 10 kilograms. This represents 50% of animal protein intake of an average Ugandan diet. Ugandan fish exports are the second largest export and foreign exchange earner. Uganda earned over 45 million dollars in 1996. It is estimated that about 1.5 million people are directly and indirectly employed in fisheries related activities. However, the industry experienced a decline after 1998 following the EU ban of fish exports to its market.

5.4 Conclusion

Uganda has very rich water resources. This is due to the fact that the country is at the centre of the Great Lakes region and it receives good rainfall. The lakes are an important source of food (fish), employment, biodiversity and water. The rivers are navigable in some parts and the water falls provide or have the potential to provide hydro power. The country has a substantial amount of underground water resources, but this resource has yet to be exploited to full potential. The country's inability to fully develop its water resource, explains the poor safe water provision in rural areas.

Questions

1. What is hydropolitics?
2. How do population activities increase harmful effects of drought?
3. How does development contribute to flooding problems in Uganda?
4. How can the effects be reduced?
5. In your village area:
    a) What are the major sources of water supply?
    b) What are some of the water supply problems?
    c) How do people travel to get water?

References and further readings

Beadle, L. C. 1981. *Lakes Edward and George, Inland Waters of Tropical Africa* (ed. Beadle, L. C.) 231-244., Longman: London.

Kaelin, A, 2002. Outline of the Path Forward In Uganda's Fisheries Sector, USAID Report.

Ministry of Water, Land and Environment. 2001. *National Rural Water Coverage Atlas*, Kampala, Uganda.

National Water and Sewarage Corporation. 2004. Annual Report.

Souther, L. 2005. Contamination of Groundwater-fed Springs in Uganda, webpage Article.

Swain, A. 2002 . *The Nile River Basin Initiative: Too Many Cooks, Too Little Broth*, SASI Review 22.2, 293-308.

The Columbia Encyclodedia, Sixth Edition. 2001-2005.

World Lakes Database. 2005. Lake Albert, www.ilec.or.jp/databese/afr.

World Lakes Database. 2005. Lake Edward, www.ilec.or.jp/database/afr.

________________ Lake George.

________________ Lake Victoria.

SECTION 2: WETLANDS RESOURCES

5.5 Introduction

Wetlands are today recognized as one of the most productive yet fragile ecosystems. They normally occupy the transitional zone between the wet and the generally dry environments, or are transitional between terrestrial and aquatic systems where the water is usually at, or near the surface. Wetland ecosystems are composed of a number of physical, biological and chemical components such as soil, soil minerals, water, plants and animal species. It is the processe among and within these elements, that make wetlands such productive ecosystems. They do not only generate a diversity of products, but also perform a wide range of ecological functions, including flood control, storm protection, regulation of the underground water table and waste water sequestration. It is the highly productive nature of these ecosystems that has attracted a diversity of land-uses and conflicts and hence the observed threat that they face globally.

Wetlands and their benefits, are now under pressure both in rural and urban areas. In urban areas, especially Kampala, wetlands were the last "free", or "cheap" areas for infrastructure development despite their designation as "green corridors". In the rural areas, small but continuous 'nibbling' at wetland edges has reduced the size of wetlands.

5.5.1 Wetlands in Uganda

Uganda is widely recognized as one of the most advanced countries in Africa and the world, in terms of wetland management. Wetlands represent one of the most vital economic resources of the country. The services and products provided by wetlands contribute hundreds of millions of US$ per year to the Ugandan economy. Without wetlands Uganda would be devastated (MWLE, 2001). Wetlands cover about 30,105km2, representing 13% of the total area of Uganda. The area under permanent wetlands is currently estimated at 7,296km2 and seasonal wetlands, 22,809km2 (NEMA 2002). At least 69% of the total area under wetlands comprises impeded drainage, while swamps constitute 30% and swamp forest, 1%.

Chapter XIII of the Constitution of the Republic of Uganda states that: "The State shall protect important natural resources, including land, water, wetlands, oil, minerals, flora and fauna on behalf of the people of Uganda." It is from here that the National Environment Statute 1995, The Local Government Act 1997, The Water Statute 1995, The Land Act 1998 and The Wetland Policy 1995 were developed to guide the people of Uganda in managing its wetlands. The Wetland Sector Strategic Plan (WSSP) whose aim is to provide a clear sense of purpose and direction and a supporting action framework for all those who will be involved in wetland management and conservation over the next ten years was developed, under the Wetlands Inspection Division (MWLE, 2001).

Although Uganda's wetlands are protected by the National Environment Statute (NES) of 1995, most of them are still being reclaimed and degraded, especially those outside protected areas. The following are some of the known forms of encroachment: conversion/ drainage of wetlands for agricultural purposes, excavation in the form of sand-mining and extraction of clay for brick making, dumping of waste, deforestation of swamp forests and rampant swamp fires. Deforestation of swamp forests for wood and other crafts is rampant in wetlands of Mukono, Masaka and Sango Bay. These forests play a vital role in protecting Lake Victoria waters from land-based sources of pollution.

The Ramsar Convention to which Uganda became a signatory in 1987 addresses issues of the loss and degradation of wetlands by designating sites of international importance and making wise use of them. The convention is being made effective to address wetland issues both at national policy level and at the level of individual Ramsar sites like Lake George.

5.5.2 Wetland usage in Uganda

The determination of land use in Uganda has been complicated by over lapping state bodies following the enactment of the National Environment Statute (NES) in 1995. Most land use zoning and permitted development in Uganda is done under the Town and Country Planning Act which predates the Environmental Statute, which empowers the National Environmental Management Authority (NEMA) to determine wetland usage. The Environmental Law, clause 47 in particular, restricts development on wetlands and this is in conformity with clause 4 of the Land Act 1998, which states that; "use of land should conform to the provisions of the law relating to the Town and Country Planning Act and other laws."

Wetlands have intrinsic attributes: they perform functions and services and produce goods. Some of these are of primarily local interests, but others have a regional, national, or international importance. Together they represent considerable ecological, social and economic value. Ecological attributes and functions are those that wetlands perform without immediate human intervention, e.g., maintenance of water table; flood control; habitat for animals and plants. Many people are unaware of the existence of wetlands, let alone the importance of their attributes and functions. The socio-economic goods and services are better known, as they involve immediate human interaction with the wetland. Human activities based on natural wetland resources generate a wide range of products that are consumed locally, or traded internationally. Many people living in communities neighbouring wetlands are to a significant extent dependent on wetland products for their well being (MWLE, 2001).

The surface area covered by wetlands is vast. Issues relating to the ownership and use of wetlands are many and complex. There remains, despite successful awareness campaigns, a high level of silent resistance to policies of wise use, where, for example, individuals perceive potential for personal gain from exploitation of wetlands. Structures and mechanisms for enforcing laws and regulations concerning wetlands are weak. This is exacerbated by the large surface area covered by wetlands and even more so, the extent of wetland edge. Contradictions and tensions still exist where policy and legislation affecting wetlands are concerned. For example, in relation to ambiguity surrounding the concept of government or local government holding wetlands 'in trust for the people' and confusion over rights and obligations of ownership, on the one hand and management on the other. There are still gaps in knowledge of wetlands-their nature and extent, their functions and values-despite great advances over the last ten years (MWLE, 2001).

5.5.3 Defining Wetlands

Much as wetlands have attracted a lot of interest globally, a universally accepted definition has been difficult to come by. The Ramser Convention of 1972 nonetheless provided a working definition that has now received international recognition. The convention defines wetlands as:

" Areas of marsh, peatland or water, permanent or temporary with water that is static or flowing, fresh, brackish or salty, including areas of marine water the depth of which at low tide does not exceed six metres."

This rather descriptive definition has served to present a tangible entity on the concept of wetlands.

In Uganda, wetlands are commonly known as swamps. They are taken to refer to areas with characteristic plant communities and soil types that have developed in association with permanent or temporary flooding. The very nature of this definition has presented considerable difficulty in demarcating wetland areas, the wetland inspection department acknowledges that wetlands and uplands can be distinguished on their ecological characteristics (soil type, vegetation and water logging) but it must be recognized that the boundary between them is mobile both seasonally and inter- annually.

5.6 Area, Distribution and Classification of Uganda's Wetlands

Wetlands in Uganda cover almost 30,000km2, or about 13% of the total area of the country. They include; areas of seasonally flooded grassland, swamp forest, permanently flooded papyrus, grass swamp and upland bog. As a result of the vast surface area and the narrow river-like shape of many of the wetlands, there is a very extensive wetland edge, which is in many places accessible only on foot (MWLE, 2001).

The aerial coverage of Uganda's wetlands is still a subject of contention. But what is clear is that a decline is occurring in almost all districts. Computations from the National Biomass 1990-1992 imagery put the total area of wetlands at 30,105km2 representing approximately 13% of Uganda's total area. Of this total, 8,840km are the central region, 8,547km in the eastern region and 7,065 km in the northern region and 5,645 km in the western region. Wetlands are widely distributed throughout the country with each district recording the existence of wetlands.

The classification of wetlands is a rather difficult task. Numerous systems exist ranging from simple typologies, with fewer than 20 or 30 categories, to complex hierarchical systems with hundreds of wetland types at the lower levels. The national wetlands management programme (1991) adopted a simple and general classification. The wetlands of Uganda are grouped into two broad categories: the lucustrine wetlands (wetlands associated with lake environments) and riverine wetlands (wetlands associated with rivers). Under the lucustrine category include;d the Kyoga/ Kwania complex, Lakes George, Edward and Albert wetlands, the Bunyonyi lake\swamp complex, the Bisina and Opeta, Wamala and many other small wetlands. The riverine wetlands are quite numerous traversing the country in almost all the districts. All rivers and streams are lined with swamps of varying magnitude. Notable among the riverine wetlands are the Okole and Kafu systems. A further differentiation of these categories based on altitude has been undertaken with the following classes emerging:

- Mountainous swamp bogs and mires occurring above 3,000m on Mt. Rwenzori and Elgon
- Valley peat swamps and upland swamps (1,900-3,000m). The valley peat swamps include; the Kabale wetlands while the upland swamps include; the swamp forests of Bwindi, the various papyrus swamps and the sedge dominated at high altitudes.
- Permanent swamps: these occur in the relatively lower and permanently flooded areas. They are characterized by *Cyperus papyrus*, segdes, *tyha,* swamp grasses and swamp forest.
- Seasonal wetlands and temporary pools: these are mainly found in the relatively flat and drier areas of the extreme East and Northern Uganda including the districts of Tororo, parts of Pallisa, Kumi, Soroti, Kamuli, Nakapiripirit, Katakwi, etc.

Another attempt to classify Uganda's wetlands was undertaken by the wetland biodiversity inventory team. The Mbeiza and Mutekanga modification of Dugans` 1990 classification recognizes 12 categories of wetlands based on dominant vegetation communities.

A) Freshwater emergent reed swamps typically dominated by a single reed species.

- The first category here are papyrus swamps which are further subdivided into numerous smaller types. Under this category are Lacustrine swamps which are floating, or are anchored on firm soils. Lakes Victoria and Kyoga wetlands fall in this category.
- A second sub-type of the papyrus swamp wetland is the riverine and associated flood plain swamps with rooted papyrus for example the River Nile and Mpologoma wetlands.
- The third subcategory of this class are the valley swamps that are occupied by rooted papyrus, include;d among these are the wetlands in south western Uganda, the Ruhuhuma swamps being one of them.

Other wetlands in this category as classified by Mbeiza include; the vossia Swamps of the Kazinga channel, Albert Nile, the Phragmites swamps of River Semliki and Lake Kanyamukali, the Cladium Swamps of lower Mobuku Valley, the wetlands near Lake George and lastly the Typha swamps, such as the Lumbuye.

B) Freshwater floating leaved but rooted vegetation which is dominated by Nymphaea and Potamogeton, e.g. the komunuo wetlands.

C) Freshwater submerged and rooted macrophytes communities dominated by *Ceratophllum* spp, *Lagarosiphon* spp, *Hydrilla* spp, *Vallisneria* spp and/or *Ottelia.*

D) Freshwater submerged but not rooted marophyte communities dominated by *Ceratophyllum* spp and *Lemna trisulca.*

E) Freshwater floating communities dominated by *Pistia stratiotes, Lemna* spp *Azolla* spp and *Eichhornia crassipes* these can be found around the Albert Nile.

F) Seasonally Flooded Herbaceous wetlands.

This category has varied species. These include; the *Echinochloa- panicum repens* and *Cynodon* swamps of Lakes Opeta and Bisina, the *Loudetia- Cynodon-Setaria* swamps of Lake Nabugabo, the *Cynodon-Setaria-Hyparrhenia-Brachiaria* wetlands around Lake George and the *Oryza* wetlands of Okole River.

G) Seasonally Flooded Wooded Grasslands

These are mainly characterized by *Acacia- Hyparrhenia,* for example, the wetlands of River Kafu and the lower Karamoja plains.

H) Freshwater Palustrine forests.

These exist in a number of subcategories and are basically swamp forests. These include; the permanent swamp forests dominated by *Mitragyna, Alchornea* and/or *Syzigium* on the Ssese Islands and in the Ishasha valley, the *Spondianthus* Sango Bay wetlands, the *Xylopia* Bukakata swamps, The *Phoenix* or *Raphia* swamps on the shores of Lake Victoria, the *Calamus* wetlands in Mabira forest, *Eleospathain* Semliki, the *Pandanus* along the Dura River and Budongo Forest. The *Uapaca* wetlands of Ssese Islands and Zika Forest and the *Ficus* of the Northern parts of Lake George. This category also include;s seasonal swamp forests dominated by *Macaranga, Croton, Pseudospondias, Sapium, Ficus or Baikeaea-podocarpus* forests of Sango Bay.

I) Freshwater riverine forest

These are known to be dominated by *Acacia, Ficus, Combretum, Zizyhus, Phoenix, Pseudospondias, Erythrina* and *Alchornea-Pseudospondias.*

J) Freshwater montane wetlands

These are commonly known as Bogs. This is a category that is dominated by a diversity of plant species including Sphagnum on the Rwenzori Bogs and Budongo swamps and Rukiga highlands, Sedges in Muchoya and Bujuku Bogso on the Rwenzori and the Echuya swamps of Kigezi.

K) Permanent saline wetlands

These are dominated by *Cyperus laevigatuand,* they include; the wetlands of Lake Katwe, Lake Bunyampaka and the Sempaya springs.

L) Seasonally flooded saline herbaceous wetlands:

These have a diversity of plant communities the most notable ones being *Sporobolus pyramidalis-Sporobolus spicatus-sporobolus* and *robustus-cynodon.* These are commonly found around Lake Munyanyange.

Using this classification, the government has attempted to provide a framework for the management of wetlands in Uganda;

However, it could be improved by adding soil types. To the criteria. This would provide better information for management and wetland conservation.

5.7 Trends and current status of Uganda's wetlands

There has been a marked decline in the coverage of wetlands, statistics computed from the Biomass imagery of 1992 revealed that approximately 2,376km about 7.32% had been converted. High rates of conversion have been witnessed in Jinja, with 43% of the original wetland area being converted. Other districts with high conversion of wetlands are Kisoro 40.3%, Kabale 36.6%, Iganga, Bugiri, Mayuge 32.7%, Tororo-Busia 32.2%, Pallisa 26.6% Kamuli 22.6% and Kampala 19.7%. With increased economic activity and population growth, wetlands are facing a very uncertain future, it is quite difficult to reverse the current trends especially in districts experiencing high population growth and rapid urbanization.

In districts like Kamuli, Kabale, Pallisa, Tororo, Iganga, Kumi Kisoro, the lack of alternative income generating activities will continue to decimate wetlands. In the rice growing areas, declining fertility on uplands has left wetlands the only productive areas and rice cultivation the only viable economic venture. In the Kigezi highlands, the combined forces of population pressure and inequitable land tenure have served to accelerate wetland degradation. Much of the wetlands have been individualized and turned into private lands. Where communal land still exists, the limited knowledge on regulations and limitations on wetland ecosystems have contributed to their misuse and subsequent degradation. In the area where some land still exists, the pressure has not yet reached alarming levels. Masindi, Hoima, Arua, Yumbe, Kitgum and Ntungamo are some of the districts with delicate ecosystems which are relatively less affected.

The recent enactment of a series of policies and legislation related to wetlands is bound to secure the future of Ugandan wetlands. The 1995 Constitution of the Republic of Uganda, the Local Government Act, the Water Statute 1995, the 1998 Land Act, the Wetland policy, the National Environment Management Policy all together are expected to guide sustainable wetland resource utilization. Added to these instruments are the current wetlands, river banks and lake shores management guidelines of the year 2000.

5.8 Wetland values, production and livelihood

Wetlands provide a number of useful goods, services and attributes. Those along the shore of L. Victoria help to maintain the quality of water in the lake. The Nakivubo wetland, for instance, provides a free service to the country through tertiary treatment of sewage effluent and run-off from industrial and residential areas of the country. However, despite the role of wetlands, there has been a dramatic change in the water quality and ecology of L. Victoria over the past 30 years. Another example is of the banks of River Mayanja; the lake shore wetlands and those along the river protect destruction from storm-water erosion, while other in-land wetlands store floodwater from nearby hills. These wetlands form a natural water distribution system and at the same time help to improve water quality. In addition, many of these wetlands are sources of water for domestic use and are particularly important in areas with no piped water. Almost every wetland in this area has a well on its fringes. As wetlands around River Mayanja continue to be degraded, the recharge and discharge capacities of wetlands are gradually being lost, leading to the drying up of adjacent wells.

Communities perceive wetlands as sources of direct benefits (craft material, sand, clay, water and agricultural land) but fail to appreciate the ecological services and other non-tangible value of wetlands. Wetlands have been identified as water sources for communities to use for domestic and livestock. However, this water is neither quantified nor given an economic value. Conversion of wetlands into agricultural land is the main threat to their survival. After a number of studies, agriculture was identified as the major activity in wetlands and all studies realized that agriculture is not fully compatible with wetland conservation.

Wetlands are no longer regarded as wastelands. Most studies on the socio-economic aspects of wetlands have mainly concentrated on the consumptive use of wetlands without an evaluation of the socio-economic sustainability of the different resource utilization practices. Furthermore, there are gender differences in the utilization of wetlands. Traditionally, women depend on wetlands more than men and therefore, the degradation of wetlands will affect women more than men. The socio-economic benefits of wetlands are better understood as they involve immediate human interaction with the wetland. Human activities in wetlands generate a wide range of products which are used locally, or traded internationally. Most of the socio-economic values are essential for the well being of local communities adjacent to the wetlands. Ugandans interface with wetlands on a regular basis and the resources in the natural wetlands contribute directly and significantly to their sustenance.

As the population and the people's expectations increase, the pressure on wetlands and their resources also increases. Where there is poverty, meeting short-term, immediate, personal needs (such as food, water, shelter, school fees, etc.) may take precedence over protecting attributes that provide long-term, indirect, general benefits (such as water storage and recharge, micro-climate regulation, biodiversity conservation). In the rural areas, small but continuous nibbling at the edges has reduced wetland areas, but this is mainly restricted to seasonal wetlands. In the east, however, almost all seasonal wetland valley bottoms fit for rice cultivation have been converted to that use and in some parts of south west, large areas of wetlands have been converted to pasture for grazing or cultivation. Nevertheless, the damage to permanent wetlands in rural areas is probably still relatively limited. Some of the natural protection from encroachment and wholesale drainage is afforded by inaccessibility and lack of suitable drainage technology (MWLE, 2001). Wetland management is a difficult and time-consuming task, which requires effective communication and co-ordination among wetland users if hydrological regimes are to be managed to support a range of activities such as crop production and reed collection.

5. 9 Threats to wetlands resources in Uganda

Wetlands in Uganda continue to face a serious threat of total destruction with high population density as one of the main causes of destruction. For example, in Kampala, it is estimated that about three quarters of the wetland has been significantly affected by human activity. In as much as the remaining wetland area is small, an accelerated decline could occur if sensible management is not undertaken. The main threats to wetlands are associated with unplanned reclamation and excessive conversion. Wetlands are also under extreme pressure due to uncontrolled development activities, the most detrimental activity being industrial and residential development, especially in urban areas.

1) Lack of coordination and planning in the allocation and development of plots. Some officials have approved development in wetlands, without regard to any plan and no environmental report. In cases where wetlands get totally blocked, storm water will spill over onto roads and other corridors, causing road foundations to collapse and culverts and underground sewage systems to get blocked. As a result of the lack of coordination, enforcement mechanisms become a problem, in that even the various existing laws and regulations usually appear contradictory, vague and lack the necessary statutory instruments for implementation. Consequently, it is often impossible to charge developers without running the risk of losing the case and end up paying damages to the developers.

2) Wetland lease procedures that are concluded now were started long before the inception of the NES and other legislative documents. Although no new leases in wetlands are supposed to be given out, the issuing of leases continues for those that were already in the pipeline. Therefore, many structural developments that are criticized today are the result of bureaucratic procedures started years ago. Such developments, although dangerous, cannot easily be stopped

due to political patronage. In many cases of wetlands abuse in Uganda, the abusers are well connected to the people at the top in government.

3) Unplanned drainage of wetlands for agricultural production. Most seasonal wetlands and permanent ones with shallow waters have been intensively drained and cultivated. This is a common practice in areas where rice cultivation has replaced the traditional cash crops in Iganga, parts of Kamuli, Tororo, Pallisa, Kumi and Bugiri districts. The permanent wetlands with shallow waters are being drained for sugarcane growing. This has affected, particularly, parts of Iganga, Mayuge, Mukono and Mpigi districts. In Kabale, Kisoro, Mbale, Sironko and Kapchorwa districts wetlands have been drained for vegetable growing and dairy farming.
4) As indicated earlier, population growth has led to increased pressure on wetlands and consequently failure to perform their major functions, namely, sustain human livelihoods and upholding the quality of the environment.
5) Drainage or modification of wetlands (especially in south west and eastern Uganda), has led to loss of valuable resources. Previous government policies allowed utilization of wetlands without clear guidelines. New schemes (e.g. rice) have tended to engender adoption of massive drainage and have led to adverse effects. In the process that policy was found unsustainable.
6) Rapid population growth and increasing rate of development require a sufficient supply of water and effluent discharge at an affordable cost. Many urban settlements including Kampala city are dependent on wetlands for water supply. These effluents cause an environmental disturbance on the wetland ecosystem.

5.10 Wetlands legislation and institutional arrangements in Uganda

Legislation is the highest form of policy articulation. It should not only echo government policy but also be a tool for implementing the policy. Wetlands have been marginalized and regarded as 'wastelands'. They, therefore, need strong government institutional arrangements and a sectoral national legislation in order to stop the high rate of degradation and ensure sustainable management. Since wetlands are a multi-sectoral resource, there is need to create and establish an appropriate institutional arrangement for their management.

Management plans ebbed in the Wetland Strategic Sector Plan (2001-2010) have include;d the local population and activities. Wetland reserves are only allowed between 30 metres -50 metres from the cover and boundaries are set for different uses such as cultivation, cattle watching, fishing, papyrus harvesting or washing.

The specific strategies that were developed for use in wetlands legislation are:

- Enacting a national law for regulating the management of wetland resources
- Encouraging district authorities to make by-laws for the proper management of wetlands
- Disseminating the broad guidelines provided herein, to district and urban authorities, as well as wetland users, researchers and academic institutions, etc.
- Establishing an inter-ministerial policy implementation institution.

Nakivubo Bay: A wetland near Kampala City

Source: NEMA

Mabamba wetland now used for brick making

Source: NEMA

Waters in Uganda

Source : NEMA

In 1995, a National Policy for the Conservation and Management of Wetland Resources was formulated, with the overall aim of promoting the conservation of Uganda's wetlands in order to sustain their ecological and socio-economic functions for present and future generations. The policy emphasized:

- No drainage of wetlands unless more important environmental management requirements are in place
- Sustainable use of wetlands to ensure that benefits from wetlands are maintained for the future
- Environmentally sound management of wetlands to ensure that other aspects of the environment are not adversely affected
- Application of environmental impact assessment procedures on all activities to be carried out in a wetland to ensure that development is well planned and managed
- Equitable distribution of wetland benefits

The implementation strategies call for stopping net drainage and destructive wetland uses and aims to ensure the utilization of wetlands such that traditional benefits are conserved in biologically diverse source water or discharge areas. Management strategies further place wetlands under public resources that are managed by the government and prohibit the leasing of wetlands. The communal use of wetlands is permitted but also regulated. Generally, the policy appears to be comprehensive in terms of conservation, though it is rigid in restricting wetland resource exploitation. No institution appears to have been mandated to enforce the wetland policy.

Since wetlands represent one of the most vital economic resources the country has, Uganda can achieve poverty alleviation through sustainable tourism development. Poverty and the prevailing high level of dependence on wetlands has forced many people who live around the L. Victoria region to over-exploit the wetlands resource base. Several laws, policies and legislation have been passed, but, many wetland issues are still multi-sectoral, hence the conventional sectoral management which is of little help to sustainable resource use. Many of these laws have created a lot of conflicts, overlaps and rivalry between and within sectors leaving some areas unattended to. Wetlands management requires a clear understanding of the stakeholders' attitudes towards the wetlands. Policies and laws therefore must reflect the aspirations of society and must be enforced in such a way as to benefit society.

5.11 Strategies for effective wetland management

1. Close collaboration is needed at all levels: national, district and local. This will ensure effective compliance monitoring of wetlands resources.
2. There should be no drainage of wetlands unless more important environmental management requirements supersede.
3. Sustainable use to ensure that the benefits of wetlands are maintained for the foreseeable future.
4. Environmentally sound management of wetlands to ensure that other aspects of the environment are not adversely affected.
5. Integration of economic, social and ecological aspects in wetland management plan. Equitable distribution of wetland benefits to all people in Uganda.
6. Implementation of environment impact assessment procedure on all development activities sited in wetlands. EIA's will be carried out to ensure that wetland development is appropriate, well planned and managed for long term sustainability.

Conclusion

Wetlands in Uganda represent very important natural resource as an ecosystem. They represent considerable ecological, social and economic value. Wetlands are recognized as fragile systems which are under pressure arising from infrastructure development, human settlements, industrialdevelopment and agriculture.

This has resulted in a marked decline in the coverage of wetlands. Wetlands in Uganda are regarded by communities to be of consumptive use without evaluating their socio-economic sustainability. Thus the damage inflicted on Uganda's wetlands requires coordinated management with special emphasis on laws and regulations that can regulate their use on a sustainable basis.

Key Terms

Term	Definition
Wetlands in Uganda	Areas seasonally flooded grassland, swamp forest, permanently flooded papyrus, grass swamp and bog
EIA	Environmental Impact Assessment
NEMA	National Environment Management Authority
NES	National Environment Statute
Wetlands Strategic Sector Plan	A document to guide all stakeholders in wetland usage and management in Uganda.

Questions

1. Discuss the various ways in which Uganda's wetlands benefit the population.
2. Make an assessment of threats to Uganda's wetlands.
3. Examine the challenges facing wetlands management and conservation in Uganda.
4. To what extent is the policy and legislation in place appropriate for sustainable use and conservation of wetlands?

References and further readings

Ministry of Water, Lands and Environment, *The Wetland Strategic Sector Plan* 2001-2010. Wetlands Inspection Division, 2001.

Sengendo, H et al. *Environmental Planning, Policies and Politics in Eastern and Southern Africa*, in M.A Mohamed Salih and Shibru Tedla (editors)

Kampala City Council, Strategic Reform Report. 1998.

Government of Uganda. 1995b. *"National Policy for Conservation and Management of Wetland Resources"*, Report of the National working Group No.2 on *Management of Water Quality and Land use including* Wetlands, Kampala.

Government of Uganda. 1995c. *"Statute No.4"*, The National Environment Statute, Kampala.

Chapter 6

POPULATION GEOGRAPHY

DEMOGRAPHIC CHARACTERISTICS AND TRENDS IN UGANDA

FREDRICK TUMWINE

6. 1 Introduction

This chapter examines the population geography of Uganda. Population geography is the study of spatial variations in the distribution, composition, migration and change of populations. Population is constantly changing because of dynamic processes such as deaths, births and population migration. Net migration is a product of in-migration less out-migration from a country or place. Birth, death and migration are called "vital" processes, since they are the means by which population replenishes itself and remains in existence. Growth, decline, or maintenance of the population makes this section of geography a dynamic one.

Population geographers focus on the following:

1. Size of the population; the number of persons.
2. Distribution; the arrangement of the population in space at a given time.
3. Structure; the distribution of the population according to sex and age.
4. Change; the growth, or decline of the total population.

The growth of population is influenced by fertility, mortality and migration. The high fertility of Uganda (6.9 TFR) is responsible for the broad age composition of the population. The population structure is broad based due to an increase in the child population (0 – 4) age (Census 2002). This creates a high dependency burden since more children need to be looked after by the relatively small "adult population".

Mortality exerts tremendous influence on age composition and structure of the population. Current data in Uganda shows drastic changes in mortality in intermediate ages due to the HIV/AIDS pandemic.

Migration is another determinant of population change. From time immemorial, populations have migrated due to various reasons. These reasons range from political, economic, to socio-cultural factors. In many cases, the youthful and energetic population constitutes the migration streams.

Migration influences the population structure of both the area of origin and that of destination. In Uganda, many people have migrated from over-populated districts in the South-Western part of Uganda (i.e. Kisoro, Kabale, Rukungiri and Kanungu) to less populated districts such as Kamwenge, Kyenjonjo, Kibaale and Masindi. Rural-Urban migration has also been pronounced in Uganda. On the other hand, many educated Ugandans have been leaving Uganda to work in many parts of world such as Europe, U.S.A. and Japan.

The direction and magnitude of the above factors can be shown by the component method which uses the equation below:

$$P_{t+n} = P_t +_n B_t -_n D_t +_n M_t$$

where;

P_t = refers to population size at given time.

"t" = e.g. Uganda had a population of 12.6 million people in 1980.

P_{t+n} = refers to population at a time "t + n" where "n" refers to the period between two population sizes usually shown in census years, Uganda had a population of 17.5 million people in 1991.

${}_nB_t$ = refers to the total number of births between the two periods where the duration of the period is denoted as "n" usually in completed years e.g. between 1980 and 1991, "n" = 11.

${}_nD_t$ = refers to the total deaths that occurred between the two periods by death between 1980 and 1991 in Uganda.

${}_nM_t$ = denotes net migration in a country between time "t" and "t+n" where "n" is the number of years. It is calculated as the difference between immigrants and emigrants, i.e, I – E during the "n" period.

Uganda's population is expected to double in every 24 years. According to Uganda Population and Housing Census of 2002, the total population was 24.7 million and it is expected to grow to 48 million in 2025 and 84.1 million in 2050. The short doubling time (24 years) is attributed to the high population growth rate of 3.4% per year. This is mainly due to the high fertility rate (6.9 TFR).

Map 6.1: Average Growth Rate by District

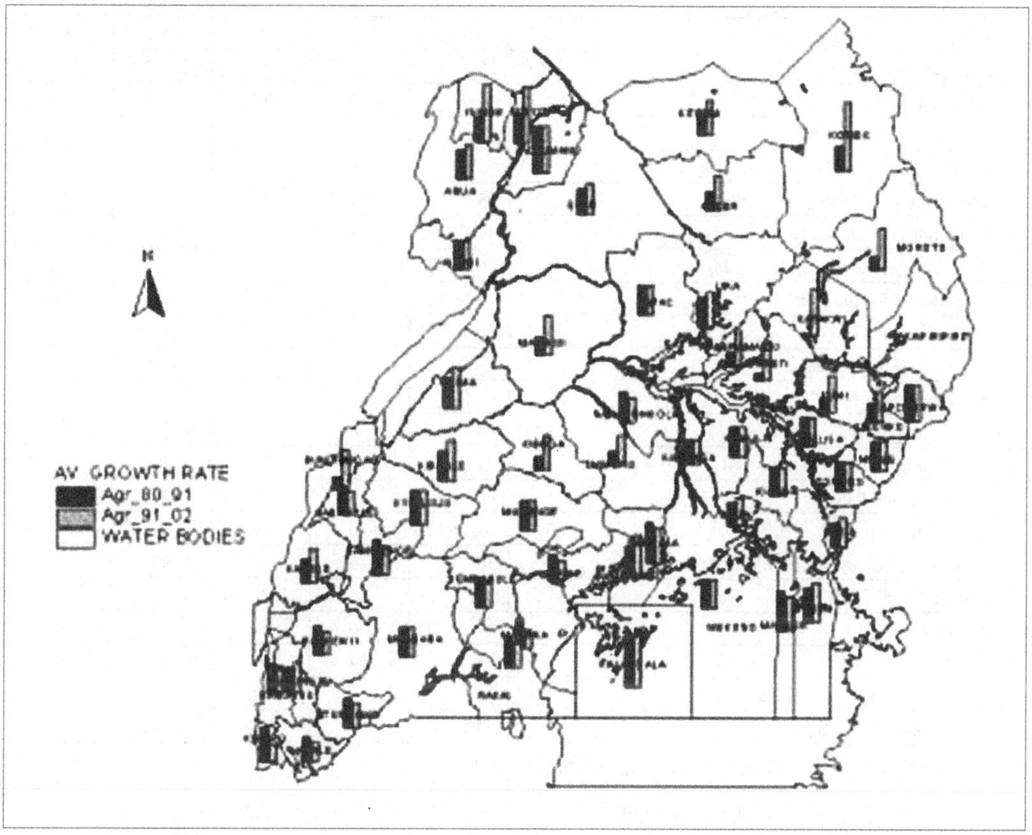

Source: Uganda Bureau of Statistics, 2004

The average growth rate is high. This is attributed to high total fertility rate (TFR) and a decline in infant and child mortality nine million in 2003.

In order to reduce the population growth rate, Uganda needs to put a lot of emphasis on modern methods of contraception. Total contraceptive prevalence rate is 23% with 18% using modern methods (Uganda Population 2004).

Mortality rates have remained abnormally high despite some slight declines. In fact general mortality declined from 122/1000 in 1991 to 88/1000 in 2002, child mortality declined from 180/1000 to 152/1000 in the same period. Maternal mortality improved from 700/100,000 in 1991 to 505/100,000 in 2002 which is still abnormally high. Life expectancy has declined from 48 years in 1991 to 43 years in 2002 mainly because of the impact of HIV/AIDS pandemic.

Table 6.1: Selected indicators of Uganda's population

INDICATOR	1990	1991	1995	2000	2002
Total Population (millions)	30.6	16.7	19.3	22.2	24.7
Female Population (Millions)	15.7	8.5	9.8	11.2	12.6
Male Population (Millions)	14.9	8.2	9.5	11.0	12.1
Population Growth Rate (%)	3.2	2.5		2.9	3.4
Total Fertility Rate (births per woman)	6.7	7.1	6.9	6.9	6.9
Maternal Mortality Ratio (per 100,000)	435	700	506	504	505
Births attended by Trained Personnel (%)	42	38	38	38	-
Infant Mortality Rate (per 1,000 live births)	76	122	81	88	88
Under 5 Mortality Rate (per 1,000)	137	180	147	152	152
Life Expectancy at Birth (in years)	50.4	48	-	43	43
Average Age at First Marriage	-	17.5	17.5	17.8	-
Average Age at First Birth	-	18.5	18.6	18.7	-
Total Contraceptive Prevalence rate (%)	24	5	15	23	23
Unmet Need for Family Planning (%)	41	52	29	35	-
Full Immunization Coverage (%)	46	31	47	38	-
HIV Prevalence Rate (%)	6.4	30	14	6.1	5.0
Pop Without Access to Safe Drinking Water	36	74	58	40	-
Stunted Children Under Five Years (%)	38	45	38	39	-
Poverty Level (%)	31	56	44	35	38
Literacy Rate (%)	-	54	62	68	-
Primary School Enrolment (millions)	-	2.3	2.6	6.8	7.1
GDP per Capita (in US$)	**370**	**251**	**330**	**350**	**350**

Poverty/level figures 56 and 44 are for 1992 and 1997 respectively.

Primary school enrolment figure 2.3 is for 1989.

Source: Census 1991, UDHS 1995, UNHS 2000 & UDHS 2000 and Uganda Human Development Report. 2000.

State of Uganda's population 2002. Ministry of Finance, Planning and Economic Development, Crane Chambers

Source: Population Secretariat MoFPED

6.2 Population Density

Table 6.2 shows that Uganda's population densities have been increasing since 1948. This increase is due to the high rate of population growth. The Central Region is the most densely populated because of Kampala which is 100 percent urban and Wakiso District which has many urban centers. The Eastern Region is the second most densely populated region followed by the Western while the Northern region is the most sparsely populated.

Table 6.2: **Population Density of Uganda, 1948 – 2002**

Index	1948	1959	1969	1980	1991	2002
Population Density (Persons per Sq Km)	25	33	48	64	85	124

Population Density by District

Population density by district is shown in Map 6.2. Kampala, Jinja, Mbale and Wasiko districts have the highest population densities. These districts have some of the highest urban populations. In the Central region, Kalangala district, the cattle corridor districts of Kiboga, Luweero, Nakasongola and Sembabule are sparsely populated.

In the Eastern region, Teso region (i.e. Kumi, Soroti, Katakwi and Kaberamaido) is the least populated. Kapchorwa is the least populated mountainous district in Uganda. The Western region as a whole is densely populated apart from Bunyoro region (i.e. Hoima, Kibaale and Masindi) and Kyenjojo District. Regionally, the southern districts have the highest densities while the northern districts have the lowest densities. High population densities in many districts are associated with problems of land fragmentation, soil erosion, landslides and encroachment on forest reserves and national parks.

Map 6.2 Population Density by District

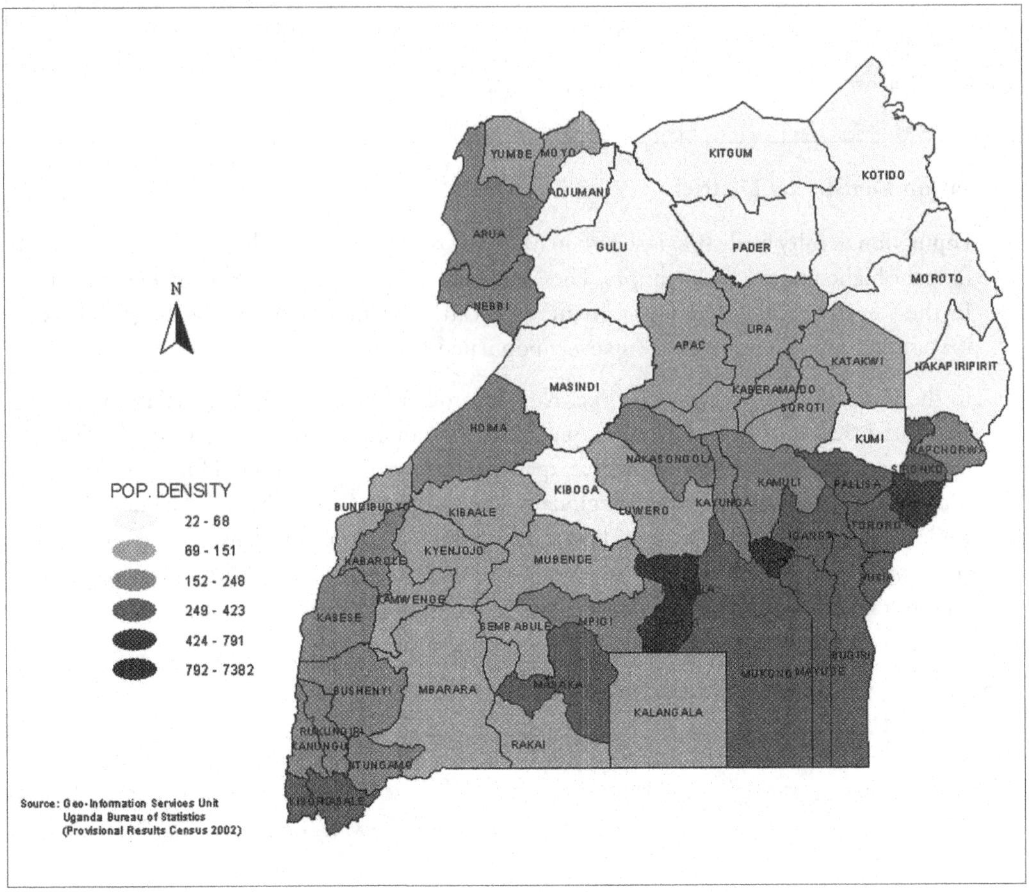

Source: Uganda Bureau of Statistics, 2005.

6.3 Demographics

6.3.1 Sex Ratio and Age

Kalangala District has the highest sex ratio 106:156 male to female (See Map 6.2). This may be partly a product of the economic activity of fishing. Fishing is primarily a male occupation. The fishermen from other districts have moved to Kalangala. This contributes to the higher male population. Kampala and Wasiko districts have the lowest ratio 83:88. This may be partly a product of urbanization and migration (BakamaNume, 1998).

The largest percentage of population in Uganda is below 20 years of age. This is an indication of the potential for the population to grow. It is also an indicator of a high dependency ratio.

Map 6.3: Sex Ration by District

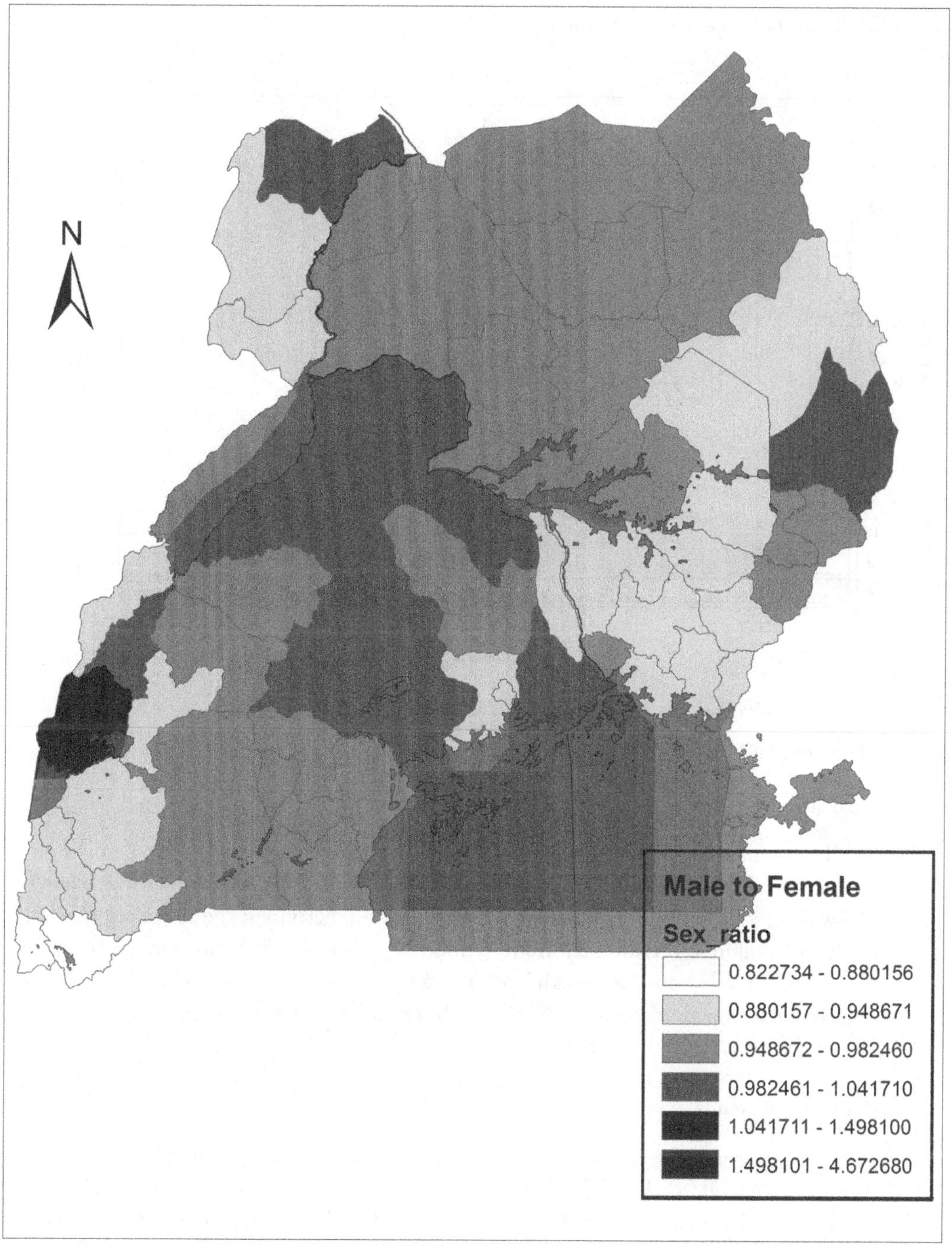

Source: Uganda Bureau of Statistics, 2005.

6.3.2 Population Pyramid

Figure 6.1: Uganda Population Pyramid

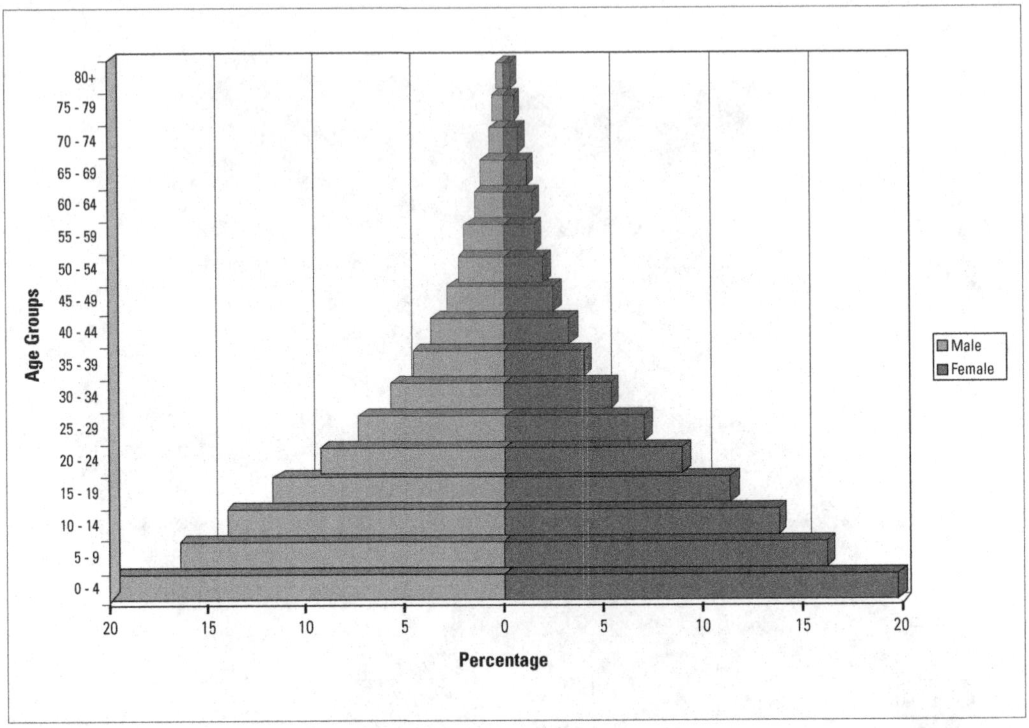

Source: Uganda Bureau of Statistics, 2004.

6.3.3 Fertility

The significance of fertility lies in the fact that it contributes to population growth. It is perceived as a major determinant of population growth and declining natural resources, which result in declining standards of a population. Fertility is the actual reproduction of live children. It is different from fecundity which is the ability to conceive. A fecund woman may choose to remain infertile by not marrying or by practising effective contraception. Some fecund women may be infertile because they may be physiologically unable to successfully complete pregnancy.

6.3.4 Fertility Trends

Table 6.3 shows the trend of fertility in Uganda using the 1969 census, the 1991 census, 1988/89 UDHS, 1995 UDHS and 2002 census. TFR remained 7.1 births per woman between 1969 and 1991 and then declined to 6.9 in 1995. The difference is so small to indicate any significant decline in fertility. It implies on average a woman produces seven children in her life time. This is very high compared to other countries in the world. For example, while Uganda and Kenya had the same TFRs in the 1970s, the situation is very different now. Kenya's TFR had gone down to 4.4 in 2002.

Table 6.3: Age-specific fertility rates from various sources

Age-specific and total fertility rates from various sources and the approximate time period to which the rates refer, Uganda 1969 – 2002

	1969 Census	1991 Census	1995 UDHS	1988-89 UDHS	1991 Census	1995 UDHS	2002 Census
Age group	1968	1990	1992-95	1984-88	1990	1992-95	2002
15 – 19	198	152	204	187	153	208	
20 – 24	341	329	319	325	331	319	
25 – 29	322	324	309	319	327	308	
30 – 34	253	275	244	273	278	235	
35 – 39	189	207	177	224	209	179	
40 – 44	87	95	89	96	97	84	
45 – 49	35	32	29	29	29	29	
TFR	7.1	7.1	6.9	73	7.1	6.8	6.9

Source: Uganda Bureau of Statistics, 2004.

Table 6.4: Selected Percentages of Uganda's Population, 1969 – 2002

Age Category*	1969	1991	2002
Population Aged Less than 5 Years	19.3	18.9	18.6
Population Aged 6 – 12 Years	22.7	22.3	22.0
Population Aged Less than 15 Years	46.2	47.3	49.3
Population Aged Less than 18 Years	51.4	53.8	56.1
Population Aged 10 – 24 Years	27.8	33.3	34.3
Population Aged 18 - 30 Years	21.7	23.6	22.4
Population aged 60 Years or More	5.9	5.0	4.5
Median Age	17.2	16.3	15.6

NB: These age categories are NOT mutually exclusive and therefore do not add to 100%.

Source: Uganda Bureau of Statistics, 2004.

The population between 0 to 12 years accounts for slightly over 40% of the total population. Over 45% of the population is 15 and below. These statistics show the growth potential for the population of the country and high dependence ratio.

6.3.5 Factors Influencing Fertility

There are several factors that affect fertility. Fertility is influenced by the socio-economic factors (such as education level, place where one lives, income) and cultural factors. Some of these factors are shown in Table 6.5.

Education

Women who received some secondary education have fewer children on average (5.2 births per woman) compared to those with primary education who have the highest level (7.1), a difference of nearly two. Education influences fertility because it increases the age at which women get married, affects decision making by women, increases income and use of modern methods of contraception.

Urbanization

Fertility levels are substantially higher in the rural areas (TFR of 7.2 children) than in urban areas (5.0). When Kampala is compared to other urban areas, TFR for Kampala is about 0.6 births lower than other urban areas. This partly explains why TFR ranges from a low of 6.3 in Central region (with big urban areas) to a high of 7.4 in the Eastern region, a difference of 1.1 per woman. However, the lower fertility level in the urban areas has a small impact on the overal level of fertility because of the urban area's small share of the total urban population (15%).

Table 6.5: Fertility by Background Characteristics

Background characteristic	TFR 1991 Paper Census	TFR UDHS 1995
Residence:		
Urban	5.6	4.97
Rural	7.3	7.17
Region:		
Central	6.9	6.28
Eastern	6.78	7.38
Northern	6.8	6.82
Western	7.9	6.98
Education:		
No education	7.09	7.04
Primary	7.36	7.12
Secondary +	5.2	5.15

Source: UDHS 1995. Uganda Population census 1991.

Contraception

Contraception use is one of the outstanding factors which determine fertility. Populations which have a high use of contraceptive measures like pills, coils, injections, implants, condoms and sterilization are likely to reduce the risks of child bearing. Such populations have lower fertility levels as compared to those populations which are not willing or lack the knowledge of using contraceptive measures to limit fertility. Table 6.6 below shows a comparison of total fertility and contraception use in different countries. It is clear from the table that countries with low contraception use have high fertility rates.

Table 6.6: Fertility and Contraception

		Percentage using contraception	
COUNTRY	TFR	TOTAL	MODERN
Egypt	3.5	56	54
Sudan	4.9	10	7
Nigeria	5.8	15	9
Kenya	4.4	39	32
Uganda	6.9	23	18
Tanzania	5.6	25	17
D.R. Congo	7.0	8	3
Botswana	3.9	42	41
South Africa	2.9	56	55
Canada	1.4	80	66
U.S.A.	2.1	76	71
Costa Rick	2.6	80	72
Brazil	2.4	77	70
India	3.2	48	43
Malaysia	3.3	33	28
Thailand	1.8	72	70
China	1.8	83	81
Japan	1.3	56	48
U.K.	1.7	72	68
Germany	1.3	85	79
France	1.9	80	74
Switzerland	1.5	82	78
Australia	1.7	67	63

Source: Population Reference Bureau 2003.

Marriage

Marriage is another factor which has a bearing on the level of fertility. Marriage legalizes child bearing and increases social frequency. Table 6.7 below shows the relationship between marital status and fertility according to the 1991 population census.

Table 6.7: Fertility and Marital Status

Marital Status	TFR 1991 Population Census
Never Married	4.09
Married	7.14
Widowed	6.21
Divorced	4.88

Source: Uganda Bureau of Statistics, 2004.

Table 6.7 shows that married women had the highest fertility followed by the widowed, divorced and then never married. Uganda's fertility is high because marriage is universal. According to the Uganda Demographic and Health Survey 1988/89, 99% of the women are married by age 50. Age at marriage also influences fertility. Early marriage lengthens the period time within which a woman is exposed to the chances of child bearing. Thus, in a population where early marriage is practised like Sierra Leone, Uganda and Yemen, fertility rates are higher than in countries like Algeria, Botswana, Japan and U.K. where the average marriage age is high as observed in Table 6.8.

Table 6.8 Fertility and Age at Marriage

Country	TFR	Average age of first marriage of women
Algeria	3.8	24
Sudan	4.6	24
Sierra Leone	6.6	17
Uganda	6.9	18
Botswana	4.1	25
South Africa	2.9	26
Yemen	6.5	17
Japan	1.3	27
United Kingdom	1.7	26
United States	2.1	25

Source: Population Reference Bureau 2003.

Age at first sexual intercourse

Although age at first marriage is widely used as a proxy for the onset of women's exposure to sexual intercourse, it is less useful in Uganda, where some women are sexually active before marriage. The 1995 UDHS shows that by the age of 15, 30% of women have had sexual intercourse and by the age of 18, 72% percent of women have had sexual intercourse whereas only 56% are married by this age. This influences the TFR and is partly responsible for child pregnancies.

6.3.5.1 Economic and Social value of children

In 1976, Caldwell argued that wealth accumulation can be used to explain change in fertility. He explained both stable high fertility and the onset of sustained fertility decline in terms of the household income level. The theory analyses the economic and social value of children. The theory suggests that in households where children make an economic contribution at an early age, through either wage employment, helping on the family farm, or freeing adults (especially women) from domestic tasks such as collecting water and fuel wood, fertility will be high. This is the case in Uganda. It was especially pronounced before the introduction of Universal Primary Education (UPE) in Uganda in 1997.

The economic benefits of children include; value attained from providing security for parents in old age. This is very important in countries like Uganda where very few people expect social security provisions in the form of pension in old age.

6.3.5.2 The economic and social status of women

Societies where women have legal and political equality with men, as well as the same social and economic opportunities as men are rare, even in the developed countries. In Uganda despite the affirmative policies by the movement government, many women cannot make their independent decisions. For example, most women cannot make decisions on their fertility. Women are also less likely to make individual decisions on female circumcision, widow inheritance, levels of education for females, employment outside the home and political positions of leadership.

6.3.5.3 Cultural factors

It has often been suggested that the persistence of high levels of fertility observed in Africa and some parts of Asia are to a large extent due to the various cultural beliefs and practices in these communities. These beliefs and practices are also believed to account for the failure of family planning programmes in these societies.

The cultural obligations and taboos connected with the sexual life of couples are based in Ankole region (Mbarara, Bushenyi and Ntungamo districts) of Western Uganda but many of them are also applicable to other parts of the country. The cultural practices fall into three main categories:

1. Resumption of sexual relations between spouses after delivery

   In pre-modern societies, Uganda inclusive, the period that couples took before resuming sex was often dictated by custom. The persistence of these customs may have important implications for fertility.

   In Uganda, two extreme cases can be referred to. The first one which encouraged high fertility is that of Ankole, Western Uganda where sex resumed four days after delivery. This was believed to prevent both the man and woman from having extra-marital sex before consummating the birth of the child which could lead to infertility or the death of a child. It was also believed that the semen has chemicals which help in the quick healing of wounds caused by giving birth.

On the other hand the Teso discouraged sex during breast feeding. This was because of the belief that if a breast feeding woman has sex it would lead to the semen contaminating the milk which would lead to the death of the child. The effect was the lowering of fertility among the Teso.

2. Obligation to have sexual intercourse

   In many societies in Uganda, customs encouraged high coital frequency because many occasions had to be consummated with sex. For example, in Ankole thirty three such occasions are identified.

3. Sexual relations with brother-in-law (Levirate)

   In some regions of Uganda there is a custom which allows the husband's brothers to have sex with the wife. In case of widowhood, the brothers are obliged to inherit the widow and look after all her needs, including her sexual ones. This custom implies that despite the absence of a husband, fertility is not greatly affected.

6.4 Mortality

As observed in Table 6.1, mortality levels in Uganda are very high. This section looks at the factors associated with high mortality rates in Uganda.

6.4.1 Factors which affect mortality:

1) Eight essential elements of primary health care

   In Uganda, like many other developing countries, eight essential elements of primary health care are either at low levels or not properly implemented.

   a) Education concerning prevailing health problems

      Due to the low levels of education, many Ugandans do not know about health problems and die from preventable diseases such as malaria, intestinal worms, measles and dysentery.

   b) Promotion of food supply and proper nutrition

      According to the UDHS 1988/89, 45% of the children in Uganda are chronically undernourished. Maternal under nourishment leads to anaemia in pregnancy with associated problems such as abortion, still births, premature births and low birth weight. Two examples can be used to illustrate poor nutrition of Ugandans. The per capita meat consumption is 5 kgs/year/person compared to Britain's 50 kgs/year/person. Milk per capita consumption for Ugandans is 40 litres/year/person compared to the 200 litres/year/person in Europe. Factors which affect nutrition in Uganda include; cultural practices; lack of knowledge; low education especially among women; large family sizes, inadequate income; lack of proper food processing and storage facilities; a poor distribution and marketing system; poor agricultural practices and poor transport system.

   c) Adequate supply of safe water and basic sanitation

      Safe water implies obtaining water from a tap, protected spring or protected well. However in Uganda, many people still use untreated water from open wells, channels, rivers, swamps and lakes which are contaminated.

d) Maternal and child health

According to the 1988/89 UDHS, only 3% of births were attended by a doctor, 36% by a nurse/midwife and 6% by a trained traditional birth attendant. An estimated 55% of women in Uganda deliver under untrained personnel or without any assistance at all. Contraceptive prevalence rate is only 23% with 18% using modern methods.

e) Immunization against the major infectious diseases

Uganda has a low immunization coverage of 38%. This low coverage is responsible for the high death of infants and children from the six killer diseases (i.e. diphtheria, tetanus, whooping cough, measles, poliomyelitis and tuberculosis).

f) Prevention and control of locally endemic diseases

Although Uganda has managed to control guinea worms and HIV/AIDS prevalence, it still faces the problem of other endemic diseases such as sleeping sickness, river blindness and bilharzia.

g) Appropriate treatment of common diseases and injuries

The government policy is to have a doctor stationed at sub-county level but some districts are far from where the doctors are stationed.

h) Provision of essential drugs

The government policy is to stock all dispensaries and hospitals with essential drugs but in many cases this has not been achieved.

2) Medical facilities

Health infrastructure in Uganda is characterized by uneven distribution and poor access to facilities, inadequate services and low per capita expenditure. According to UDHS 1988/89, over 50% of the hospitals were in the urban areas where 11% of the population lived. In 1992, 76% of the doctors, 80% of the midwives, 72% of the nurses and 64% of medical assistants were working in the urban areas.

3) Percentage of GDP spent on health

In 2002 only 12% of GDP was spent on health in Uganda (UBOS, 2004). This is an equivalent of only 5.65 US dollars per capita/year or about 50% of the absolute minimum level of 12 US$ for proper care.

4) Urban and Rural Influence

The environment in which people in Uganda live may influence the health of the population. In urban areas the poor people live in pathetic conditions. According to 1991 population census, 58% lived in one-roomed houses and 4.7% did not have toilet facilities. In the rural areas, some houses are poorly constructed to an extent that mosquitoes enter freely. 54.5% of the households had roofs made from grass and banana leaves in the 1991 population census. High temperatures in Uganda create ideal conditions for disease vectors of bilharzia, sleeping sickness, malaria and river blindness. The tetanus virus thrives in humid and wet conditions.

Agriculture has been known to provide breeding ground for vectors such as mosquitoes which spread malaria.

5) Education

Education affects use of modern methods of family planning which reduce maternal mortality through the four toos:

Too young	- giving birth before age 20.
Too old	- giving birth after age 35.
Too often	- giving birth every year.
Too many	- giving birth to many children, for example as many as 15.

Education also affects mortality through influencing age at first marriage, decision making, income, demand of children and the environment in which the child grows and lives.

6) Food situation and eating habits

Eating habits influence our health. Many people in Uganda are poorly fed not because of lack of food but because of poor transport and belief in quantity rather than quality. Some people take one meal a day and others do not have balanced diets. For example, in Ankole, Western Uganda, a family can take a big plate of bananas flavoured with only salt for supper.

7) Culture

Despite the education, culture still plays a role in influencing mortality. There are some negative cultural practices which raise mortality levels. For example, widow inheritance has been known to influence the spread of HIV/AIDS.

6.5 Conclusion

Uganda's population growth has been phenomenal and it is also a cause for concern. The high fertility rate is a concern to many population researchers and policy makers. The country has to reduce its fertility. Efforts to educate the population on family planning should be instituted. Child mortality is also high. Immunization and improvement in basic health services should be implemented to reduce the high child mortality.

Key Terms

Birth rate	Contraception	Death rate
Dependence ratio	Fertility	Growth rate
Mortality	Population pyramid	Population density
Sex Ratio		

Questions

1. What is the difference between birth rate and fertility rate?
2. What factors influence fertility?
3. How does fertility rate differ in different places in Uganda?
4. What factors influence mortality?
5. What is a population pyramid? Construct a population pyramid for the counties in your district.
6. What factors influence population migration?
7. Where is the major destination of migrations?
8. What is migration selectivity?

References and further readings

Bakama B. BakamaNume. 1998. *Inter-District Migration in Uganda,* Unpublished Paper.

Louis Henry. 1976. *Population: Analysis and models.*

Jones Huw R. 1987. *A Population Geography,* 2nd Edition, New York, The Guilford Press.

Shyrock Henry, Siegel S, Jacob and Associates .1976. *The Methods and materials of Demography,* London, Academic Press.

Barrett H.R. 1995. *Population Geography,* New York, Pearson School Press.

Statistics Department, Ministry of Finance and Economic Development. 1996.

Uganda Demographic and Health Survey 1995, Government Press, Entebbe.

Uganda Demographic and Health Survey(UDHS), 1988/89.

Population Reference Bureau. 2002. *World Population Data Sheet,* New York.

Population Reference Bureau. 2000. *The World's Youth 2000,* New York.

Ntozi J.P.M. 1993. *The role of men in determining fertility among the Banyankore of South-Western Uganda,* Unpublished Paper, Makerere University.

Ntozi, J. P. M et al. 1991. *Some determinants of fertility among Banyankole: Findings of the Ankole* fertility survey.

Ntozi, J. P. M et al. 1990. *Some aspects of determinants of fertility in Ankole,* Uganda: Findings of Elders survey.

Uganda Bureau of Statistics. 2003. Ministry of Finance, Planning and Economic Development

Chapter 7

URBAN GEOGRAPHY OF UGANDA

Hannington Sengendo

7.1 Introduction

Urban geography is the study of how urban phenomena are organized in space. As such, urban geographers are concerned with analyzing the evolution of urban societies, the urbanization processes, spatial pattern of internal structure of cities and the cause and effect relationships associated with urbanization. Geographers are also concerned with city growth, decline, the dynamic interaction, functional relationships between cities and their environments in an urban system. An urban geographer, therefore, has to examine questions such as how do cities or urban areas evolve and what form and structure do they assume? What are the major concerns of the urbanization process? How have urban societies changed through time? The answers to these and many other questions may be found in a wider context of social, economic and political organization of society.

7.2 Evolution of Urban Societies

During the onset of the nineteenth century, most populations were rural except in countries such as Britain, which experienced urban growth during the pre-nineteenth century. Thus, urban societies have evolved over time and this evolution has been extremely rapid during the nineteenth century. Urban societies have evolved from settlements forms, which were influenced by a number of factors such as landscape, building patterns, sociability and immigrations. These societies developed into class structures such as the gentry, upper middle class, lower middle class and lower classes. This classification was based on status, land ownership, prosperity and family size among others.

Segregation between classes and the erosion of a sense of community villages led to formation of neighbourhoods. The high and middle class became more conscious of squalor and unhealthiness of exploding towns and took refuge in the emerging suburbs. Seen in a broader perspective therefore, one can say that urban societies have evolved through patterns of inequality and spatial differentiation and this socio-economic differentiation is arguably the most important cleavage within contemporary cities (P.L Knox 1991).

7.3 The Internal Structure of Cities

Anyone who lives in or visits towns or cities is aware of variations in land uses. However, in cities of developing countries, these variations do not show any standard patterns as may be observed in cities of the developed world. There are different land uses, for example, commercial retail and whole sale, industrial, agricultural, residential (high, medium and low class areas), transportation and recreation among others. The land uses create visible patterns and regions. The patterns and regions are responsible for the formation of internal structures of cities.

Classical theories of city formation stem from ideas of Burgess. They are collectively known as the Chicago school. They describe an urban environment where the inner city had largely been vacated by rich populations who had moved outward to the suburbs in rings and wedges from the town centre. The migration of population to the outer rings left the decaying inner city to the most disadvantaged groups. The Chicago school saw the internal spatial organization of cities as an outcome of ecological competition for riches between social classes who behaved like different species in terms of their endorsements and wants and who would compete for different land uses, with the strongest groups taking the most desirable positions and the weaker groups occupying residual spaces. As society and transport technology changed and the circumstances of the groups altered, or housing became inadequate, they would vacate particular areas leaving them for new immigrants or social groups who would occupy them.

Figure 7.1: The Burgess Spatial Scheme for Chicago

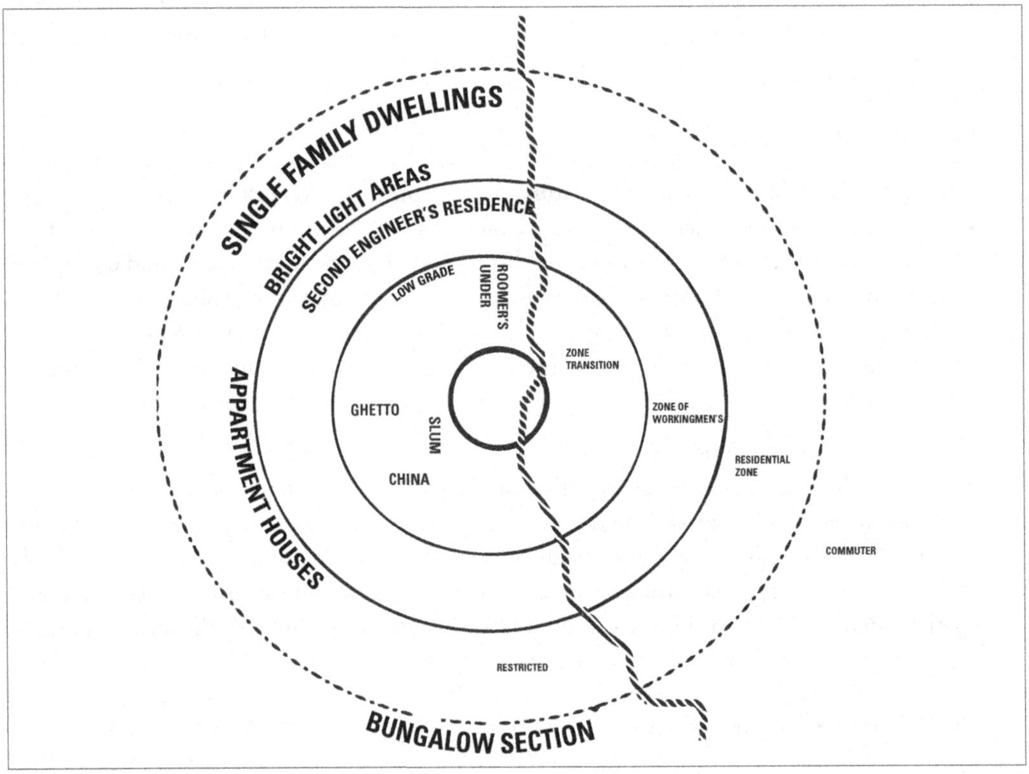

The zones of interest in the diagram are those designated as workers' housing and zones in transition. Earlier generations of working men's houses were slowly being taken over by warehouses, immigrants and the urban poor, as better off households vacated for the suburbs. These zones in transition were the ghettos, slums and bright light areas. The diagram further shows that all the zones shown were at one time in the early history of the city included within the circumference of the inner zone, the present day centre. However, as the city grew, those uses, which needed a central location, came to monopolize the centre, buying up and replacing or converting buildings previously used for other purposes. Such uses were shops and offices hence there developed what Burgess called the Central Business District (CBD) a term which is widely used.

7.4 Urban Form in cities of Developing Countries

Older European or North American cities grew in an environment where marked norms and feudal landholding systems had been well established and dwellings could be readily traded by owners for alternative uses. They also grew initially, during a time when most work was centrally located and people walked. With improved transportation cities expanded to suburbs along revile corridors, filling between these corridors as personal motor transport became universally available.

However, the situation of cities that have emerged as substantial centres in the developing world during the past 50 years is often very different from that of succession of land uses described by Chicago school. Their business centres have often been in the historic centre, but have been purposely built and multinucleated with access to airports and the residential zones of the more affluent. The shape of the city has been determined not by centralized rail networks but by minibuses and private cars.

The cities we find in many parts of the developing world do not usually show the classic Chicago pattern of cities with a decaying, possibly partly rejuvenated core, surrounded by rings of garden suburbs. The most common ones in Africa and South Asia include colonial-style cities, with well-built formal core, surrounded by large areas of informal settlements. Others include planned ethnic separations, which is an extreme example of urban segregation. For example, during the apartheid years, Soweto and other ethnic satellite cities of Johannesburg were made possible by cheap subsidized daily bus transport for workers to the centre.

It is also important to note that cities in third world countries exhibit dual characteristics whereby the existing native cities were partly colonized at the time of occupation. The Europeans for example had little desire to live in close contact with the indigenous population, partly for cultural reasons and partly for security reasons. The response, therefore was to build a new European city removed from, but usually adjacent to the native city hence creating a dual character.

The first element of dualism in third world cities is the *formal economy* of the international sphere of manufacturing, trading, business and commerce. This is epitomized by the modern towers of concrete and glass commonly located in major cities of the world. The second element is the *informal economy*, which is characterized by marginal squatter settlements. It consists of a great multitude of minor, but important activities, which support the low-income groups.

Cities also demonstrate a strong contrast between residential areas in terms of high and low status areas, as well as peripheral location of low status areas. Usually, the high-income residential areas are central in location although this is now changing. The structure of third world cities therefore falls between modernization and westernization, which have induced social, cultural and economic changes. The most general interpretation of this process of change therefore is that of superimposition of those processes and their physical consequences, which have created the western city, onto the underlying, indigenous form of urban development. Along with that goes the additional feature of the creation of peripheral shanty areas, which are a consequence of urbanization.

7.5 Urbanization in Uganda

7.5.1 Uganda's urban geography

In order to understand Uganda's urban centres, the urban geography of Uganda should not be studied in isolation of other urban geographical models and theories. The main reason for this is to relate the original ideas for urban form, urban structure and internal organizations of cities in the past and those in the present.

In Uganda, many urban centres have evolved and emerged because of several factors. The urban landscapes are continually being transformed through different processes. The signs of significant change in many urban landscapes in Uganda include extensive redevelopments, increase in run down areas, settlements on edges of the existing urban areas. Good examples of these are found on Housing Estates of Nakawa in Kampala and Walukuba in Jinja.

Uganda's urban areas have evolved overtime and many of them emerged during the time of colonial occupation, mainly as administrative and commercial centres. In some cases, they also developed where traditional leaders had their headquarters. Administration at these centres was on behalf of colonial masters through local colonial officers, while commercial centres were mainly bases for exporting locally produced raw materials and importing goods from colonial centres.

Uganda's major urban centres are Kampala, Jinja, Mbale, Entebbe, Tororo, Gulu, Masaka and Mbarara. They owe their growth, development and importance to a number of functions they played including administration, commerce, manufacturing, seats of traditional leaders and finance functions.

In addition to the administrative and commercial functions of these areas, there were a few other functions. The urban areas acted as educational centres particularly in smaller administrative centres such as Kabale and Gulu, financial centres, landing sites such, as Entebbe town council, which later on became an administrative center and a seat of colonial government. Currently, Entebbe serves as the main gateway to Uganda because of the International Airport.

Other urban centres evolved as manufacturing centres notably Kampala, Jinja, Tororo and Mbale. Towns like Masaka developed because of its location in a pleasantly fertile and densely populated coffee growing area and a collecting center for hides and skin. All urban centres play administrative, commercial, industrial and manufacturing and finance roles, though Kampala is more active in the above roles compared to the other towns.

Changing skyline of Kampala

Source: Uganda Tourism Board

Urbanization is the change in the proportion of the population that lives in an urban area. In Uganda, urban areas have been gazetted according to the number of persons. For instance, in the 1992 population and housing census, trading centres with 1,000 or more persons were referred to as urban although they were un-gazetted. The 2002 population and housing census included only those gazetted towns, cities and municipalities with 2,000 persons and above. The urban population as enumerated in 1969, 1980, 1991 and 2002 is given in Table 7.1 below.

Table 7.1: Urbanization in Uganda, 1969-2002

Index	1969	1980	1991	2002
Number of towns	58	97	150	74
Urban Population	634,952	938,287	1,889,622	2,921,981
Proportion (%)	6.6	7.4	11.3	12.2
Urban growth rate	8.17	3.93	6.35	3.73
% in Capital city	53.9	47.9	41.0	40.7
% in 20 largest towns	87.4	80.4	74.4	76.6

Source: UBSO

The data for 1969, 1980 and 1991 are as per the 1991 definition, while those from 2002 are as per the 2002 definition and hence should not be compared.

It should be noted that urbanization in most developing countries has not been accompanied by other essential structural changes. Consequently, some of the positive externalities which are associated with the urbanization process such as emergence of modern technology and rural agriculture, have not manifested themselves significantly in the economies under discussion.

The urbanization process in Uganda has created two types of cities formal and informal. The informal city consists of extra illegal housing and unregistered activities. Some of the activities include agriculture involving cultivation, livestock and poultry keeping, household activities-cooking, beer brewing, metal works, carpentry and petty commodity trade. Studies have shown that there is a close correlation between poverty, informal housing and informal income generation. Furthermore, as population influx continues to swallow the urban enclaves, a number of socio-economic activities also spring up, mainly as survival strategies for the urban dwellers. The extent of the impact of these activities on environmental quality seems to have a direct relationship with population increase in urban areas. This situation is further aggravated by the minimal control of these activities by urban authorities (Nuwagaba, A. and Mwesigwa, D. 1997).

7.5.2 Trends of Urbanization in Uganda

In Uganda, urban growth started in the first half of the last century with the establishment of economic and administrative centres all over the country. This triggered the influx of people to the urban nuclei development into full-grown district urban centres. The influx however later outpace urban facilities such as housing, utilities, land and other infrastructure, a phenomenon that is continuing. Table 7.2 below depicts the magnitude of urbanization of major towns in Uganda since 1959.

Table 7.2: Magnitude of Urbanization of Major Towns in Uganda 1959-1991

Major Urban Centre		Total Population			
No		1959	1969	1980	1991
1	Kampala 1	111,483	330,700	458,593	773,463
2	Jinja 2	29,807	47,872	45,080	60,779
3	Masaka 4	4,782	12,987	29,123	49,070
4	Mbale 3	13,569	23,544	28.039	53,634
5	Fort Portal	8,317	7,974	26,806	32,627
6	Mbarara	3,844	16,078	23,255	40,385
7	Entebbe	9,941	21,096	21,289	41,638
8	Gulu	4,770	18,170	14,958	42,841
9	Lira	2,929	7,340	9,122	27,143
10	Kasese	1,564	7,213	9,917	18,554

Source: UNEP 1992 Strategic Resource Planning in Uganda Vol. X Human Settlements

Map 7.1: Location of major towns in Uganda

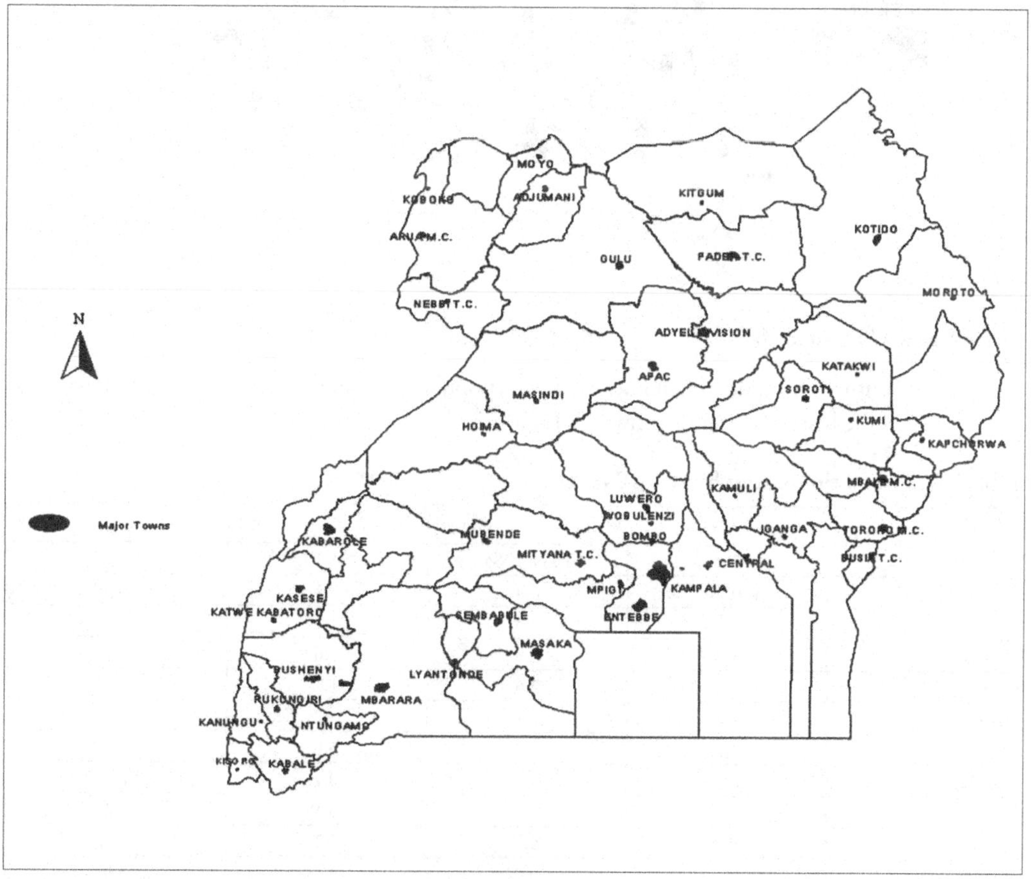

Source: Population Reference Bureau 2003

Map 7.2: Uganda - Rural/Urban Population by District

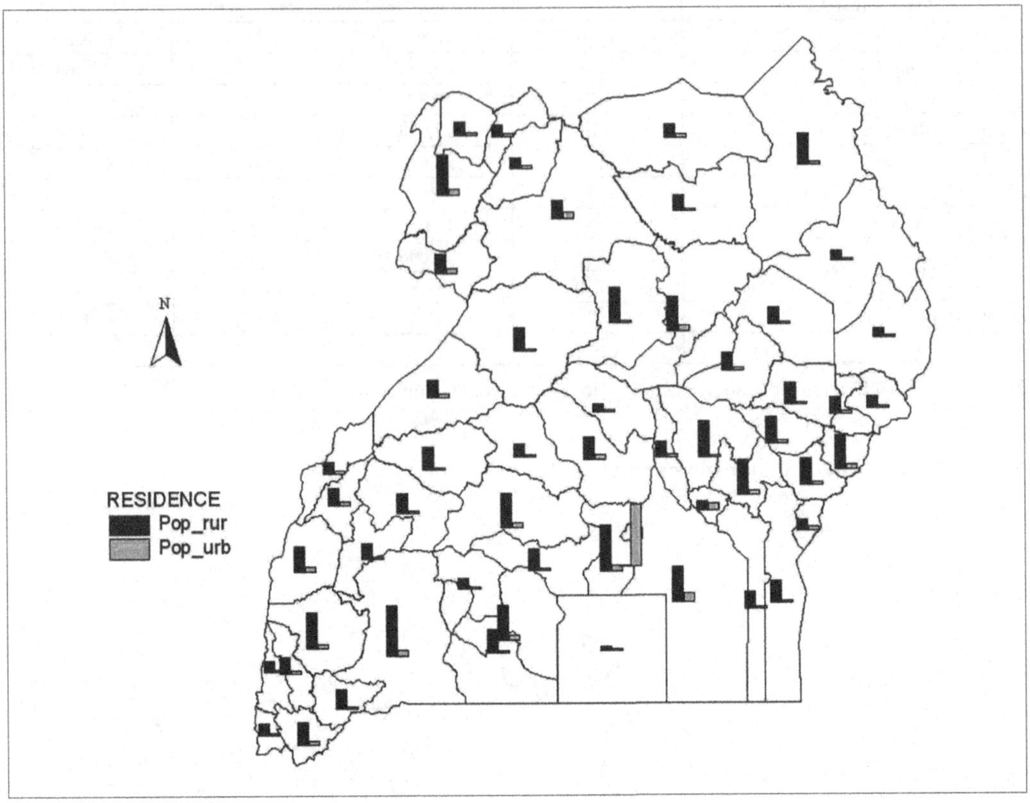

Source: Uganda Bureau of Statistics, 2004.

Southern Uganda has a higher urbanized population than northern Uganda. This reflected in the historical trends shown in Table 7.3.

Table 7.3: Population Growth rates 1948-1980

Major Urban Centres	Growth rate(years)			
	1948-1957	1959-1964	1969-1970	1980-1991
Kampala	10.5	6.8	3.1	4.9
Jinja	4.7	2.0	-0.6	2.8
Masaka	10.0	8.9	0.6	2.8
Mbale	5.5	3.6	1.7	6.1
Fort Portal	0.5	5.7	11.7	1.8
Mbarara	14.3	8.8	3.5	5.1
Entebbe	7.5	3.7	0.1	6.3
Gulu	13.4	5.6	-1.9	9.4
Lira	9.2	5.6	2.1	10.4
Kasese	15.3	9.0	3.1	5.9

Source: UNEP 1992 Strategic Resource Planning in Uganda Vol. X Human Settlements

According to the 1959 census, 276,211 persons were enumerated in 35 urban areas that existed in Uganda at the time. Later censuses show that in 1969, there were 58 urban centres with a total population of 634,592. By 1980, urban population had increased to 938,287 in 96 urban centres and by 1991, 1,889,622 people lived in 150 urban centres (Uganda Population and Housing report 1991).

Urbanization in Uganda however has certain unique features. The pattern of urban centres for instance assumed linear settlements along major routes and road functions. During the 1960's Uganda's policy on industrialization was aimed at developing urban centres, creating employment opportunities to absorb the rural-urban migrants and empower them economically. Many of the urban centres in Uganda were built without proper planning. The continuing influx population to these urban centres has caused many problems.

A comparison of levels and trends of urbanization between Uganda, Kenya and Tanzania, reveals that Uganda's level of urbanization (12.3%) is lower than in Kenya (20%) and Tanzania (22%) (Uganda Population and Housing Census 2002). Nevertheless, the urban population in Uganda has increased rapidly from less than one million people in 1980 to three million by 2002 representing a more than three-fold increase. The urban population in Uganda has been characterized by a high growth rate of 5.1% between 1991 and 2002. This can be explained by the persistent rural poverty that has caused people to migrate to urban areas with the hope of improving their livelihood.

The increase in urban population in Uganda has led to challenges in provision of housing, water and other amenities for many low-income urban dwellers. This has further culminated in the growth of squatter settlements or slums with inadequate services such as sanitation and water services. A closer look at Kampala and its environs shows that several unauthorized settlements as well as commercial and industrial buildings are located in wetlands. Wetlands have been drained to give way to construction and other socio-economic activities. These developments for example can be observed in Ndeeba, Bwaise, Kasubi and Kawaala among others. In Kampala and other urban areas of Uganda, there has not been strict enforcement of urban planning regulations resulting in all sorts of illegal developments.

7.6 Challenges of Urban Growth in Uganda

7.6.1 Economic and Social problems

Since the 1970s when the then President (Idi Amin) expelled Asians, the majority of urban centres in Uganda have been experiencing severe economic problems. These include unemployment due to decline in manufacturing sector, for example Jinja, Mbale, Tororo have experienced factory closure and deindustrialization. Jinja (the former industrial center of the country) lost all the economic advantages of industrial manufacturing.

The process of deindustrialization has further displayed a pronounced social dimension with mainly middle class residents moving to the suburbs or beyond, while the low-income people move into the towns. This has led to unprecedented development of informal settlements and increase in informal sector activities. The growth and increase of informal settlements or slums has had its impact on the environment of urban centres as will be explained later on in this chapter.

While there are considerable similarities in the urbanization process between the developed and developing countries, one notable difference is that the large cities in the developed countries emerged only after the countries had achieved high levels of income and intermediate urbanization. In developing nations, particularly those of sub-Saharan Africa, large cities have appeared despite low levels of both income and urbanization. Migration to cities is largely driven by the "urban illusion"- the promise of a job and a better life that may never materialize. This has tended to create an urban crisis because the continuous influx of migrants creates a ripe atmosphere for criminal activity, intensifies pressure on available services and resources and exposes the inability of governments to deal with the situation.

7.6.2 Urbanization and the Environment

Urbanization is manifested in a number of varying phenomena in Africa, which can be generalized as environmental issues. A large proportion of Africa's urban populations live in un-serviced settlements and a large percent of this population find their livelihood in informal economy. In some cities, up to 90% of the new housing stock has been provided informally and more than half of the work force is in unregistered employment. Most cities have severe environmental problems mainly because of emissions from transport and industry. These environmental problems are also linked to inadequate housing.

In Uganda, just like other Third World Countries, there is a close relationship between urbanization and environmental decadence. Urban environment decadence is a derivative of structural issues critical among which is the lopsided development policy that tends to favour particular areas. The skewed distribution of socio-economic opportunities has exacerbated the rural-urban interface in the form of heavy influx of rural populations to Uganda's urban centres. This phenomenon has culminated into an environmental crisis in Third World cities. Recognizing that urbanization is inevitable and irreversible, the real challenge then is to devise strategies for its more efficient and effective management so as to fully realize the negative impacts on population and society and at the same time try to minimize these negative consequences.

Nowhere perhaps are the effects of ongoing environment degradation less understood or more in need of understanding than in cities and particularly Kampala where rapid population growth will continue for many years to come, creating confining pressure on the environment. Environment in an urban context has only recently begun to be seriously addressed in literature and by the full range of stakeholders who are involved in the urban development process. Urban areas, as already shown, have become environment "hot spots", which urgently require special attention not only in terms of project specific environmental assessments, but also in terms of environment planning both at municipal and regional level. The impact of urban areas on resources immediately adjacent to and far removed from their actual locations, is becoming increasingly apparent.

In spite of the increasing benefits to individuals and society, the process of urbanization has given rise to a number of environmental problems, which many urban authorities in Uganda have failed to contain. The main forms of urban environmental problems in Uganda include uncontrolled construction, solid waste dumping, inappropriate agriculture, severe sanitation problems, residential occupation of hazardous areas, industrial activities and emission, clearance of vegetation, rapid increase in motor vehicles, indoor pollution from charcoal use, water, air and noise pollution. The impacts associated with environmental problems and urbanization in Uganda include land pollution from garbage disposal, health problems, water borne diseases, malaria which is on the increase, lack of urban planning, increase in air pollution and destruction filtering capacity due to clearance of papyrus swamps among others. (Kampala First Urban Study 1993).

In Kampala, environmental quality is associated with a number of social economic activities that have sprang up due to population influx in the city. The following activities contributed tremendously to the deterioration of the city. The activities include agriculture, involving cultivation and rearing of animals, informal economic activities such as open vehicle repair, metal fabrication and carpentry and petty commodity trade among others. The visible effects of these activities on environment quality are accumulation of garbage in places of residence and along the roads, overcrowding, unplanned poor housing structures, sanitation and polluted environment. The impact of these activities on environmental quality seems to have a direct relationship with the population growth in Kampala.

There are several sanitary urban problems in Uganda. Sengendo's study (1992) revealed parts of Kampala where 77% of the inhabitants were using shared latrines. The study also found that 2.5% did not have latrines, or toilet facilities at all and 12.5% did not have any bathing facilities. The study further revealed that 32.5% and 4% of the inhabitants collected water from springs and natural streams respectively. These are sources of water, which collect filthy water from other sources and thus tend to be a source of diseases prevalent mainly in high-density settlements.

Environmental deterioration is clearly manifested by many indicators including, wetland degradation, poor solid waste management, poor housing and related facilities such as sanitation, overcrowding air, water and soil pollution among others. For the last twenty or so years, Uganda has undergone a series of events and developments that have greatly altered her ecological balance and eroded her natural resource base. Consequently, Uganda today finds itself with a deteriorating environment as evidenced by pollution, improper waste management and recurrent drought appearance of desert like conditions, indiscriminate wetland destruction, wildlife destruction, urban blight, sprawl, decay and congestion. While human abuse and over use of natural resources is not a new phenomenon in

Uganda, what is new is the realization of the effects of rapid deforestation, inappropriate agricultural practices and careless disposal of domestic and industrial wastes, excessive fishing and human encroachment into ever marginal and sensitive environments.

7.6.3 Urban Agriculture

Urban farming is becoming a complex and dynamic feature of urban landscapes and a socio-economic reality in Uganda. It involves crop cultivation and animal husbandry. There are various theoretical considerations for this type of agriculture. The Marxian view suggests that it is a survival strategy of the urban poor. This type of agriculture is seen as a means through which the poor are forced to bear the social costs of capitalist development. City planners claim that it is a rural cultural artifact of a recently urbanized population-remnant of bush life. Informal sector advocates see it as a form of market rationale, micro entrepreneurship, responding to incentives in the local economy. It is regarded by some as an adaptive strategy by city dwellers in the face of economic adversity, while others claim that urban agriculture does not constitute resourcefulness but rather symbolizes decay.

Although urban agriculture is emerging as an important economic activity within urban informal activities, few planners and decision makers assume that urban agriculture is a worthwhile legitimate activity. Urban agriculture is looked at as a contradiction of the common image of the city, some kind of activity which detracts from images of the ideal city - a planned city. As a result, many people view urban agriculture negatively.

In Kampala and other urban areas in Uganda, urban agriculture has intensified since the early 1970s. The territorial expansion of Kampala for instance, brought peri-urban areas of Kawempe, Nakawa and Mengo into the city boundaries. These areas boosted farming activities in Kampala despite the prohibitive urban by-laws which restrict cultivation within the urban confines with the exception of vegetable gardens and flowers. But evidence already adduced shows that such restrictions are hardly based on environmental concerns. Urban settlements sometimes produces kitchen waste compost to improve its plots.

Urban agriculture in Uganda

Source: Hannington Sengendo

7.6.4 Solid Waste Management

Urban wastes are a dangerous to human health and the environment. The urban poor, especially the slum dwellers and squatter settlers, are a high risk group because they live close to waste dumps and lack potable water and basic sanitation. Hence, they rely on open canals and holes in the ground as a regular means of waste disposal. Consequently, there is a high incidence of sanitation-related diseases among the urban poor because they are exposed to epidemics such as cholera, typhoid and dysentery.

The increasing quantity of waste is creating a serious environmental problem in the urban areas in the developing countries. The status of waste management in our major cities is unhygienic and unsatisfactory. In Dar-es-Salaam, (Tanzania) the level of refuse collection is only 16%. In Nairobi, (Kenya) less than 50% of the refuse generated is collected while the rest remains scattered and rotten in open spaces, open drains and road sides. In Kinshasa, (Democratic Republic of Congo) collection of household waste and cleaning are not carried out in a coherent way, with most of it put on the road, illegal dumps, in storm water drains, or buried in plots. In Kampala, a maximum of 15% of total household waste is collected by the city council.

Garbage left to rot by the roadside

Source: Hannington Sengendo

According to the United Nations Report (1992), it is argued that a deteriorating urban environment is the enemy of sustainable development and that protecting the environment is a precondition of efficient economic development. Yet, the economic activities prevalent in the urban centres of Uganda have been detrimental to sustainable development.

In Kampala, there has been an increase in urban economic and industrial activities, which have been accompanied, by a population increase. These have led to an increase in the quantity of solid waste generated.

Informal settlements (a view of Nsambya slums)

Source: Hannington Sengendo

7.7 Evolution of Kampala and Jinja as urban centres

7.7.1 Kampala

Kampala started as an administrative capital of Buganda Kingdom in the 1800s. Between 1856 and 1890, the palace site ("*Lubiri*") moved over ten times from one hill to another. The site accommodated up to 3000 people. It was regarded as *"Kibuga"* where the palace (*"Lubiri"*) was located. Beside the administrative role, religious groups played a vital role in the development of Kampala. In 1877, the Church Missionary Society (CMS) established the First European Mission at Natete. The Original French Catholic White Fathers later followed it, when they built their mission at Kitebi between Rubaga and Lake Victoria. Following the death of Kabaka Mutesa 1 in 1884, the Lubiri moved for the last time from Rubaga to Mengo. In 1885, Roman Catholics were given a site (vacated by the Lubiri) on the southern slopes of Rubaga Hill, before final movement to the hilltop in 1891. The Arab (Moslem) settlement was established at Natete in 1892 following the relocation of the Roman Catholic mission.

In 1893, Uganda was declared a British Protectorate with its headquarters in Entebbe. However, this was short lived because in 1902, Captain F.D Lugard of the Royal Geographical Society (RGS) arrived and established a fort on the present Old Kampala hill. Lugard was also an agent of the Imperial British East Africa Company (IBEAC). The Fort was used as an administrative centre by the colonial government and at the same time acted as a nucleus for the eventual growth of the city. In 1895, British Mill Hill Catholic mission was granted a site at Nsambya Hill for their Diocesan headquarters.

In 1903, the Uganda township ordinance was passed. The ordinance provided the first legal framework for urban growth. It empowered the Commissioner to declare any place in the protectorate to be a township. In 1905, Kampala Township was transferred from Old Kampala to Nakasero Hill and in 1906 Kampala was gazetted as a township. An administrative boundary at a radius of three miles from Nakasero was established. In 1908, the surveyed area of Kampala included Natete in the west, Mulago in the north and Kibuli in the east and Kabowa in the south. The first planning scheme for Kampala was prepared in 1912. The population of Kampala stood at 2,850 and the township occupied about 1,400 acres. By 1928, the township was 3,200 acres. The addition of Makerere, Wandegeya and Mulago to Kampala occurred in 1938 and by 1944; the township covered 4,600 acres (City Council of Kampala, 1991).

Kampala was declared a municipality in 1949. The 1959 population census showed its population at 46,735 people over an area of 8.25 square miles (5,376 acres). Modern urbanization in Uganda started in the early twentieth century, with the establishment of economic and administrative centres all over the country that increased tremendously in the 1950's (Strened, 1994). The economic centres created "pressure zones", a phenomenon punctuated by heavy demographic shifts from rural areas to urban areas. The urban migrants later overran existing facilities leading to increased pressure on housing utilities, land resources and other infrastructure.

By the time Uganda gained political independence in 1962, Kampala city had a population of 50,000. In 1968 Kampala area increased to 75 square miles with the inclusion of Kawempe, Lusanje, Kisasi, Kiwatule, Muyenga, Ggaba and Mulungu, the hills such as: Makerere; Mulago; Kibuli; Namirembe; Kololo; Mbuya; Muyenga; Kawempe; Nsambya; Kasubi and Naguru with an average altitude of 1,200 metres are flat topped lateritic capped which are chemically stable to support tall and heavy buildings. Most of these hills are overlooking Lake Victoria. The valley bottoms of the hills are wide and shallow. These areas however, were not populated at the time, but as populations grew through urbanization, they have become settlement areas with dense populations.

The different hills have led to the expansion of Kampala as many of them started as nuclei serving different functions. Makerere started as a centre of learning, Old Kampala as an Islamic center, Namirembe a Protestant headquarters, Rubaga as centre for the Catholics, Mulago as a health headquarters and Kololo, Muyenga and Mbuya as residential areas for the mushrooming urban workers.

Good drainage has also contributed to the continued urbanization of Kampala. Water flows to two main drainage basins namely: the Lake Victoria basin and the Lake Kyoga basin through a number of streams. Lugogo and Kitante streams join Nakivubo stream (now Nakivubo channel) before empting into Lake Victoria. Other streams, which drain into Lake Victoria, are Lufuka, Vubyabirenge and Kajjansi. Nsooba, Nabisasiro and Nalukolongo streams drain into Lake Kyoga through Lubigi swamp and river Mayanja. Wakongolo, Kakerere, Katonga, Walufumbe and Nyajerade rivers, drain into Lake Kyoga through river Sezibwa. Apart from the valley bottoms that are prone to flooding, many areas are suitable for housing development to accommodate the ever-growing urban population.

Kampala lies in the Lake Victoria climatic zone. It receives an average rainfall of 1,850 mm per year. Rainfall occurs throughout the year with two peaks in March-May and October- December. The high rainfall favoured agriculture of both food crops such as bananas and cash crops especially coffee, tea and cotton. The rich agricultural region has made it an important collecting and marketing centre for the products. Fallers (1964) noted that Buganda had become the most economically developed part of Uganda. This was especially manifest in the area within 40 kilometres of Lake Victoria. From initial basis in the establishment of cotton in the early years of the twentieth century, this changed to coffee in the 1950's and to dairy farming in the late 1960's.

Baker (1963) noted that the population migration into the economic opportunity region of the Lake Victoria basin of Buganda and Busoga played a significant role in the growth of Kampala. In-migration was partly from inside Uganda (internal) and partly, as in the case of the Banyarwanda and the Luo, from adjacent countries (international). According to the 1959 census, there were 828,027 non-Baganda Africans in Buganda as compared with 1,006,101 Baganda. The immigrants comprised almost half the total African population. In the intercensal period 1948-59, the African population of Buganda increased by 531,966 with non-Baganda accounting for more than two-thirds of the increase. Everywhere in the south and central Buganda there was an appreciable proportion of non-Baganda.

Kampala is located close to the centre of the country. According to central place theory, the actual centre of Uganda is Bombo a few miles north of Kampala. This has contributed to the city developing as a transport centre. Roads radiate from Kampala to other towns all over the country making rural-urban migration easy. In 1931, Kampala became a railway terminus as the railhead was extended from Kisumu to Kampala and now is connected to Mombasa. Through Port Bell, Lake Victoria steamers serve Kampala to Bukoba, Kisumu, Mwanza and Jinja ports.

Kampala's central location has made processing activities possible. Industrial development led to population increase of the city through both rural-urban and urban-urban migration. Fallers (1964) observed that it was in this zone, within 40 kilometres of Lake Victoria that most industrialization and urbanization took place and upon which the growth of the capital city had the most impact. After 1986, Kampala overtook Jinja as the most industrialized town in Uganda with the emergence of new industrial zones namely: Nakawa, Natete and Kawempe.

The political instability in the country has also contributed to the growth of the city. Between 1981 and 1985 during the fighting in the Luwero Triangle, there was a large influx of population from the area to Kampala for safety. Later in 1987 and throughout the years of fighting in the north, people have been forced from Acholi region to Kampala in search of safety.

Uganda's urban growth rate was high in the 1970s. The usurpation of political power by the military and the subsequent political crisis led to the death of the formal economy especially the industrial sector. This resulted in serious economic disequilibria which resulted in the development of the informal economy that promised vibrant employment opportunities in activities such as welding and metal fabrication, hawking, petty commodity trade, mobile markets and urban farming. This persistently triggered the rural-urban influx and the subsequent high rate of population in the urban areas.

Between 1959 and 1969, the urbanization rate in Uganda was in the region of 7.5% per annum, over twice the rate of growth of the total population (Langlands, 1974). According to the 1969 census results, half of the population of Jinja and Kampala was born outside the towns. This was a sign of in-migration and immigration. The result was that in 1969, about 7% (668,264) of Uganda's population lived in 33 urban centres with over 2,000 people each, compared with 4.3% (280,000) in 1959. Almost exactly half of these lived in Kampala. This was a slightly higher degree of urbanization than Tanzania (6.2%) but considerably less than Kenya, which even by 1962 was 7.8% urban and by 1973 it was over 10%.

Hirst (1975) attributed Kampala's doubling its population between 1959 and 1969 largely to in-migration. He further asserted that persons born in Northern Province displayed a stronger tendency towards clustering and were characterized by six distinct foci at Kawempe, Kiswa, Lubiri, Nsambya, Mbuya and Luzira. This reflected segregation by occupation coupled with the provision of official housing quarters by the police, armed forces, prisons and railways in the last four neighbourhoods, as well as the voluntary choice to live together, in Kawempe and Kiswa. Occupational stratification of migrants was reportedly important in Kampala and was predicted to have an important bearing upon the degree of residential segregation. In case of northerners, this was expected to lead to marked residential clustering.

Throughout the period 1969 to 2002, Kampala has remained a primate city with a contribution of more than 80% of the total urban population in the central region in 1991. The central region constitutes the highest level of urbanization as indicated by 15.0%, 16.6% and 24.4% in 1969, 1980 and 1991 respectively.

The 1991 Uganda population and housing census reported Kampala's population as 774,241 people occupying 190 km^2. The 2002 census reported the population at 1,189,142 people occupying an area of 198 km^2. The proportion of the population of Kampala city to the total urban population declined from 54% in 1969 to 40% in 2002. This is because of the fixed geographical boundaries of Kampala city as well as the growth of many urban centres, especially with the creation of new districts. More than half (54%) of the urban population of Uganda lived in the Central region, while the other regions had almost equal proportion of the urban population (Northern 17%, Western 14% and Eastern 13%). The high proportion of urban population in the Central region is attributed to Kampala city, which had 40% of the total urban population in 2002.

If Kampala is excluded from the Central region, the proportion of the urban population for Central region drops to 15%, which would compare well with the other regions (2002 Uganda Population and Housing Census). Nine urban centres i.e. Kampala, Jinja, Masaka, Entebbe, Njeru, Busia, Iganga, Lugazi and Mukono out of the twenty-five are located in the Lake Victoria basin. Kampala has remained the largest urban area in Uganda since independence. Currently, Kampala's functions are primarily commercial, administrative, cultural and industrial.

Planning of Kampala

Early planning efforts of Kampala and indeed of Uganda were initiated during the colonial period. It was established as a township in 1903 and it was structured under the policy of creating exclusive and defensible enclaves by means of careful planning under the recommendations of Professor Simpson. This resulted in the segregation of Europeans and Asians from African enclaves where it was emphasized that Europeans should not live in proximity to natives in order to avoid infections and to safe guard against fires (from huts) which were so common in native quarters. Furthermore, segregation was advocated for because it would remove the inconvenience felt by Europeans, whose rest was disturbed by drumming and other noises dear to the natives.

Since that time, planning efforts were drawn with emphasis on health and water services provision, location of land uses through zoning, control of rodents and drawing of swamps.

Map 7.3: Structure plan for Kampala since 1972- 2004

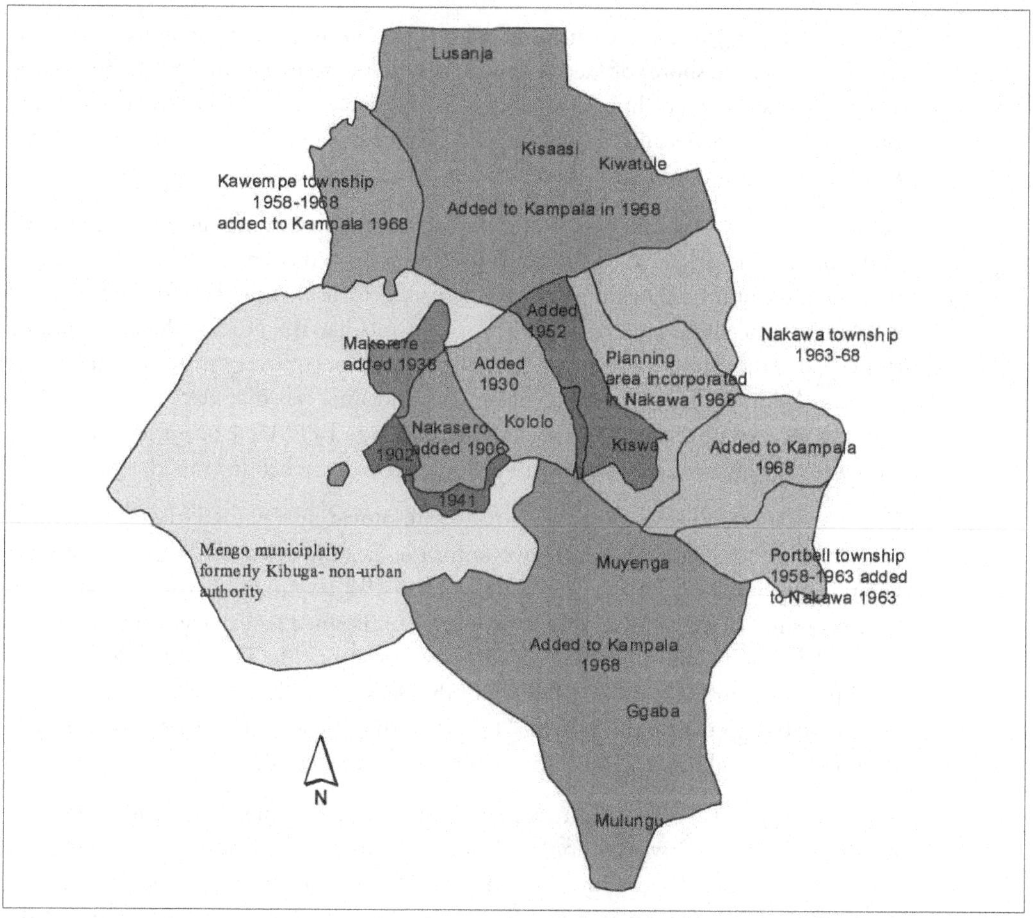

Source: Kampala City Planning Department

It should be noted that most of the planning schemes that have operated in Kampala city bear the hallmarks of colonialism with the planning systems and expertise brought in from outside. Until independence in 1962, Kampala was a colonial city and that shaped its economic and land use distribution. This is mainly so because all businesses have been and are still based within Kampala, though Jinja was originally the country's industrial city. The land area for Kampala has increased 150 times between 1959 and 1999, whereas the population has increased just 20 times in the same period. This population growth pattern is not uniformly spread in the city, with high densities found close to the city while low densities are found on the urban periphery (Figure 7.1). The land area has increased as a result of the city's annexation policies of surrounding urban areas and townships, with the approval of new planning schemes. Over the years, new planning schemes have given Kampala a role of annexing small townships and villages.

7.7.2 Jinja

Jinja lies in the south east of Uganda, 87 kilometers (54 mi) east of the capital, Kampala. It is located on the shores of Lake Victoria, near to the source of the White Nile River. The city is the capital of Jinja District and is considered the capital of the Kingdom of Busoga. Nearby towns and villages include Njeru Buwenda (2.8 nmi/5.2 km/3.2 mi), Kimaka, Mpumudde, Masese, Walukuba and Bugungu.

Before 1906, Jinja was formerly a fishing village that benefited from being on long-distance trade routes. The origin of the name 'Jinja' comes from the language of the two tribes (the Baganda and the Basoga) that live either side of the River Nile in the area (http://www.jinjatowncentre.com). In most of Africa, rivers like the Nile hindered migration, this explains the ethnic boundaries along the Nile as one moves north from the river's source at Lake Victoria. However, the area around Jinja was one place where the river could be reached due to the large rocks near the Ripon Falls. Here on either bank of the river were large flat rocks, where small boats could be launched to cross the river.

These rock formations were also accredited with providing a natural moderator for the water flow out of Lake Victoria. However for the locals it was a crossing both for trade and migration and as a fishing post. This explains why despite this barrier the two tribes have very similar languages and the more powerful Baganda had an enormous influence on the Basoga. The area was called the 'Place of Rocks' or 'The Place of Flat Rocks'. The word for stones or rocks in the language of the Baganda is "Ejjinja" (Plural "amayinja") and in the Basoga dialect this became "Edinda". The British used this reference to name the town they established - 'Jinja' (http://jinjatowncente.com).

When the Ripon Falls were submerged with the building of the Owen Falls Dam (later renamed Nalubaale Power Station) in 1954, most of the 'Flat Rocks' that gave the area its name disappeared too. However a description of what the area looked like can be found in the notes of John Hanning Speke, the first European to find the Source of the Nile:

"Though beautiful, the scene was not exactly what I expected, for the broad surface of the lake was shut out from view by a spur of hill and the falls, about twelve feet deep and four to five hundred feet broad, were broken by rocks; still it was a sight that attracted one to it for hours. The roar of the waters, the thousands of passenger fish leaping at the falls with all their might, the fishermen coming out in boats and taking post on all the rocks with rod and hook,

hippopotami and crocodiles lying sleepily on the water, the ferry at work above the falls and cattle driven down to drink at the margin of the lake, made in all, with the pretty nature of the country—small grassy-topped hills, with trees in the intervening valleys and on the lower slopes—as interesting a picture as one could wish to see." (J H Speke notes, 1954).

The town was founded in 1907 by the British, as an administrative centre for the Provincial Government Headquarters for Busoga region. This was around the time that Lake Victoria's importance in transport rose due to the Uganda Railway linking Kisumu, a Kenyan town on the lake, with Mombasa on the Indian Ocean, 900 miles (1,400 km) away. Cotton-packing, nearby sugar estates and railway access all enabled it to grow in size. By 1906 a street pattern had been laid out and Indian traders moved in from around 1910. The Indians were Catholic Christians and English-speaking and originated in the former Portuguese colony of Goa on the west coast of India.

The town remained the capital of Busoga region and in 1956, it was granted municipality status. At that time, it was the industrial heart of Uganda between 1954 and the late 1970s - supported by power from the hydroelectric Nalubaale Power Station at the Owen Falls Dam, which was completed in 1954. The dam meant that Jinja enjoyed clean, potable water on tap and an unwavering electricity supply throughout the 1960s. There was also a new and highly efficient drainage system leading into capacious sewers that emptied directly into the River Nile. Cars began to appear in the 1960s, often as taxi services.

Manchester-based Calico Printers Association, in association with the Uganda Development Corporation, constructed a large vertical textile mill ("Nyanza Textile Industries" or more popularly "Nytil") in the mid 1950s. This utilized hydro-electric power from the Owen Falls Dam. By 1973 Nytil employed about 3,000 people and exclusively used Uganda cotton to spin, weave and dye or print, to sell via its own retail chain, Lebel, throughout Uganda and Kenya. Genuine Nytil fabric was recognized by the "Silver Shilling" - a foil piece resembling a shilling which was inserted at one yard intervals along the edge of every cloth length produced. As Jinja grew, new roads were constructed, serving local taxis and the many people who lived outside the town. Each morning in the 1960s there would be a line of two-wheel traffic heading for the 'sokoni' or marketplace with cargoes of bananas or sacks of charcoal.

Jinja in the 1960s, like all the towns in Uganda, was subtly segregated, with little mixing of white, East Indian and black neighbourhoods. The white area was by the lakeside, with houses affording large gardens, near a lakeside club with golf, yachting, a rugby pitch and swimming pool. White children studied at the Victoria Nile School and were then sent to be schooled at Nairobi or the United Kingdom. The East Indians were the commercial and business class and lived in the rest of the town and they greatly valued education: in 1968, the huge Jinja Secondary School had one white student and about half a dozen blacks, while the remaining 500 students were all Asian.

East Indians were expelled from Uganda by Idi Amin in 1971 and 1972. Under Idi Amin's bloody rule, it is said that so many bodies were dumped in Lake Victoria that they often blocked the hydroelectric intake channels at the Owen Falls Dam. Much of Jinja's architecture is Indian-influenced, although the detailed shop-fronts and buildings were poorly maintained after the Indians left. Local industrial concerns also collapsed after the Indians were expelled. Many of the East Indians who are now returning to Uganda are choosing to set up businesses in Jinja.

Map 7.4: Structure plan for Jinja

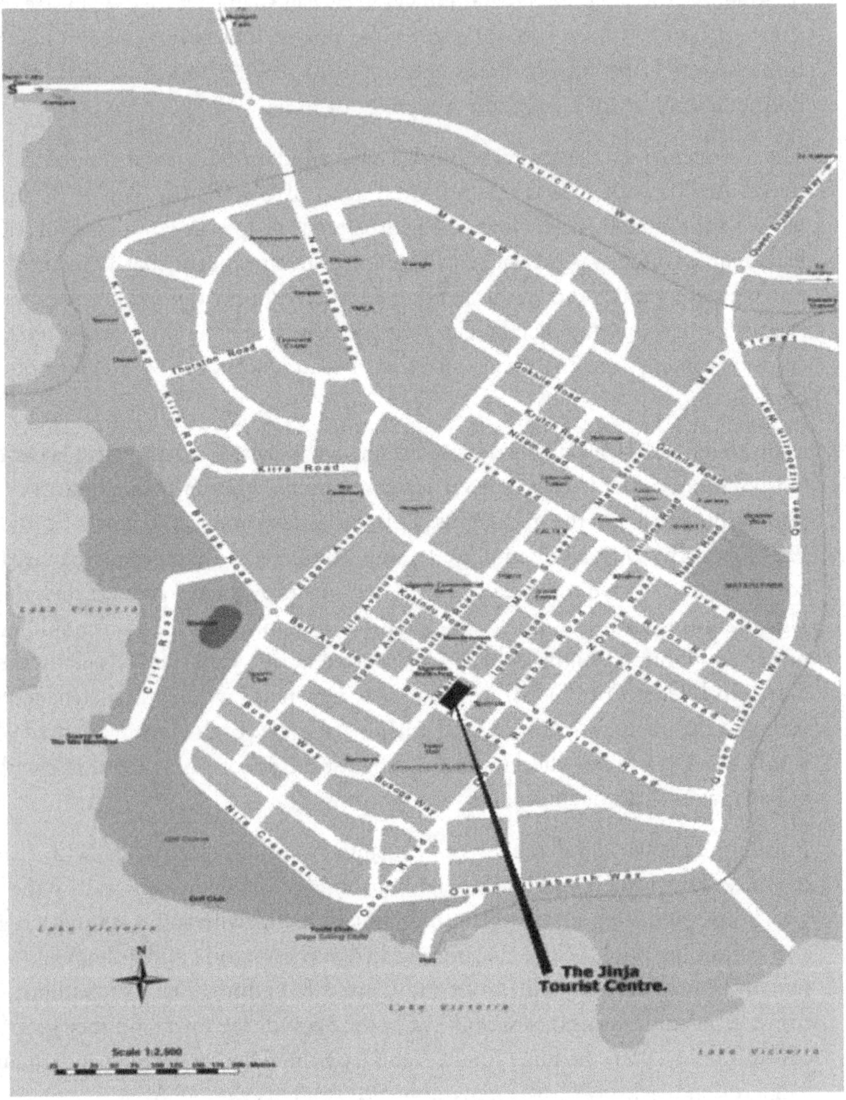

*Source:*Jinja Municipal Council

Map 7.5: Divisions of Jinja

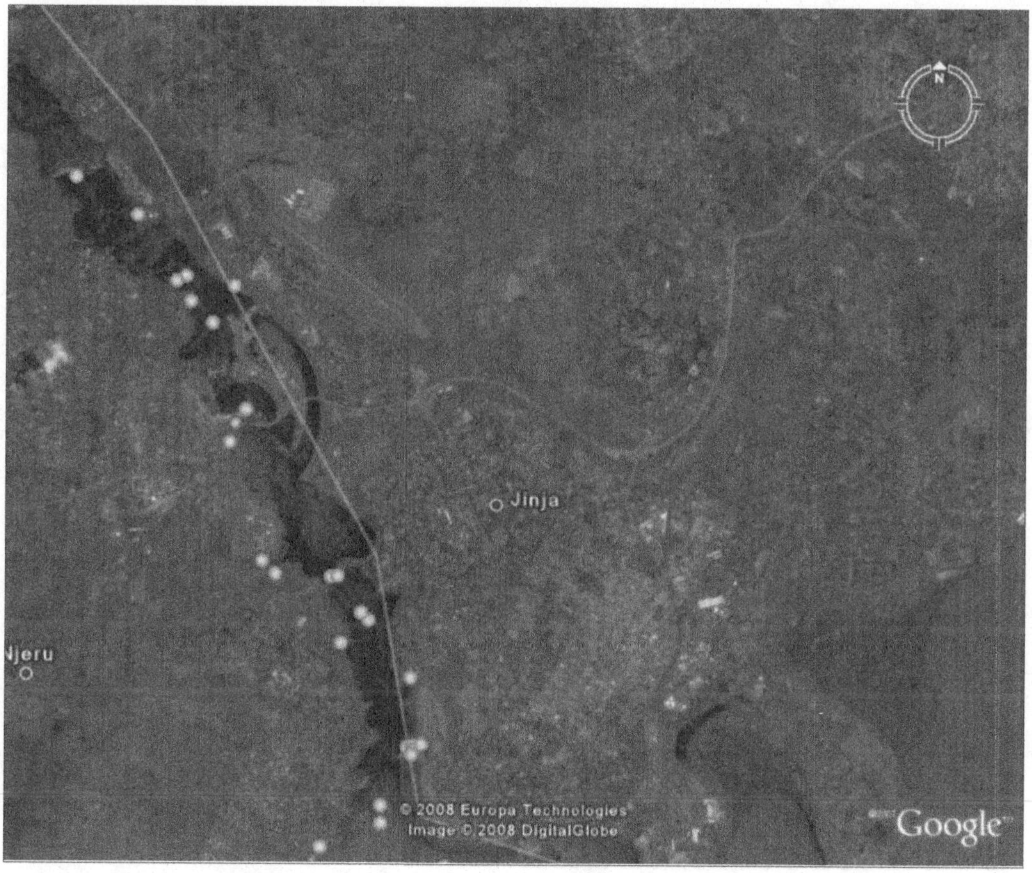

Source: Jinja Municipal Council

Map 7.5: Divisions of Jinja

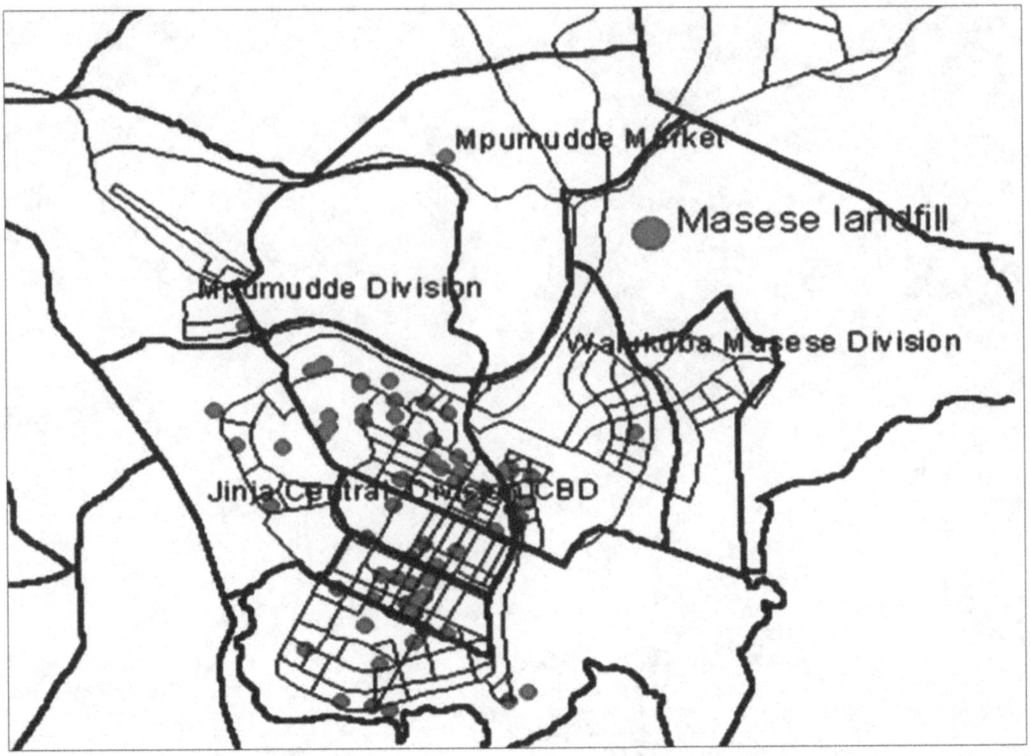

Source: Jinja Municipal Council

7.8 The Future of Urbanization in Uganda

Available evidence shows that urbanization in Uganda will continue to grow as long as Uganda towns continue to act as centres of innovation, diffusion for new agricultural methods, centres of multinational companies and new centres which have the basic service provisions in the form of facilities and infrastructure. This urbanization process will have direct impact on the urban environment as long as the towns continue to receive rural-urban migrants. Profound effects will also be felt by the existing urban dwellers. A combination of these two will exert a lot of pressure on existing facilities which are not commensurate with population numbers.

It is therefore, suggested that considerations for the future of urbanization in Uganda should look at the ways economic prosperity can be achieved without causing environmental damage. The ever increasing numbers of people may mean high costs per household for the provision of piped water, the collection and disposal of human wastes, health facilities, education etc. This, therefore, will require urban authorities to promote environment revolutions where intermediate technologies could be applied to reduce freshwater use, in the city and towns, recycling of waste and effective sanitation systems.

Poor drainage systems

Source: Hannington Sengendo

The support here should be addressed to the needs and priorities of the city's and town's low income people, especially the vulnerable groups—women and children. The question which must not be left unanswered however is "how does such support address the needs and priorities of these poor people?" In Uganda, if cities have to be looked at in future as engines of economic development, which are experiencing the impact of rapid urbanization, the immediate solution will be for the policy makers and planners to incorporate preventive and curative measures in response to the complex issues of urbanization, urban poverty and environmental degradation.

7.8.1 Towards Effective Government Interventions

It is not always clear whether there is a pro-active public policy position towards urbanization by most governments in developing countries particularly Sub-Saharan Africa. There are of course other discussions and statements of intent in planning and public policy documents. Urban policies therefore, tend to be more reactive except those that relate to zoning and housing standards.

7.9 Conclusion

The urbanization perspective shows that there are strong links between urbanization, urban social economic activities and urban environmental quality. Further still, there are also strong links between urban population growth, environmental deterioration and environmental management. Urban households manage or fail to manage environmental resources as part of sustaining their daily lives, for instance in form of urban agriculture and other formal activities which involve securing an income, maintaining social relations and investing in the life of their children.

These include the pace and intensity of the urbanization process, population growth patterns and city size, the complexity of urban ecosystems, economic sectors and political jurisdictions, the spatial scale at which particular urban based environmental problems occur, urban land use and environmental interrelationships and a multitude of key public and private sectors.

In Uganda, until recently, policy makers failed to evolve appropriate policies regarding urbanization and environmental deterioration. There was no regard for policies related to addressing urban environmental problems in relation to what can be termed as INTERNAL OR EXTERNAL factors in regard to environmental crisis in urban locations. This resulted mainly in unprecedented social and economical problems. For instance, this is evidenced by improper waste management, urban plight, decay and congestion, urban agriculture to supplement incomes of the urban poor or used as a means to offset the social costs arising from the Structural Adjustment Programme.

For the urban environmental problems to be effectively dealt with it is essential to address the underlying political, institutional and local conditions perpetuating urban environmental degradation and government inaction. The most important issues that need to be resolved therefore include: poor urban governance including weak institutional capacity in planning as well as the managerial and operational aspects of urban service delivery, jurisdiction complexity and the lack of public education about environmental issues. In addition to the above are, insufficient community and private sector participation in planning and implementing of urban environmental infrastructure and services, the pervasive lack of public awareness of environmental problems and the unwillingness of government authorities to adequately address them.

Another issue which needs to be addressed is the ill-conceived regulatory and economic policy that encourages environmental degradation and which fails to provide a necessary basis for comprehensive environmental management. The National Environmental Policy for Uganda is currently taking this into consideration.

Key Terms

1. Urban structure

   Refers to the standard pattern that a city takes according to different land uses depending on historical, economic, social, political and physiographic factors

2. Urban form

   Refers to a degree of regularity from changes that take place with time. This change may be due to economic and social conditions which modify its morphology to a definite pattern of its internal organization.

3. Urbanization

   Refers to the process of becoming urban or relative concentration of a territory's population in towns and cities. This process can be accompanied by economic development, change of urban boundaries and change of administrative status among others.

4. Urban agriculture

   An activity that produces food and other products derived out of land, water, livestock and poultry in urban and peri-urban areas.

References and Further Readings

Bigstem, A. and Kayizzi, M. 1992. *Adaptation and Distress in the urban Economy:* A study of Kampala Households, *World Development*, Vol. 20, No.10.

Government of Uganda, (1993) Background to the Budget 1993/1994, Government Printer, Entebbe, Uganda.

Government of Uganda. 1991. *Uganda Population and Housing Census*

Government of Uganda. 2002. *Uganda Population and Housing Census*

Gutkind P. C. W. 1963. The Royal Capital of Buganda, The Hague, Mouton and Co.

Kampala City Council. 1993. *Kampala First Urban Study*

Hauser, P.M.1965. *The Study of Urbanisation*, Wiley, New York.

Knox P. L. 1991. "The Restless Urban Landscape: Economic and Social Cultural Change

Transformation of Washington DC", Annuals of Association of American Geographers 81, 2.

Little, K.L. 1974. *Urbanisation as a Social Process*, Routledge, London.

Maxwell D. and Zziwa S. 1992. *Urban farming in Africa; The case of Kampala Uganda*, ACTS Press. Acts Studies, Nairobi, Kenya.

Mougeot L. 1993. *Urbanisation and Urban Environment Management in Canada.*

Nuwagaba A and Mwesigwa D. 1997. *Environmental Crisis in Peri-Urban Settlements in Uganda; Implications for Environmental Management and Sustainable Development.* Makerere Institute of Social Research Kampala.

O'Connor, A. 1986. *The African City*. Hutchinson and Company Publishers, London.

Tevera, D. 1993. *Urban Waste Management in Central and Eastern Africa In Urban Environment Management, IDRC Workshop Proceedings,* Ottawa, Canada.

Savio, C.J.1993. *Urban Agriculture: Eastern and Central Africa*, IDRC Workshop Paper, Ottawa, Canada.

Sengendo H. 1992. *The Growth of Low Quality Housing in Kampala between 1972-1989.* Thesis submitted to University of Nottingham UK in partial fulfillment of a PHD.

Sengendo H. 1994. *Urban Poverty, Shelter and Urban Governance*, Makerere University, Kampala.

Southall Aw nand Gutkind P. C. W. 1957. Townsmen in the Making, *Institute of Social Research East African Studies No. 9*

Speke J H. 1954. *Notes on Description of Source of the Nile*, Government Archives Entebbe, Uganda.

Stren R. and White R. 1989. *African Cities in Crisis: Managing Rapid Urban Growth*, Boulder, Westview.

Townsend, A. 2004. *Ubiquitous Information Technology, Envisioning the Ubiquitos City*, http:urban.blogs.com.

United Nations Report. 1992. UNEP. 1992. Strategic Resource Planning in Uganda Vol. X Human Settlements World Bank Report. 1991.

http://jinjatowncentre.com

Van Nostrand Associates. 1994. *Processes of Urbanization and Their Environment Impacts.*

Wally N'Dow. 1994. *Population, Urbanisation and Quality of Life*, United Nations Centre for Human Settlement (Habitat), Nairobi.

World Bank. 1991. *First Urban Project Kampala City Council*, Kampala, Uganda.

Wratten. 1995. *"Conceptualising Urban Poverty"* in *Environment and Urbanisation*, Vol. 7, 1, London.

Chapter 8

MEDICAL GEOGRAPHY OF UGANDA

BAKAMA BAKAMANUME

8.1 Introduction

Medical geographers attempt to explain the geographical variation associated with medical and health issues (Meade et al. 2001). This chapter, however, focuses on four health issues disease distribution, mortality causes, distribution of medical facilities and HIV/AIDS and malaria prevalence. The chapter demonstrates how health care issues can be illustrated and analyzed using geographic and cartographic techniques.

In seeking to explain disease distribution, medical geographers attempt to answer four related questions: *who* is affected with a disease, *what* disease(s), *where* is the highest prevalence and incidences and *why?* This relationship is studied within the general theme called the disease ecology. The disease ecology identifies three aspects, host, vector and agent.

Figure 8.1: Disease Ecology

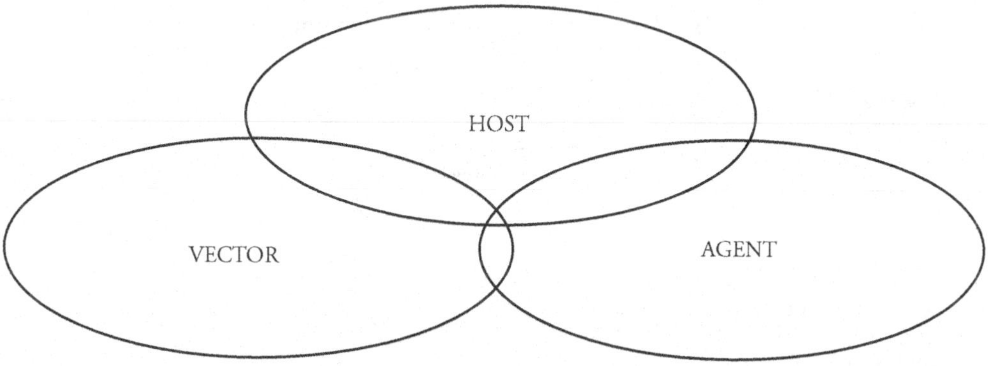

Medical geographers examine disease agents (pathogen, germs, parasites), disease vectors (transmitters of disease, for example, mosquitoes, fleas) and hosts (organism affected by disease). They use the concepts and techniques of geography to investigate health related issues. The term "medical geography" was introduced by Dr. John Snow, who was working on the cholera disease in London in 1854.

8.2 Measures of Frequency of Health Events: Rates, Incidence and Prevalence

Medical geographers focus on the frequency of disease or risk factors. In order to map and compare the frequency of disease in different groups of population and regions, it is necessary to calculate rates. A *rate* is calculated by dividing the numerator (number of cases) by the denominator (the population at risk). It is expressed as a certain number per 1,000 (10,000 or 100,000).

Disease frequency or occurrence is measured by incidence and prevalence. *Incidence* is the number of reported new cases, during a particular period of time, for a given population. Incidence rate (incidence density) considers the time/ duration each individual in the risk group remained under observation.

Prevalence is the number of people infected with a particular disease. The terminology is used to determine presence of disease in a population at a particular time.

COMPONENTS OF A RATE: NUMERATOR DENOMINATOR

Disease rate:

There are 10 sick persons in one year in a population of 1,000.

$$\text{Disease Rate} = \frac{10}{1{,}000 \times 1 \text{ year}}$$

Incidence:

$$\frac{\text{Numerator is new cases}}{\text{Total population at risk at the given time}}$$

Prevalence:

$$\frac{\text{All cases}}{\text{Total population at risk}}$$

DEFINITION OF IMPORTANT RATES IN TERMS OF NUMERATOR AND DENOMINATOR

Birth rate

$$\frac{\text{Number of live births during the year}}{\text{Total Population}} \times 100000$$

Death rate

$$\frac{\text{Number of deaths during the year}}{\text{Total Population}} \times 100000$$

Fertility rate

$$\frac{\text{Number of live births to women 15-49 years}}{\text{Number of women 15-49 years of age in the population}} \times 100000$$

Cumulative incidence

$$\frac{\text{Number of new cases in a specified time}}{\text{Population at risk and the specified time}}$$

Incidence rate

$$\frac{\text{Number of new cases of a disease during a given time}}{\text{Total Population at the time}}$$

Figure 8.2: Incidence and Prevalence

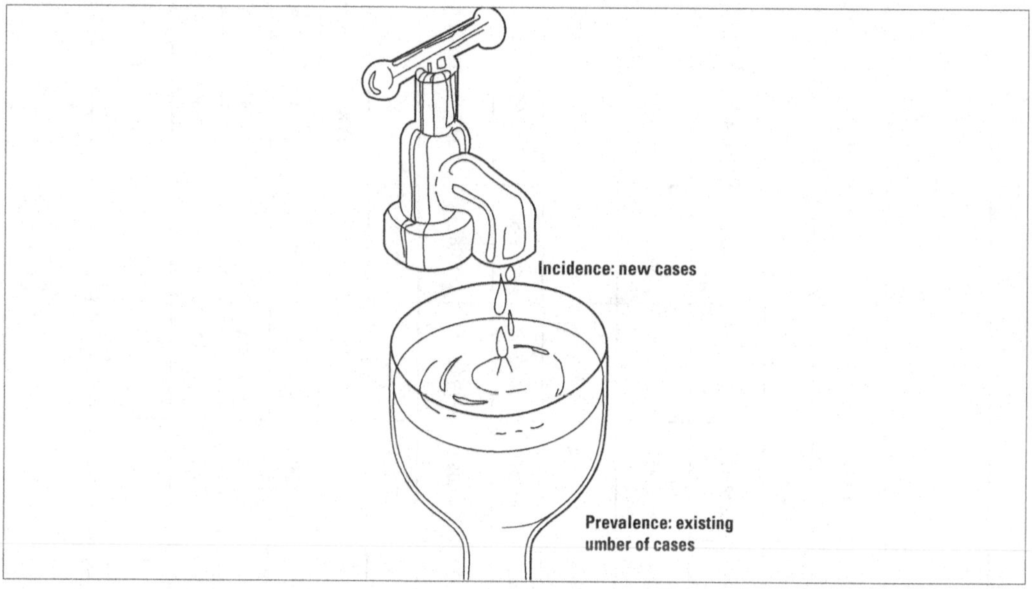

Source: Kibei M A, Wagstaff LA, 1995.

Incidence is similar to running water. New cases are added to existing cases. Existing cases are similar to water already in the glass.

8.3 Disease Distribution

This section discusses the distribution of some of the leading diseases in Uganda. The diseases responsible for the largest proportion of morbidity and mortality are Malaria, Acute Respiratory Infections HIV/AIDS, Tuberculosis, Malnutrition, Maternal and Prenatal conditions, Cardiovascular conditions and Trauma/ accidents (Ministry of Health, 2004). The data and trends on disease morbidity and mortality are shown in Table 8.1 and Figure 8.3 respectively. The data and graph show that malaria accounts for the largest proportion of morbidity and mortality. This trend promises to continue.

Table 8.1

MORBIDITY TRENDS FOR 1991-2002

TYPES OF DISEASE	1991	1992	1993	1994	1995	1997	1998	1999	2000	2001	2002
Malaria	2,708,118	2,446,659	1,470,662	2,220,402	1,444,352	2,317,840	2,845,811	2,845,811	2,923,620	5,622,934	5,725,866
Lower Respira-tory	606,078	587,910	354,153	1,198,866	828,072	1,147,752	1,222,380	1,239,642	1,272,064	2,024,452	2,343,912
Upper Respira-tory	157,1471	141,813	937,921	526,634	598,452	655,890	563,557	426,691	549,064	789,239	815,367
Intestinal worms	1,005,,006	986,297	634,268	735,783	531,054	733,818	744,917	784,708	778,463	1,235,399	1,564,773
Diarrhoea	896,066	749,962	456,459	539,885	367,203	572,950	556,557	565,055	580,250	696,184	723,241
Complication from pregnancy	97,784	79,511	55,158	72,322	54,052	84,820	84,618	103,733	111,421	140,423	179,123
Tuberculosis	14,679	14,249	20,628	15,003	17,134	19,329	22,892	33,522	21,133	30,809	35,289
Ear Infection	247,464	222,962	122,832	128,380		125,713	134,072	148,887	142,799	191,426	257,018
Eye Infections	434,834	439,102	254,664	342,353	226,381	311,809	240,331	148,887	244,145	282,316	0
Dental Diseases	149,635	141,368	109,600	108,838	87,487	124,061	145,781	170,697	174,488	257,495	0
Malnutrition	114,206	123,045	60,651	81,360	66,699	64,412	79,345	69,360	65,864	63,539	0
Whooping Cough	14,088	9,210	6,302	5,306	2,073	0	0	0	0	0	0

Source : Ministry of Health Statistical Abstract 2004

Figure 8.3 : Morbidity for Ten Major causes of Illness

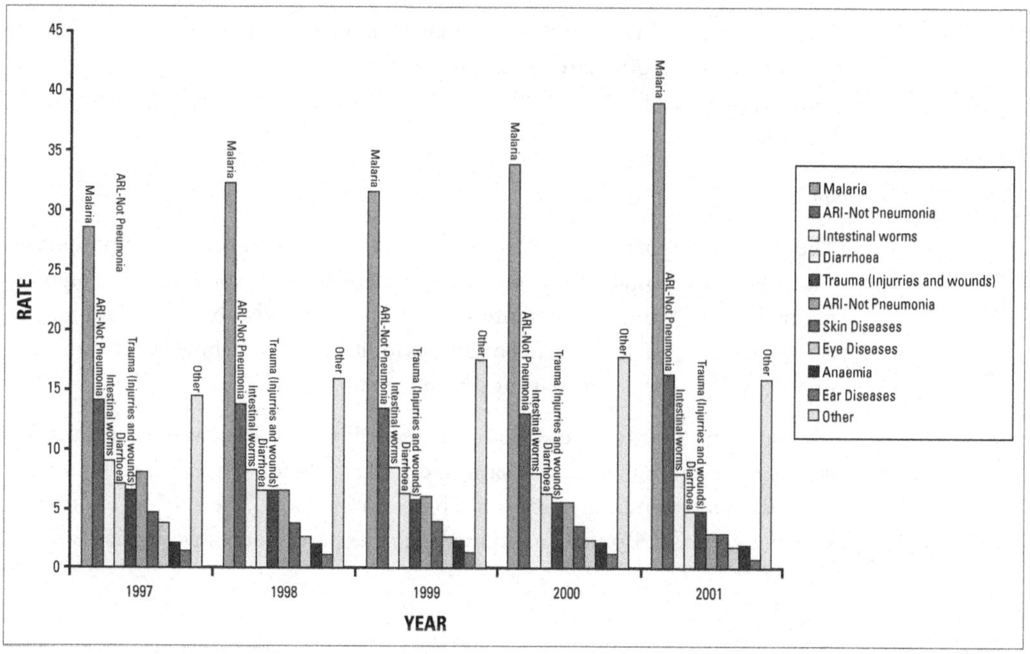

Source: Uganda Bureau of Statistics, 2004

Malaria is the leading cause of morbidity. It is followed by ARI Not Pneumonia and Intestinal worms.

8.3.1 Malaria

Malaria is very common and widespread in Uganda. Historically, malaria has been a serious health problem and currently poses a great threat to the population (MOH, 2003). One out of five deaths (23.4%) in Uganda is caused by malaria (New Vision, 2000). A recent report puts the figure at 40,000 people every year (Omaswa, 2003). In terms of resource burden, Malaria is responsible for 25-40% of outpatients, 20% of hospital admissions, 9-14% of in-patient deaths and it is a major cause of death among children under five years, refugees and Internally Displaced Persons.

There are four species of the human malaria parasite in Uganda (McCrae, 1975). *Plasmodium falciparum* (malignant tertian) is the leading type. It causes severe forms of illness (e.g. cerebral malaria). It is also the leading cause of mortality (Gongora and McFie, 1959; Ministry of Health Report, 2003). It is associated with anemia in pregnancy and low birth weight (Jelliffe, 1966). *Plasmodium malaria* (quartan) is the second most prevalent. It often combines with *Plasmodium falciparum. Plasmodium ovale* is widespread but not very common. *P. ovale* and *Plasmodium vivax* (benign tertian) account for about 4% of malaria infections.

8.3.1.1 Endemicity of Malaria

In 1975, Hall and Langlands stated that Malaria is endemic in most parts of the country below 1,500 metres. Currently, malaria cases have been reported in areas above 1,500 meters (WHO, 2002). This is often referred to as highland malaria. (See Map 8.1).

Vectors

The major vectors of malaria are *Anopheles gambiae* and *Anopheles funestus* mosquitoes. The *A. gambiae* mosquito is encouraged by muddy pools and puddles of a temporary nature especially during the rainy season. The vector also benefits from human activities which create the right conditions for its breeding. The *A. gambiae* is most prevalent during the early part of the rainy season.

The breeding sites for the *A. funestus* are shallow bodies of water with vertical emergent vegetation, such as swamps, calm lakes, pool margins and fish ponds. The seasonality conditions are favourable to both vectors. The first half of the rainy season is good for *A. gambiae* and the later drier part is more favourable for the *A. funestus*.

Map 8.1: Malaria Rates in Uganda

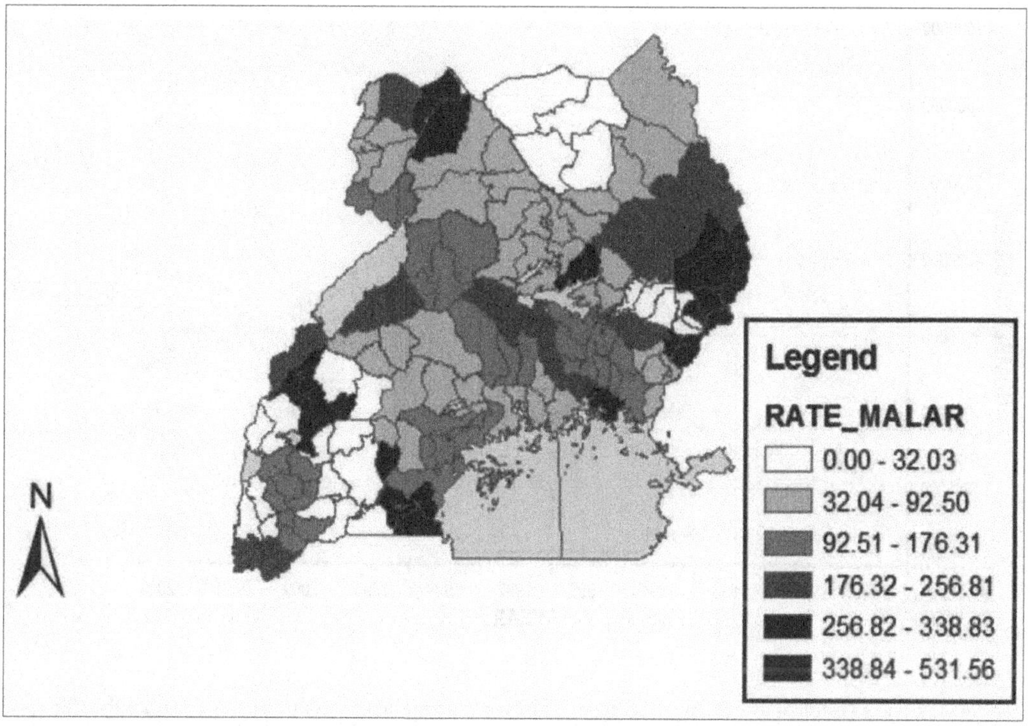

Source: Uganda Ministry of Health 2004

High rates of malaria are found in the lake basin regions of Uganda (Lakes Victoria, Kyoga, George and Edward) and the Nile river valley in the Northwest.

8.3.1.2 Prevalence of Malaria

Malaria in Uganda shows a prevalence trend that is closely associated with seasonal rainfall. The months of May and June have the highest reported cases. The South-western region has the highest number of malaria cases related to changes in weather. The most affected districts in the southwest are Ntungamo, Kabale, Rukungiri, Mbarara and Bushenyi. The districts with the highest number of malaria cases are Ntungamo, Tororo, Rakai, Wasiko and Mbale (Ministry of Health Report, 2004). Since 1995, the number of cases has increased rapidly. This trend is shown in Figure 8.4. The graph shows a general continual increase in the number of malaria cases except for the years 1993 and 1995. The number of malaria cases declined in 1993 and 1995. The decline shown may be a product of reporting rather than interventions.

Figure 8.4: Malaria Infection Trend

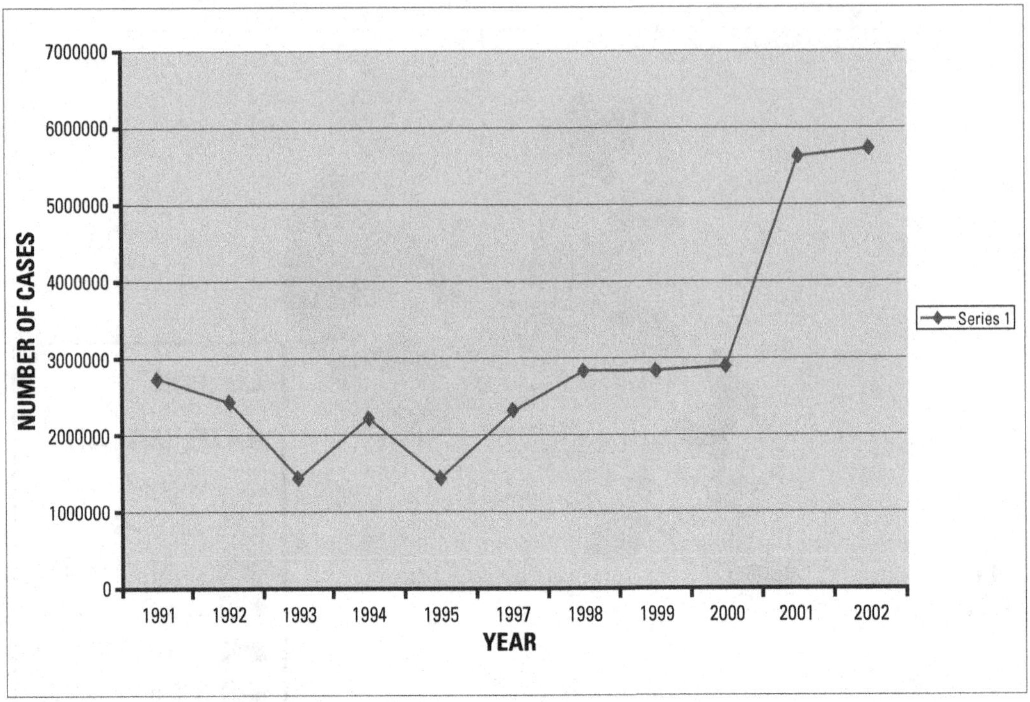

Source: Ministry of Health Statistical Abstract, 2004/ The Malaria trend shows a continual increase since the 1990s.

8.3.1.3 Some of the Public Health Impacts of Malaria

1. Major cause of mortality among children
2. Causes childhood anaemia
3. Reduces growth and causes mental retardation among children
4. Causes still and premature births among pregnant women
5. Responsible for 60% of the miscarriages in the country
6. Responsible for 23.4% of the deaths in the country

Discussion Box 1

1. Discuss the economic impacts of malaria.
2. What are the socio-cultural impacts of malaria.

8. 3.1.4 Control and Eradication of Malaria

Uganda is turning back to DDT pesticide. This is a pesticide that was banned worldwide because of its environmental impact. The former minister of health, Honorable James Muhwezi, is on record as a strong supporter of the idea/project to use DDT to eradicate mosquitoes. Even though the Malaria Pre-Eradication Programme was established in 1964 by WHO, over the years, the major vectors have become resistant to pesticides used, hence the call to return to DDT. DDT has a drastic effect on the environment and its use was stopped all over the world. A recent report indicates that DDT has been identified as one of the contaminants of Lake Victoria (Monitor, August 2005).

8. 3.1.5 Summary

Malaria is a leading cause of ill health, mortality and poverty in the country. Thus, controlling malaria will improve human development and enhance the fight against poverty. Malaria transmission is increasing, partly due to deforestation, cultivation of wetlands, poor environmental sanitation, other human impacts such as fish ponds and constructions sites. These conditions provide the breeding grounds for mosquitoes. It has also been demonstrated that malaria is more widely spread today than in 1970s.

Discussion Box 2

1. What are some of the likely environmental effects of DDT?
2. Discuss the advantages and disadvantages of using DDT

8.3.2 HIV/AIDS Related Diseases

Acute respiratory infections (both lower and upper) are closely associated with HIV/AIDS prevalence in Uganda (MOH, 2003). The first AIDS case in Uganda was reported in 1982 (Uganda AIDS Commission, 2003). In 1983, 17 more cases of "Slim" disease were reported. Slim was the name given to the new disease because it caused great loss of weight before death. In 1984, "slim" disease was confirmed as AIDS. The epidemic spread from major towns to rural populations following major transport routes (BakamaNume, 1997; Smallman, 1991). To date, all districts have reported AIDS cases. HIV/AIDS prevalence reached a peak in 1992. Since then, the prevalence has leveled off and it continues to decline (UNAIDS, 2004). The decline is attributed to the openness approach in the country. However, it should be noted that while prevalence has continued to decline, the number of HIV/AIDS cases (incidence) has continued to increase.

8.3.2.1 Prevalence of HIV/AIDS

HIV/AIDS prevalence rates in pregnant women have been declining in both rural and urban areas (Okware etal. 2001). Sentinel surveillance have shown remarkable decline among pregnant women (AIDS Control, 2003). For example, rates in Nsambya (see Table 8.2 and Figure 8.5) have declined by over 50% between 1992 and 2002. Regionally, the rates are slightly higher in the north than the south and urban areas have generally higher rates than rural areas. The northern region has seen declines from 27.1 in 1993 to 12.3 in 2000. This decline is mainly among the young age cohorts.

Table 8.2: HIV Infection Rates (%) at Selected Antenatal Sentinel Sites in Uganda

Site	1989	1990	1991	1992	1993	1994	1995	1996	1997	1998	1999	2000	2001	2002
Nsam-bya	24.5	25.0	27.8	29.5	26.6	21.8	16.8	15.4	14.6	13.3	12.3	11.8	9.5	8.5
Ribaga			27.4	29.4	24.4	16.5	20.2	15.1	14.8	14.2	10.5	10.7	10	8.3
Mbarara	21.8	23.4	24.3	30.2	18.1	17.3	16.6	15	14.5	10.9	11.3	10	10	10.8
Jinja	24.9	15.8	22	19.8	16.7	16.3	13.2	14.8	11	10.5	10.8	8.3	7.4	5
Tororo		4.1	12.8	13.2	11.3	10.2	12.5	8.2	9.5	10.5	4.5	4.7	7	6.3
Mbale	3.8	11	12.1	14.8	8.7	10.2	7.8	8.4	6.9	6.3	5.7	5.5	5.6	5.9
Kilembe					7.0	16.7	11.1	10.4	8.5		7.5	4.2.5	2.1	4.2
Pallisa				7.6	5.1	1.2			3.2	2.6	3.2	3.8	3.7	
Soroti					9.1		8.7	7.7	5.3	7.7	5.2	5	5.0	4.6
Matany					2.8	7.6		2	1.6	1.3	.9	1.9	1.7	.7
Hoima								12.7	9	5.4	3.5		5.3	4.6
Kagadi									10.3	11.5	11	10.5	7.4	6,4
Mu-tolere		4.1	5.8		4.2		3.6	2.6		2.5	2.3	2.1	4.1	1.5
Moyo					5.0		3.1			2.3	5.2	2.7	2.7	4.3
Arua					4.4						5.2	5.2	4.8	5.2

Source: Uganda Ministry of Health 2004

Figure 8.5 shows that HIV rates at the five selected sites have declined since 1992. The rate initially rose to a peak in 1992. In 1992, some of the sites had a 30% infection rate. The rate has declined to 10.8%.

All the urban sites shown in the graph have experienced a decline in HIV rates. The median HIV prevalence for the six sites, was 7.2% in 2002. Okwale et al (2001) attribute the decline in HIV/AIDS prevalence to a number of factors.

1. National/ government effort
2. Institutional capacity building
3. Public Education for Behavioral change
4. STD management
5. Care and support for people with HIV/AIDS
6. Surveillance systems for HIV/AIDS

Figure 8.5: HIV Infection Prevelence Rates at Five Intenantal Sites

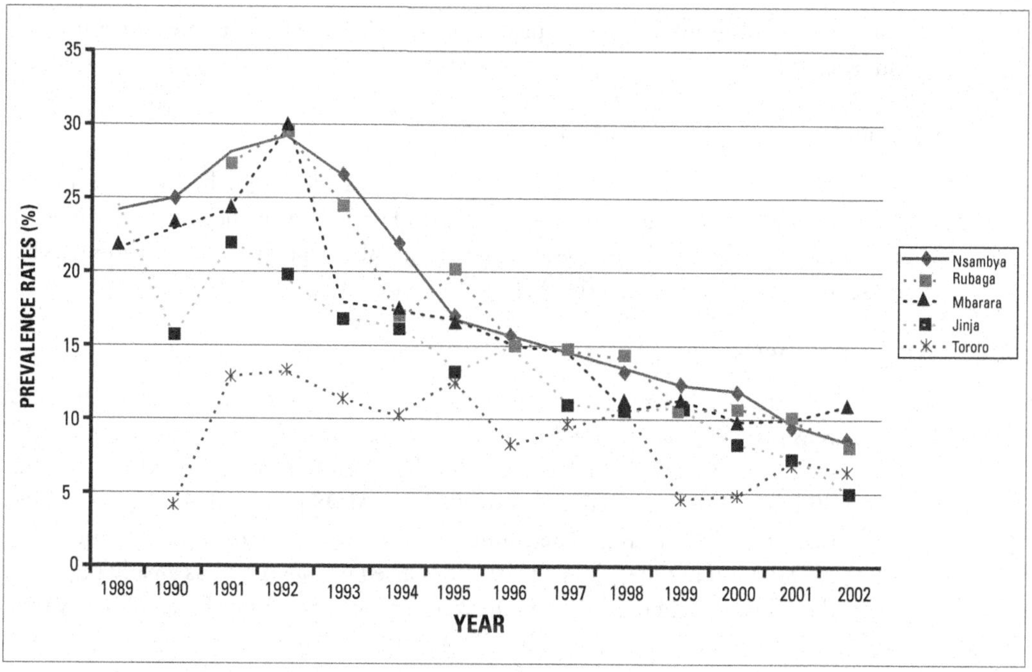

Source: Uganda Ministry of Health 2003

1. National Effort

The government took a sustained response to create an environment where openness about HIV/AIDS was encouraged. The Uganda AIDS Commission was established in 1992 to coordinate all efforts to fight HIV/AIDS.

2. Institutional Capacity Building

Institutions have been set up at regional and national levels to provide services, conduct research and train personnel in monitoring HIV infection. The institutions are both government and non-government, for profit and non-profit.

3. Public Education and Behavioural Change

Population based surveys in several districts in Uganda have shown that there is a high level of awareness of HIV/AIDS, increase in levels of knowledge of protection from HIV/AIDS, increase in condom use and late participation in first sex. Public education programmes in schools, at higher institutions of learning, through the media and drama have contributed to awareness of HIV/AIDS.

4. STD Management

The ministry of health has put in place a reasonable STD programme which involves surveillance.

5. Care and Support for People with HIV/AIDS

The AIDS Support Organization (TASO) is one of the original Non Government Organizations (NGO) that began the fight against HIV/AIDS and helping people infected with the disease. Its model has been replicated in other countries. The government has also put in place programmes that help people with HIV/AIDS.

8.3.2.2 Summary

Life expectancy has declined since the HIV/AIDS onset. According to the 2002 census, life expectancy is 43 years. However, there have been some minor gains in the last two years (Health Report, 2003). The health service sector is slightly better off but still understaffed. Some of the HIV/AIDS patients have access to drugs that prolong their lives. The country as a whole has almost no stigma on HIV/AIDS (MOH, 2003). HIV/AIDS is discussed openly. This has greatly improved the efforts to fight the disease in the country. Uganda is much better off today than in 1992.

Map 8.2: HIV/AIDS Rates

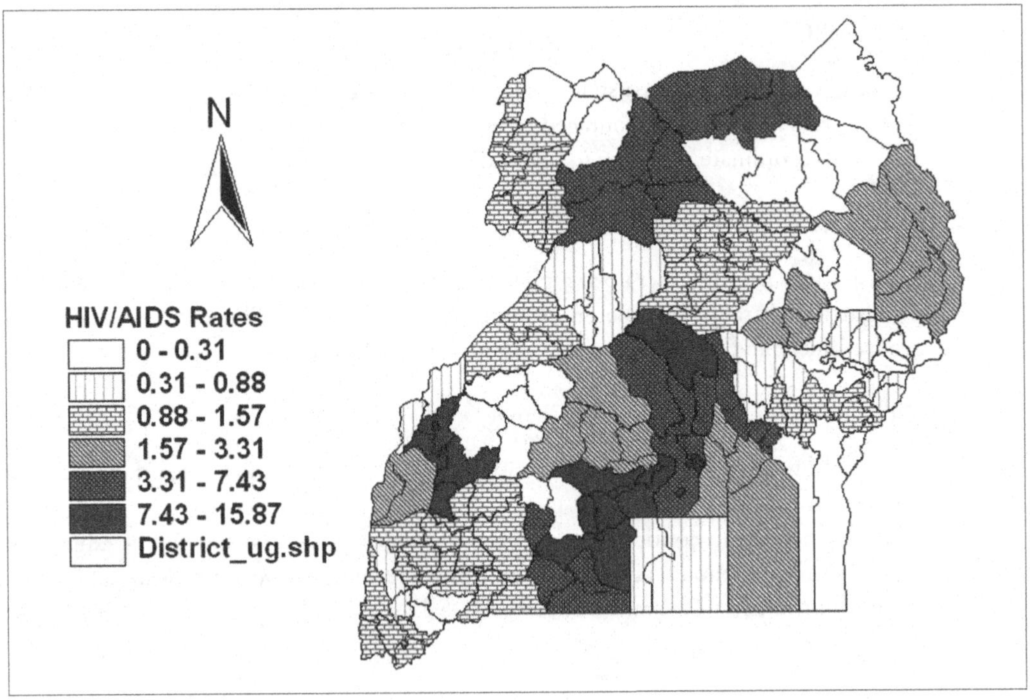

Map 8.3: TB Rates in Uganda

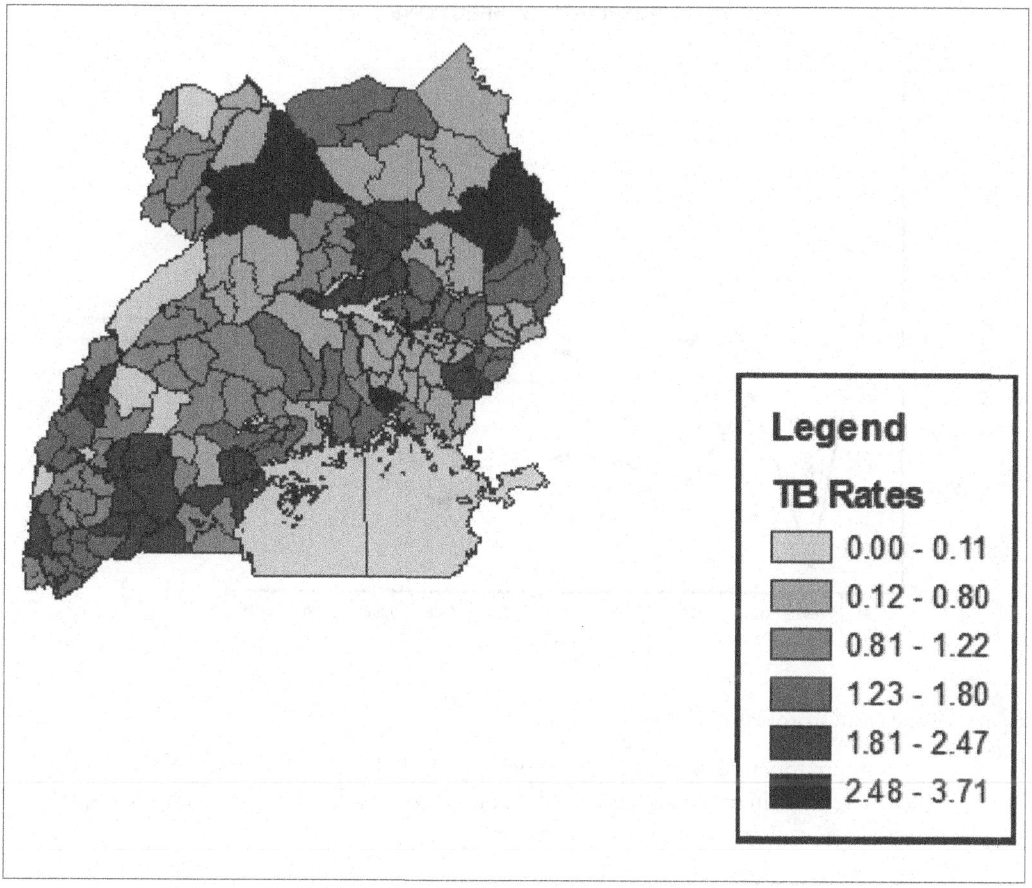

Source: Uganda Ministry of Health 2004

HIV/AIDS rates are highest in the urban region of the south and north and in Kasese Area. TB rates show some close association with HIV/AIDS rates. The Northern and Southern Regions especially Rakai District have the highest rates.

8.3.3 Other Diseases

8.3.3.1 Respiratory Infections

Respiratory infections (lower and upper) are the second largest proportion of morbidity and mortality in the country after malaria. These infections are commonly associated with HIV/AIDS prevalence. Lower respiratory cases have continually been on the increase (see Figure 8.8). This could be attributed to the rise in HIV/AIDS cases.

Figure 8.8

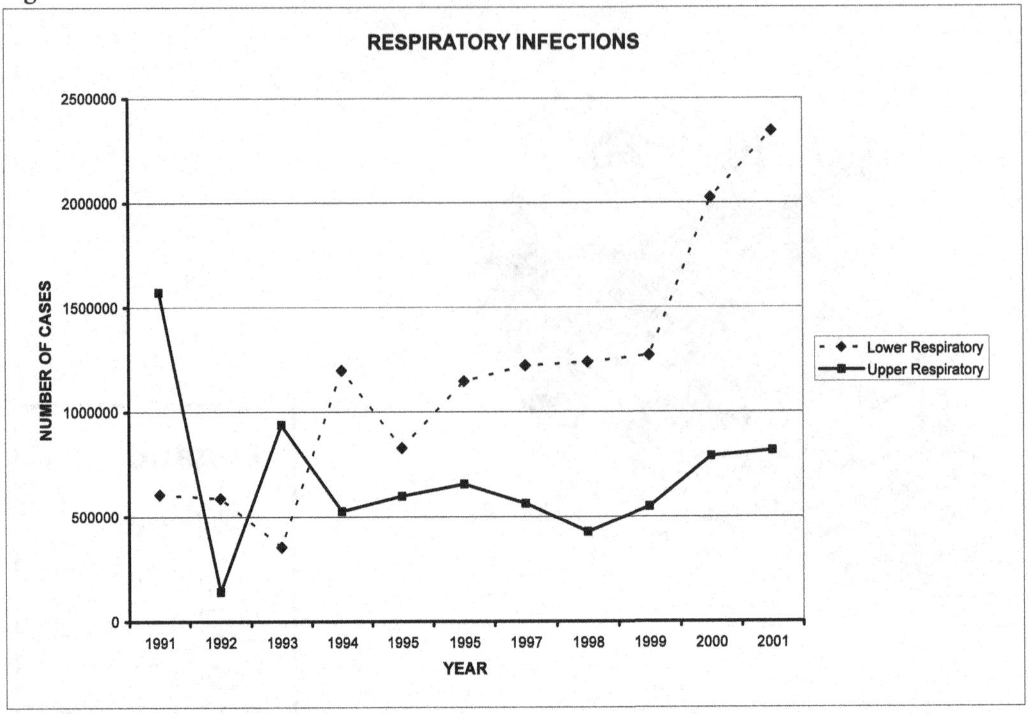

Source: Uganda Ministry of Health 2004

Upper respiratory infections were at the peak in 1991. This was followed by sharp decline in 1992. The rates increased after 1992 but have leveled off since 1994. Lower respiratory infections have risen since the decline in 1993.

8.3.3.2 Tuberculosis (TB)

HIV and Tuberculosis (TB) have posed a serious challenge to health care delivery in Uganda (Ministry of Health, 2003). TB prevalence is associated with HIV/AIDS. There are two types of tuberculosis in Uganda (Morrow, 1975). Pulmonary tuberculosis is common and outpatients and inpatients returns show an increase in prevalence. Myco tuberculosis is also present among the population. TB rates are shown in Map 8.3. The map shows no clear pattern. Four-five areas have relatively high rates. The areas are Northwest central, Northeast, South central (Rakai) and urban areas.

8.3.3.3 Tetanus

Tetanus is a major medical problem in Uganda. It occurs throughout the country from infected wounds and neonatal tetanus due to infection of the umbilicus (Bennet F. etal. 1975). Tetanus is a problem in the country, with rates of nearly 10 per 10,000 people. In Busoga region (Jinja, Kamuli, Iganga, Mayuge, Bugiri), tetanus was the second cause of hospital deaths in childhood in 1960s and early 1970s (Bennett et. 1975). Currently, the situation is better, but tetanus still remains a problem in the country.

8.3.3.4 Brucellosis (Malta fever)

Brucellosis in Uganda has been historically associated with keeping goats and drinking raw goat milk (Watts, 1975). There have been few cases reported of brucellosis in humans in the recent years. This is partly because of the aggressive approach to vaccinate calves and to educate the people on dangers of raw milk. The cases in humans are noticeable among cattle keepers especially in Ankole area. Boiling milk effectively eliminates the risk.

8.3.3.5 Guinea Worm (Drancunculiasis)

In 1991, Uganda had a severe guinea worm infection problem. This prompted the establishment of the Uganda Guinea Worm Eradication Programme (UGWEP). The UGWEP was established as a joint venture of the Uganda Government and USAID. UGWEP has made tremendous achievement in the fight against the disease. There has been a 99% reduction in the number of cases since 1992 (See Figure 8.9). Guinea worm infection has declined since 1992.

Figure 8.9: Guinea Worm Eradication in Uganda

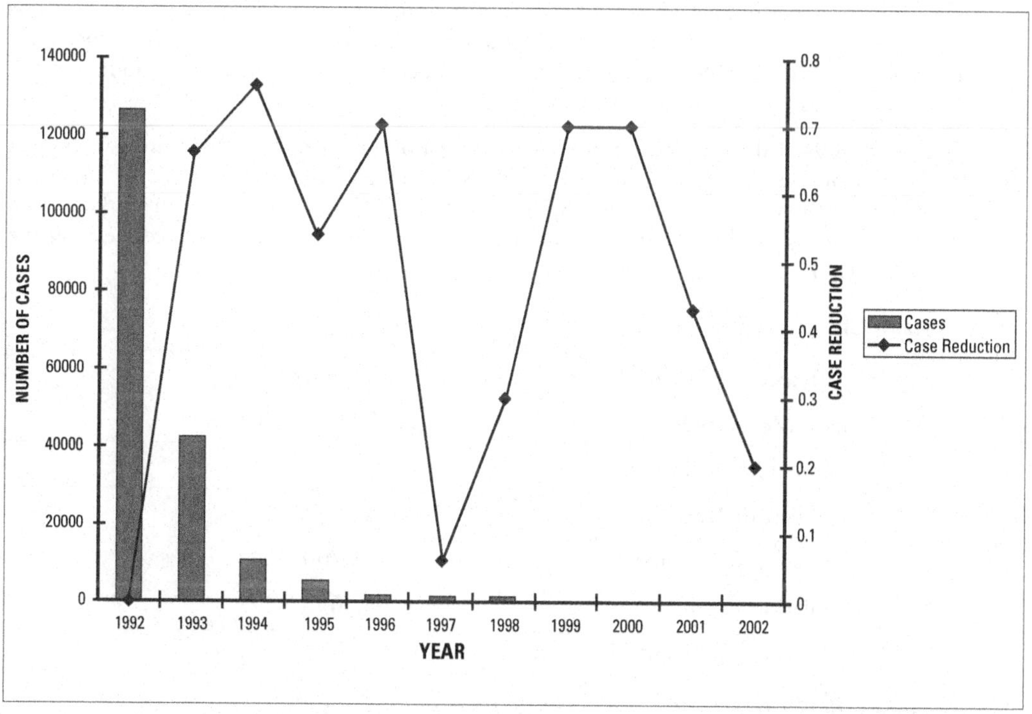

Source: Uganda Ministry of Health 2004

8.3.3.6 Accidents and Trauma

In June 2004, a road accident claimed the life of the attorney general of Uganda. There is a growing concern about the rising trend in mortality and morbidity from road traffic accidents in developing countries due to their effect on health care resources and budgets. Traffic accident injuries account for high medical care costs and loss of productivity (Murray, et al., 1996). In many developing countries, traffic accidents are the second, or the third leading cause of mortality. Yet, there is scant literature on the epidemiology of road traffic accidents in Africa and in Uganda in particular. The role of urbanization and transport changes in causing a shift in the relative importance of various sources of mortality are important elements in epidemiologic analysis. An improvement in road conditions is often promoted, in part, as a tool to improve traffic safety.

EXERCISE

Look up the traffics accidents data and plot graphs of fatality and serious accidents. Using any Geographic Information System (GIS) software, design a map of traffic accidents. What is the relationship between accidents and road miles, urbanization levels and population concentrations? Where do most accidents occur?

8.4 Distribution of Medical Services/ Health Care System

This section deals with four aspects: the number of medical professionals by district, the number of hospital beds per district, the number of health clinics and immunization in Uganda. The country has poor health indicators and a burden of preventable diseases. The life expectancy has declined since the 1980s, infant mortality is still comparatively high and population per doctor, as well as population per nurse ratios are poor.

More than 90% of the districts do not have a dentist or a pharmacist. Other areas with great need are mental health services. The country has fewer than twenty psychiatrists (Basangwa, 2004). The population per doctor of 18,700:1 is too high. There are few health workers, with an ever increasing workload. At the same time, medical professionals have been leaving the country in large numbers in search of better working conditions.

Health Characteristics

Life Expectancy (World Bank, 2005)	47 years
Infant Mortality Rate	88 per 1,000
Maternal Mortality Ratio	504
Total Fertility Rate	6.9
Population per doctor	18,700
Population per nurse	3,065
Population per hospital bed	870

Source: UBOS, 2005

8.4.1 Common Sources of Data for Health Indicators

- Vital Events Registers e.g. birth and death records
- Census
- Epidemiological Surveillance
- Sample Survey
- Disease Registers
- Routine Health Service Records

Map 8.4: Uganda Health Care Workers by District

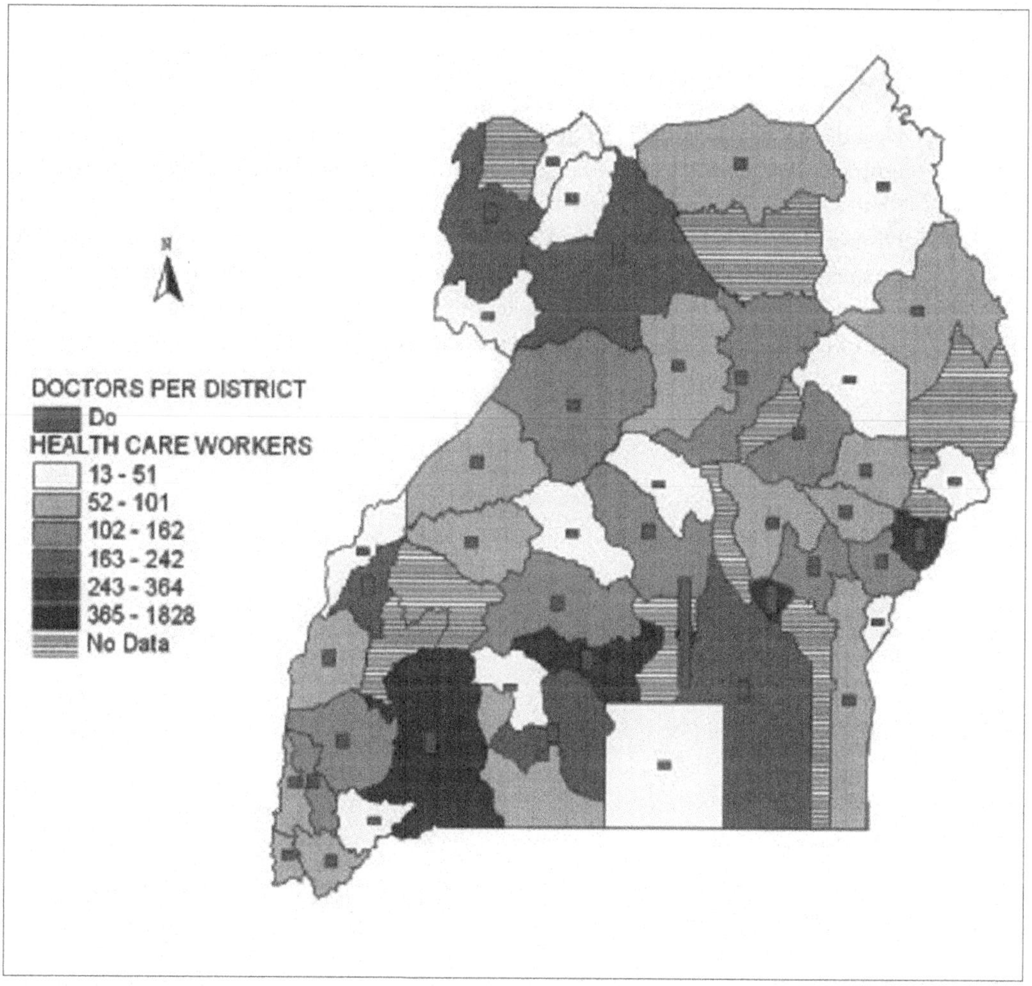

Source: Uganda Ministry of Health, 2004

The map shows doctors and health workers per district. The southern part of the country is better served than the northern and the urban districts have more health workers than rural districts. Since 80% of Uganda's population is rural, the country is poorly served. The districts with the least number of health workers are Adjumani, Kalangala, Kapchorwa, Katakwi, Moyo, Nakasongola, Ntungamo and Sembabule.

Table 8.3: Health Units by Ownership

Type	Government	NGO	Total	Percentage NGO
Hospitals	55	49	104	47
Health Centres IV	143	16	159	10
Health Centres level III	614	173	787	22
Health Centres level II	781	1,244	2,025	61
All Units	1593	1,482	2,075	48

Source: Uganda Ministry of Health 2004

Uganda has 104 hospitals, 159 health centres and 2,075 health units. The government hospitals are classified in three categories: national referral, regional referral and district/ rural hospitals. The national referral hospitals are Mulago and Butabika, both in Kampala District. There are 230,769 persons per hospital, 11,566 persons per health unit and 870 persons per hospital bed. These are very poor indicators. The provision of hospital and health care services in Uganda has not kept up with the growth of the population. Today's rate of persons per hospital bed is below the rate of 1980s.

Map 8.5: Uganda - Population per Hospital Bed by Districts

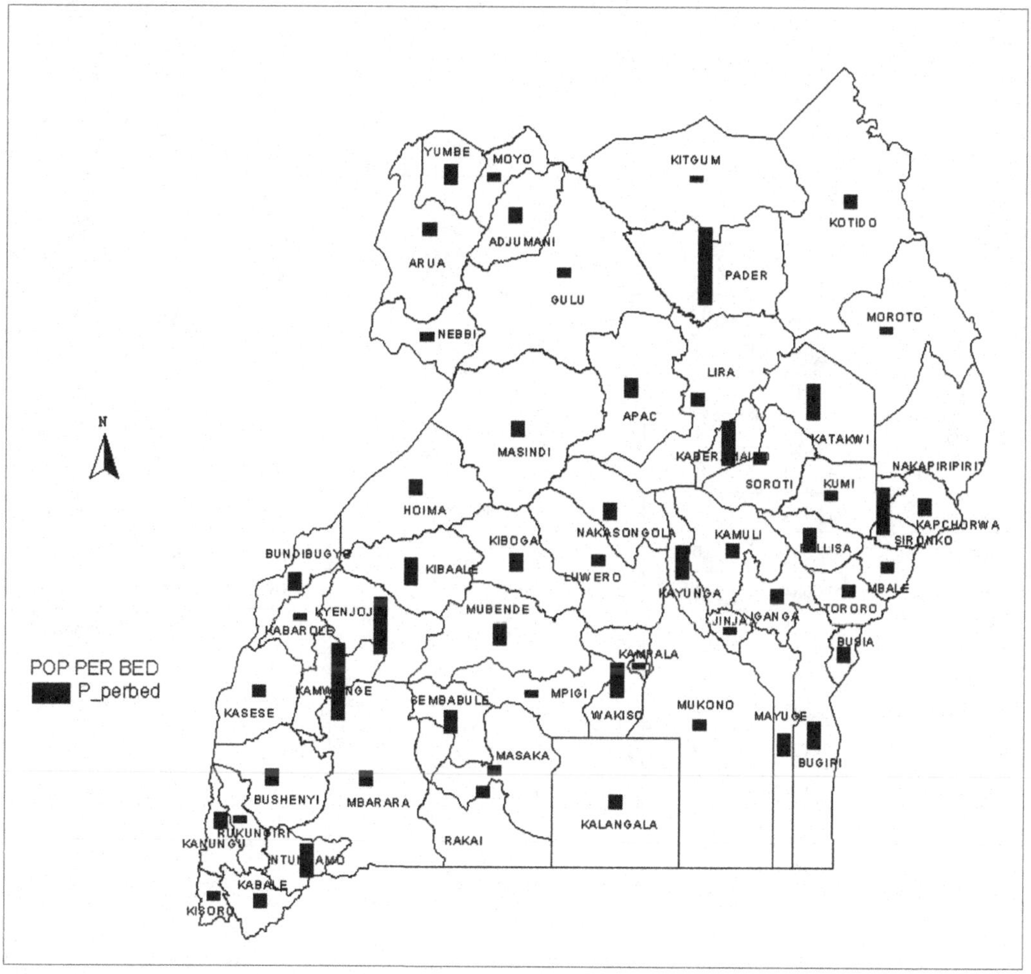

Source: Uganda Ministry of Health 2004

Table 8.4: Health Units by District

District	Population	Hospital	HCIV	HCIII	HCII	Total HC	Beds	Pop/Bed
Adjumani	201,493	1	0	8	23	32	165	1,221.17
Apac	676,244	2	3	13	22	40	425	1,591.162
Arua	855,055	3	5	30	24	62	919	930.42
Bugiri	426,522	1	2	10	29	42	183	2,330.72
Bundibugyo	212,884	1	1	2	16	20	149	1,428.752
Bushenyi	723,427	3	5	15	43	66	571	1,266.947

Busia	228,181	0	2	8	10	20	182	1,253.742
Gulu	468,407	4	3	16	28	51	835	560.97
Hoima	349,204	1	1	16	19	37	290	1,204.15
Iganga	716,311	1	4	21	67	93	648	1,105.42
Jinja	413,937	3	4	16	45	68	859	481.88
Kabale	471,783	1	5	14	48	68	433	1,089.57
Kabarole	359,180	3	2	11	12	28	911	394.27
Kaberamaido	122,924	1	1	7	2	11	30	4,097.47
Kalangala	36,661	0	2	4	3	9	30	1,222.03
Kampala	1,208,544	13	5	30	837	885	3563	339.19
Kamuli	712,079	1	5	17	31	54	598	1,190.76
Kamwenge	295,313	0	2	7	10	19	40	7,382.82
Kanungu	205,095	1	1	9	19	30	150	1367.3
Kapchorwa	193,510	1	1	13	14	30	161	1,201.92
Kasese	532,993	3	5	30	27	65	553	963.82
Katakwi	307,032	0	3	11	20	34	93	3,301.42
Kayunga	297,081	1	2	6	7	16	100	2,970.81
Kibaale	413,353	1	2	14	17	34	175	2,362.02
Kiboga	231,718	1	2	14	17	34	164	1,412.91
Kisoro	219,427	2	2	9	7	20	367	597.89
Kitgum	286,122	2	2	10	6	20	905	316.16
Kotido	596,130	2	1	6	37	46	529	1,126.9
Kumi	388,015	3	2	16	10	31	545	711.95
Kyenjonjo	380,362	0	3	18	16	37	70	5,433.74
Lira	757,763	2	6	5	36	49	720	1,052.45
Luwero	474,627	3	4	27	27	61	659	720.22
Masaka	767,759	3	8	16	33	60	1152	666.46
Masindi	469,865	2	2	13	30	47	391	1,201.7
Mayuge	326,567	1	2	3	21	27	164	1,991.26
Mbale	720,925	2	7	21	16	46	922	781.91
Mbarara	1,089,051	2	7	39	26	74	853	1,276.73
Moroto	170,506	2	0	2	12	16	442	385.76
Moyo	199,912	1	1	12	14	28	328	609.49
Mpigi	414,757	2	2	17	21	42	977	424.52
Mubende	706,256	2	7	17	41	67	387	1,824.95
Mukono	807,923	4	5	21	21	51	950	850.45

Nakapiripirit	153,862	1	2	3	10	16	30	5,128.73
Nakasongora	125,297	1	1	5	17	24	97	1,291.72
Nebi	433,466	1	3	17	24	45	668	648.9
Ntungamo	386,816	1	3	7	15	26	127	3,045.79
Pader	293,679	1	1	7	11	20	40	7342
Palisa	522,254	1	3	28	10	42	255	2,048.05
Rakai	471,806	2	2	22	68	94	534	883.53
Rukungiri	308,696	2	1	11	16	30	767	402.47
Semabule	184,178	0	2	6	10	18	93	1,980.41
Sironko	291,986	0	3	21	7	31	70	4170
Soroti	371,986	1	3	15	6	25	444	837.81
Tororo	559,528	4	3	16	17	40	609	918.76
Wasiko	957,280	4	5	27	39	75	300	3191
Yumbe	253,325	1	0	8	10	19	150	1,688.83

Source: Ministry of Health Statistical Abstract 2004

HCIV Health Centre Level Four

HCIII Health Centre Level Three

HCII Health Centre Level Two

Many districts have a ratio of persons per hospital bed that is higher than the national average of 870 persons per bed. Furthermore, the majority of the districts do not have dentists, pharmacists and eye doctors. The population per doctor ratio is too high, but it is better than before. Over the past 15 years, there has been some effort to restore the health sector.

Disease control programmes have been re-established and Primary Health Care has improved (Ministry of Health, 2003; WHO). Infant mortality rates have declined and utilization of health services has increased. The gains are, however, over-shadowed by high prevalence of communicable diseases, rising incidence of non-communicable diseases and causes of mortality (for example accidents), rapid increase in demand for services because of population growth, HIV/AIDS impact and resource constrains. Furthermore, the health sector has been losing professionals. At least 30% of the doctors leave the country for greener pastures every year (Ministry of Health, Vision January, 2005).

8.4.2 Immunization Rates and Ambulance Services

Immunization of the population against diseases is one of the areas of preventive medicine. The rates are shown in Table 8.5. The country wide immunization rates for antigens such as measles, polio, BCG and DPT3 have greatly improved over the years. The only exception is immunization for tetanus toxoid in non-pregnant women (Uganda Bureau of Statistics,

2004). The national average immunization rate is 67%. Thirty four districts are above the national average. The immunization rate is a good indicator of the service utilization. It is slightly below what is expected (75%+). Ambulance service is very poor in all regions/ districts. This is an area of concern for all health services. The government has to improve this sector if lives are to be saved.

Table 8.5: Immunization of people against diseases

District	Population	Infants 2002	Number of Children	DPT3/HepB/ Hib%
Adjumani	201,493	8,398	5,895	70
Apac	676,244	34,514	14,016	41
Arua	855,055	41,713	26,875	64
Bugiri	426,522	20,739	14,591	70
Bundibugyo	212,884	10,581	5,641	53
Bushenyi	723,427	36,918	27,066	73
Busia	228,181	11,392	11,147	98
Gulu	468,407	23,975	11,530	48
Hoima	349,204	17,152	11,770	69
Iganga	716,311	35,732	21,012	59
Jinja	413,937	19,565	17,963	88
Kabale	471,783	23,089	19,463	84
Kabarole	359,180	18,040	17,963	100
Kaberamaido	122,924	6,596	6,551	99
Kalangala	36,661	1,724	1,525	88
Kampala	1,208,544	60,978	32,744	54
Kamuli	712,079	35,341	25,538	72
Kamwenge	295,313	13,365	10,112	76
Kanungu	205,095	10,295	8,507	83
Kapchorwa	193,510	9,765	7,092	73
Kasese	532,993	26,501	18,404	69
Katakwi	307,032	13,365	12,135	91
Kayunga	297,081	14,805	10,622	72
Kibaale	413,353	20,639	14,933	72
Kiboga	231,718	11,562	5,515	48
Kisoro	219,427	11,084	8,496	77
Kitgum	286,122	14,232	6,753	47
Kotido	596,130	30,266	9,276	31
Kumi	388,015	19,664	21,704	110
Kyenjonjo	380,362	19,027	14,533	76
Lira	757,763	37,556	26,675	71
Luwero	474,627	23,996	13,084	55
Masaka	767,759	38,633	27,134	70
Masindi	469,865	23,310	12,755	55
Mayuge	326,567	16,341	9,327	57

Mbale	720,925	36,062	43,504	121
Mbarara	1,089,051	54,669	31,431	57
Moroto	170,506	9,739	5,889	60
Moyo	199,912	10,115	5,863	58
Mpigi	414,757	20,726	16,631	80
Mubende	706,256	34,847	20,628	59
Mukono	807,923	39,417	28,909	73
Nakapiripirit	153,862	7,758	4,315	56
Nakasongora	125,297	6,406	4,550	71
Nebi	433,466	21,726	16,771	77
Ntungamo	386,816	18,990	15,958	84
Pader	293,679	15,594	4,077	26
Palisa	522,254	26,112	18,339	70
Rakai	471,806	23,361	18,224	78
Rukungiri	308,696	13,906	9,857	71
Semabule	184,178	9,103	5,689	62
Sironko	291,986	0	3	21
Soroti	371,986	18,506	17,267	93
Tororo	559,528	27,205	19,129	70
Wasiko	957,280	45,706	19,745	43
Yumbe	253,325	12,720	6110	48

Source: Ministry of Health Statistical Abstract 2004

8.5 Conclusion

The medical sector faces enormous challenges. The government and people will need total commitment to effect change in this sector. There are several steps that need to be taken.

1. Stopping or reducing the flow of medical professionals leaving the country.
2. Providing better service in the government hospitals.
3. Encouraging the private sector to invest in the health sector.
4. Continuing the decentralization process – decentralize health activities to sub-district and community levels.
5. Providing more resources for the health sector at all levels – central, district and community.
6. Having the political will to effect these policies.

Exercise

The Ministry of Health has identified eleven diseases termed notifiable diseases. These diseases are dysentery, cholera, guinea worm, malaria, meningitis, measles, plaque, rabies, tetanus, typhoid fever, yellow fever and AFP/ polio. The ministry provides a weekly epidemiological bulletin for these diseases. Using the weekly epidemiological reports from the local paper (New Vision), plot the prevalence trends of five diseases for at least five districts, or places. What is the trend observed? Explain the trend using information obtained from the medical library.

Key Terms

Crude death Rate	Crude birth Rate
Infant Mortality Rate	Incidence
Life expectancy	Maternal Mortality
Morbidity	Mortality
Prevalence	Total Fertility Rate

Questions

1. What are the components of disease ecology?
2. What is prevalence and incidence?
3. Explain the following rates: birth rate, fertility rate, cumulative rate and incidence rate.
4. What are morbidity trends? What are the major causes of morbidity in Uganda?
5. What factors influence the spread of a disease?

References and further readings

BakamaNume, B. 2004. *Road Traffic Accidents in Uganda: Epidemiological and Transport Policy Implications, African Social Science Review* (ASSR), Vol. 3: 3.

BakamaNume, B. 1996. Spatial Patterns of HIV/AIDS Infection in Uganda, 1987-1994, *African Urban and Rural Studies*, 3, 2: 141-161.

Bennett, F, Farsy, S and Hutt M. 1975. *Bacterial Infections*, in *Uganda Atlas of Disease Distribution* edited by Hall and Langlands.

Curtis, S. and Taket, A. 1996. *Health & Changing Perspectives*, New York, Arnold.

Hall S and B.W. Langlands editors. 1975. *Uganda Atlas of Disease Distribution*, East African Publishing House, Nairobi.

McCrae, A. 1975. Malaria, in S. Hall and B. Langlands editors *Uganda Atlas of Disease Distribution*, East African Publishing House, Nairobi, 30-36.

Meade, M., Florin, J. and Gesler, W. 2000. *Medical Geography*, New York: The Guilford Press.

Ministry of Health. 2003. Health Policy State 2001/2002: Executive Summary, in the fight against malaria, Internet Article www.health.go.ug/policies.htm

Ministry of Health. 2001. The Burden of Malaria in Uganda: Why all should join hands in the fight against malaria, Internet Article www.health.go.ug/malaria.htm

Ministry of Health. 2001. Incidence of Meningitis, Internet Article www.health.go.ug/meningitis.htm

Miti, J. and A. Ssenkabirwa. 2005. High rate of epidemics blamed on poor funding, *The Monitor*, March 2005.

Morrow, R. 1975. Tuberculosis, in S. Hall and B Langlands. 1975. *Uganda Atlas of Disease Distribution*, 60-63.

Nsangi, K. 2005. DDT Contaminates Lake Victoria, *The Monitor*, August 17, 2005

Okwale, S, Opio, A, Musinguzi, J and Waibale, P. 2001. Fighting HIV/AIDS: Is success possible?, *Bulletin of the World Health Organization* 79(12) 1113-1120.

Rickett, T, Savitz, L, Gesler, W and Osborne, D. 1994. *Geographic Methods for Health Services Research*, New York: University Press of America.

STD/AIDS Control Programme. 2003. STD/HIV/AIDS Surveillance Report, Ministry of Health, Kampala.

Watts, T. 1975. Brucellosis, in S. Hall and B. Langlands, *Uganda Atlas of Disease Distribution*, 17-18

Chapter 9

POLITICAL GEOGRAPHY OF UGANDA

Bakama BakamaNume

9.1 Introduction

Political geography is the study of organization and distribution of political phenomena from place to place (Hartshorne, 1954). Another definition emphasizes political phenomena in their areal context (Jackson, 1971). Taylor (1980, 2001) defines political geography as a sub- discipline of Geography that examines spatial attributes of political processes in their areal expression. These political processes include administrative structures, geo-politics of administrative units and geo-elections. The focus of this chapter is to discuss the evolution of administrative units in Uganda and the administrative system in Uganda and to examine the electoral geography in Uganda. Why administrative units? This is because administrative units are vital for planning purposes, political representation, infrastructure building and for effective governing.

9.2 History of Administrative Units in Uganda: The Evolution of Administrative Boundaries

By 1900, the British had instituted the Uganda Agreement. The agreement provided the basis of the demarcation of present day Uganda. Later on, in 1926, significant border adjustments were made between Uganda and its neighbours. Uganda lost territory to most of its neighbours (Langlands, 1973). It lost territory to Kenya, Sudan and Congo. From 1945 or earlier until after independence, Uganda was divided into four provinces and the provinces were subdivided into districts. The district was going to become the main local administrative unit. The criteria for demarcation of provinces included among others population, physical features and cultural-ethnic factors.

Table 9.1: Provinces and Population 1943-1969

Province	1948	1959	1969	Area (km²)	Capital
Buganda	1,317,705	1,881,149	2,667,332	61,609	Kampala
Eastern	1,514,428	1,902,697	2,817,066	63,018	Jinja
Northern	945,104	1,249,310	1,631,899	57,320	Gulu
Western	1,177,939	1,503,375	2,432,550	54,913	Masindi
4 provinces	4,955,176	6,536,531	9,548,847	236,860	

Source: Langlands, 1975

Table 9.1 shows the populations of the provinces according to the censuses of 1948, 1959 and 1969. The Eastern region had the highest population and the Northern region had the least. The Eastern region was the largest in terms of size and the Western province the smallest. Note that Kampala was listed as the capital of Buganda. During the colonial period Entebbe served as the capital of the nation.

In 1960, the status/name of the provinces changed to regions but the district remained the primary administrative unit. However, the events that happened from 1960 to 1980s marked the beginning of change in Uganda's administrative units. Just before political independence in 1962, Mbale territory, with an area of 25 km^2 was created from a territory disputed between Bugisu and Bukedi districts. The town of Mbale was named the capital of both districts (Langlands, 1973; GE-29). Throughout, the colonial era and immediately after independence, Mbale Town remained a very unique exception. It was a town which served as a capital for two districts (Bugisu and Bukedi) and it was also a territory on its own and capital of Mbale territory.

On October 9, 1962, Uganda gained political independence. This historical event did not however, result in the change of administrative units. It was a change that marked African leadership. At the time of independence, the total number of administrative units was 15 districts, three kingdoms and one territory (See Table 9.2). The districts were further subdivided into Masaza (counties), which were subdivided into Magombola (sub-counties), which were in turn subdivided into Mirukas (parishes) (Langlands, 1975; GE-29).

Four years after independence, Uganda underwent drastic changes on the political level. The 'political crisis of 1966' brought two significant changes. The political crisis has been defined as the period when President Obote made Uganda a Republic. Two major events took place. Firstly, kingdoms were abolished. Secondly, Buganda kingdom was divided into four districts: Bombo, Masaka, Mpigi and Mubende (Langlands, 1975, GE-29). The districts that curved out of Buganda, were named after their capitals. This trend was to be repeated in 1986 when the National Resistance Movement took power and established a grass-root based system of administration.

Table 9.2: Districts and Capitals

District	Type	Capital
Acholi	District	Gulu
Ankole	District	Mbarara
Bombo	District	Bombo
Bugisu	District	Mbale
Bukedi	District	Mbale
Bunyoro	Kingdom	Hoima
Busoga	District	Jinja
Karamoja	District	Moroto
Kigezi	District	Kabale
Lango	District	Lira
Madi	District	Moyo
Masaka	District	Masaka
Mbale	Territory	Mbale
Mpigi	District	Mpigi

Mubende	District	Mubende
Sebei	District	Kapchorwa
Teso	District	Soroti
Toro	Kingdom	Fort Portal
West Nile	District	Arua

Source: Langlands, 1975

The changes in 1967 are attributed to the political events of 1966. In 1967, a new constitution was promulgated. The constitution instituted several major changes. Firstly, it abolished the kingdoms, which became districts. Secondly, Mbale territory was combined with Bugisu. Thirdly, the capital of Bukedi was moved to Tororo and there were two name changes. Bombo was renamed East Mengo and Mpigi was renamed West Mengo (Langlands, 1975; Government of Uganda).

By the 1969 population census, there were 19 districts in Uganda (see Table 9.3). Acholi district was divided into two – East and West Acholi and Karamoja was divided into South and North Karamoja.

Table 9.3:1969 Districts

District	Population	Area (km²)	Capital
East Acholi West Acholi	463,844	27,853	Gulu
Ankole	861,145	16,182	Mbarara
Bugisu	421,433	2,546	Mbale
Bukedi	527,090	4,553	Tororo
Bunyoro	351,903	19,609	Hoima
Busoga	949,384	14,047	Jinja
East Mengo	851,583	23,440	Bombo
Karamoja	284,067	27,213	Moroto
Kigezi	647,988	5,218	Kabale
Lango	504,315	13,740	Lira
Madi	89,978	5,006	Moyo
Masaka	640,596	21,300	Masaka
Mubende	330,955	10,310	Mubende
Sebei	64,464	1,738	Kapchorwa
Teso	570,628	12,921	Soroti
Toro	571,514	13,904	Fort Portal
West Mengo	844,198	6,559	Mpigi
West Nile	573,762	10,721	Arua
TOTAL	9,548,847	236,860	

Source: Uganda Bureau of Statistics

In 1971, President Obote was overthrown in a military coup led by Major General Idi Amin. This political event was to reshape the administrative boundaries for the next nine years. In 1974, President Idi Amin reorganized the districts into provinces. The number of administrative units was increased to thirty eight. These were grouped into 10 provinces (See Table 9.4). The province became the most important administrative unit. These administrative units were led by Governors appointed by the President Amin.

Table 9.4: 1974 Provinces

Province	Population	Area(km.^{2})	Capital	Former
Busoga	1,221,872	13,340	Jinja	Busoga
Central	1,117,648	6,270	Kampala	West Mengo
Eastern	2,015,530	22,260	Mbale	Bugisu, Bukedi, Sebei, Teso
Karamoja	350,908	26,960	Moroto	Karamoja (North and South)
Nile	811,755	15,730	Arua	Madi, West Acholi (part), West Nile
North Buganda	1,554,371	27,010	Bombo	East Mengo, Mubende
Northern	1,261,364	41,520	Gulu	Acholi (East and part of West), Lango
South Buganda	905,754	15,970	Masaka	Masaka
Southern	1,963,428	21,280	Mbarara	Ankole, Kigezi
Western	1,427,446	30,980	Fort Portal	Bunyoro, Toro
	12,630,076	221,320		

Source: Uganda Bureau of Statistics

In 1979, President Idi Amin was forced out of office and the country by the Uganda Liberation Army and Tanzania Defence Forces. The period between 1979 and 1986 was marked by seven different political governments but no changes in administrative units. In 1979, the name Buganda Province was changed to Central Province. Uganda was also reorganized from 10 provinces into 33 districts. Since the 1990s, the government has been redistricting the country. The provincial administrative units have been abolished in favour of districts. The redistricting, according to the government, is necessary for decentralization of the government. Indeed, the district is smaller than the province and district headquarters are close to the population served. But as for whether or not the new administrative units are more efficient than the old large units is a question yet to be answered.

The process of redistricting has continued into the 2000s. Some of the changes are listed here.

- Kalangala district split from Masaka in 1990
- Kibale was made a district by separating it from Hoima in 1991
- 314 km.^{2} of territory were transferred from Masindi to Hoima
- Kiboga was separated from Mubende
- Kisoro was separated from Kabale
- Pallisa split from Tororo

In 1996, Ntungamo district was formed by taking parts of Bushenyi and Mbarara.

By the beginning of 1997, the redistricting process had produced 39 districts (see Table 9.5) and appatite for district was set in motion.

Table 9.5: 1996 Districts

District	Population	Area(km.²)	Area(mi.²)	Former
Apac	460,700	6,488	2,505	Northern
Arua	624,600	7,830	3,023	Nile
Bundibugyo	116,000	2,338	903	Western
Bushenyi	734,800	5,396	2,083	Southern
Gulu	338,700	11,735	4,531	Nile, Northern
Hoima	197,800	5,492	2,120	Western
Iganga	944,000	13,113	5,063	Busoga
Jinja	284,900	734	283	Busoga
Kabale	412,800	1,827	705	Southern
Kabarole	741,400	8,361	3,228	Western
Kalangala	16,400	5,716	2,207	South Buganda
Kampala	773,500	238	92	Central
Kamuli	480,700	4,348	1,679	Busoga
Kapchorwa	116,300	1,738	671	Eastern
Kasese	343,000	3,205	1,237	Western
Kibale	219,300	4,718	1,822	Western
Kiboga	140,800	3,774	1,457	North Buganda
Kisoro	184,900	662	256	Southern
Kitgum	350,300	16,136	6,230	Northern
Kotido	190,700	13,208	5,100	Karamoja
Kumi	237,000	2,861	1,105	Eastern
Lira	498,300	7,251	2,800	Northern
Luwero	449,200	9,198	3,551	North Buganda
Masaka	831,300	10,611	4,097	South Buganda
Masindi	253,500	9,326	3,601	Western
Mbale	706,600	2,546	983	Eastern
Mbarara	929,600	10,839	4,185	Southern
Moroto	171,500	14,113	5,449	Karamoja
Moyo	178,500	5,006	1,933	Nile
Mpigi	915,400	6,222	2,402	Central
Mubende	497,500	6,536	2,524	North Buganda
Mukono	816,200	14,242	5,499	North Buganda

Nebbi	315,900	2,891	1,116	Nile
Ntungamo	----	----	----	Southern
Pallisa	356,000	1,919	741	Eastern
Rakai	382,000	4,973	1,920	South Buganda
Rukungiri	388,000	2,753	1,063	Southern
Soroti	430,900	10,060	3,884	Eastern
Tororo	554,000	2,634	1,017	Eastern
	16,583,000	241,038	93,065	

Source: Uganda Bureau of Statistics

While other redistricting exercises have created more districts for decentralization, it has been argued that the government rewards members of parliament by creating districts. Such a process, if true, is similar to **gerrymandering**. Gerrymandering is a redistricting which favours one political party. Redistricting after 1997 is shown in Table 9.6. Here are some of the changes:

- 1997: Adjumani district separated from Moyo; Bugiri separated from Iganga;
- Busia separated from Tororo; Katakwi split from Soroti;
- Nakasongola split from Luwero; Sembabule split from Masaka.

Table 9.6: 1997 Changes in Districts

District	Population	Area (km.^{2})	Area (mi.^{2})	Reg
Adjumani	96,264	2,888	1,115	N
Apac	454,504	5,887	2,273	N
Arua	637,941	7,595	2,932	N
Bugiri	239,307	1,453	561	E
Bundibugyo	116,566	2,097	810	W
Bushenyi	579,137	3,827	1,478	W
Busia	163,597	705	272	E
Gulu	338,427	11,560	4,463	N
Hoima	197,851	3,563	1,376	W
Iganga	706,476	3,370	1,301	E
Jinja	289,476	677	261	E
Kabale	417,218	1,695	654	W
Kabarole	746,800	8,109	3,131	W
Kalangala	16,371	432	167	C
Kampala	774,241	169	65	C
Kamuli	485,214	3,332	1,286	E
Kapchorwa	116,702	1,738	671	E
Kasese	343,601	2,724	1,052	W

Katakwi	144,597	4,647	1,794	E
Kibale	220,261	4,208	1,625	W
Kiboga	141,607	3,872	1,495	C
Kisoro	186,681	620	239	W
Kitgum	357,184	16,136	6,230	N
Kotido	196,006	13,208	5,100	N
Kumi	236,694	2,457	949	E
Lira	500,965	6,151	2,375	N
Luwero	349,194	5,360	2,070	C
Masaka	694,697	3,214	1,241	C
Masindi	260,796	8,458	3,266	W
Mbale	710,980	2,504	967	E
Mbarara	798,774	9,733	3,758	W
Moroto	174,417	14,113	5,449	N
Moyo	79,381	1,780	687	N
Mpigi	913,867	4,514	1,743	C
Mubende	500,976	5,949	2,297	C
Mukono	824,604	4,594	1,774	C
Nakasongola	100,497	3,179	1,227	C
Nebbi	316,866	2,781	1,074	N
Ntungamo	289,222	1,981	765	W
Pallisa	357,656	1,564	604	E
Rakai	383,501	2,317	895	C
Rukungiri	390,780	2,584	998	W
Sembabule	144,039	3,889	1,502	C
Soroti	285,793	3,879	1,498	E
Tororo	391,977	1,631	630	E
45 districts	16,671,705	236,036	91,134	

Source: Uganda Bureau of Statistics

The new districts have been named after their capitals. This is a trend that started in 1966 with the creation of Bombo, Mpigi and Masaka districts. The only exceptions are Kabarole and Sembabule.

Between 2001 and 2005, more districts were created. The list includes: Yumbe separated from Arua; Wakiso separated from Mpigi; Kayunga separated from Mukono; Mayuge separated from Iganga; Sironko separated from Mbale; Pader separated from Kitgum; Kamwenge separated from Kabarole; Kyenjojo separated from Kabarole; Kanungu separated from Rukungiri; Kaberamaido separated from Soroti; Nakapiripirit separated from Moroto. In July 2005, more districts were created and most recently, in 2008, more districts were formed. The list includes Mayuge, Kaliro, Namutumba and many more. The redistricting process is an interesting topic for future research.

Map 9.1: Districts of Uganda

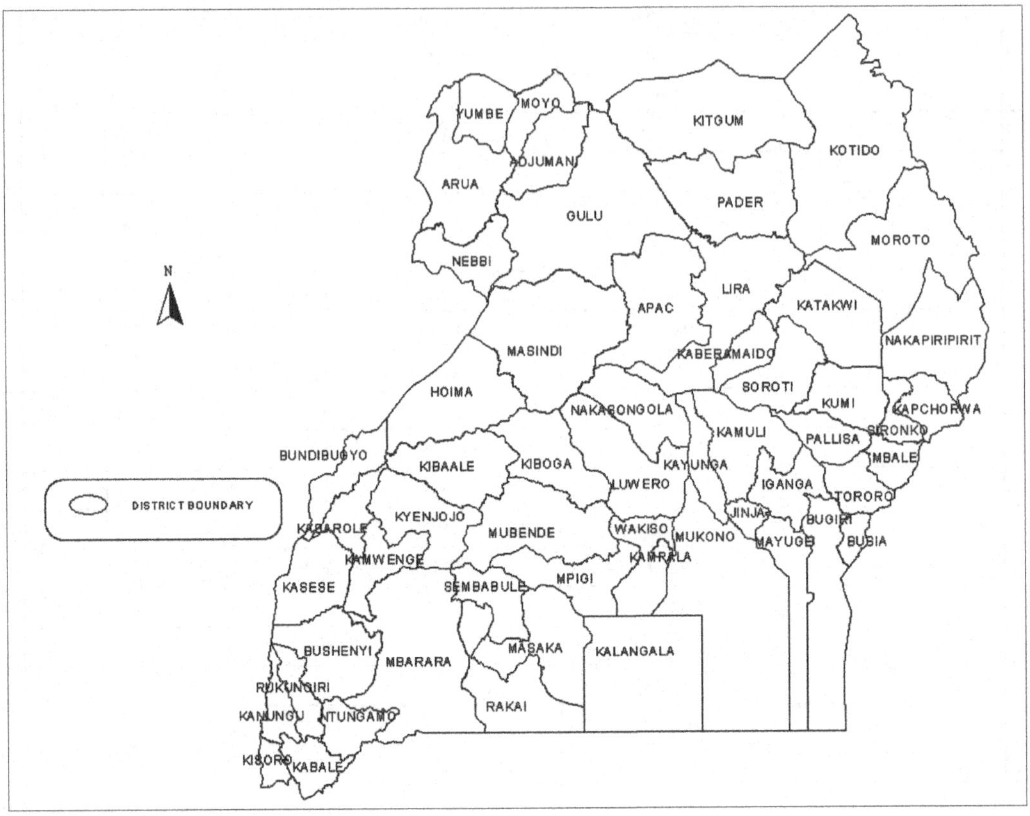

Source: Uganda Bureau of Statistics 2004

There are currently more than 111 districts in the country. More districts are being proposed.

9.3 The Uganda Administrative System

Uganda has an administrative system that has been described as partly centralized and partly decentralized (Local Government Act 1997). There are other views that regard the system as centralized in guise of decentralization. The central and local governments have non-conflicting powers and functions. These are exercised in area of jurisdiction. The central administrative system is based on Parliament – the national legislative body. The local government system is based on the district council as the administrative unit. Below the district, there is the lower local administrative body (Constitution of Uganda, 1995; Local Government Act 1997). The city is on the same level as the district and the divisions are the same as the sub-county.

Table 9.7: Administrative Units

Administrative Unit	Legislative Body
Country	Parliament
District	District Council
Sub-County	Sub County Council
City	City Council, City Division Councils
Municipal	Municipal Council, Municipal Division Councils
Town	Town Council

Source: Uganda Bureau of Statistics

9.3.1 Central Government

The central government administration includes Parliament which is the legislative body, Cabinet – the executive body and ministries – which are the administrative units. There are 279 Members of Parliament (MPs) or representatives in parliament and the Speaker. The Speaker presides over the sessions. The cabinet ministers are appointed by the president and approved by parliament. The cabinet is mostly drawn among the members of the National Assembly and is headed by the Prime Minister who also acts as head of government business (Ogoso-Opolot, 1998; Constitution of Uganda, 1995). In July 2005, the portfolio of prime minister was among the amendments in the constitution in which it was legally established.

9.3.2 District/Urban Local Governments

The governments here include district and city councils. There are currently 80 such governments. Since the number of districts is likely to increase these local governments will increase too. The district/city has an elected chairperson, councilors who represent sub-counties or divisions of the city, councilors for youths and councilors for persons with disabilities. All districts have an executive committee (district cabinet). The committee comprises of the chairperson, vice chairperson and five councilors (Local Government Act, 1997).

9.3.3 Lower Local Governments

The lower local governments are the councils in sub-county/ city division, municipal, municipal division and town councils. The Uganda Bureau of Statistics (2004) lists 13 municipalities and 59 towns. The head of the lower local government is the elected chairperson (LC 5). Under the chairperson are councilors representing each parish or its equivalent, a councilor for youths and councilor for person with disabilities (Local Government Act, 1997).

9.3.4 Local Administrative Units

These include a county, parish and village. Each one of these administrative units has a council. The chairperson of the council is the assistant chief administrative officer (ACAO) and town clerk.

9.4 Conclusion

There are more administrative units today than there were in 1980s. The creation of more districts has been gradual and considered necessary for effective administration and decentralization of services. This process is manifest through an evolution of administrative units in Uganda, a system that has changed over time. Finally, there is political representation based on administrative units and to a certain extent, there has been some degree of democracy in electing the representatives.

9.5 Electoral Geography of Uganda

Electoral geography is a sub branch of Political Geography. Geographers attempt to understand spatial variation of representation, political participation and voting behaviour. The three main themes: *Geography of voting; Geographic influences on voting; and Geography of representation.*

The district/administrative unit provides the basis for representation and representation is accomplished through the electoral process. This section examines electoral geography in the recent past. It is a case study of only the 1996 presidential election. This is because of the controversy surrounding the 2001 and 2006 elections. Critics indicate that the 2001 and 2006 elections were rigged just like the 1980 elections. Therefore, these three elections (1980, 2001 and 2006) may not represent a fair electoral process. It is further argued that the 2006 elections were further compromised by the constitutional change to remove term limits. This leaves the 1996 elections as the logical and fair choice for analysis.

9.5.1 Background

Most recent elections in African countries have been characterized by external persuasion by donor countries in which the push for democracy has been tied to foreign aid. The election process, in such cases, often favours and/or legitimizes the incumbent. Uganda's 1996 election is a good case study to test this hypothesis and explain the election within the global context. Uganda experienced political instability in the 1970s and 1980s. Since its independence in 1962, the country has been ruled by nine different governments, of which only three were elected. This section examines the nature and implication of the elections. It attempts to provide some insight to the following questions: Did the election process uphold democracy or peace? Did the process legitimize a reign? What factors influenced voting behaviour patterns? The 1996 elections gave President Museveni a first/second term in office. His elected government was the ninth government and paved the way for continuation of his non-elected first term of ten years.

The gun rather than the ballot, has dominated Ugandan politics since 1971, when a military coup led by General Idi Amin overthrew President Milton Obote's government. The Amin reign lasted nine years and he was forced out of power by a combined force of exiled Ugandans backed by the Tanzania army in 1979. Dr. Yusufu Lule became the first president in the post Amin era. After three months in office, Lule was replaced by Godfrey Binaisa, a former Attorney General in Obote's first government. Binaisa came into power through what is often referred to as an in-house coup.

In early 1980, a Military Commission overthrew President Binaisa after eleven months in office. The military commission later facilitated the manipulation of the electoral process to return Obote to power in what is believed to have been rigged elections. In 1985, Obote suffered a second coup at the hands of the army led by General Tito Okello. Okello's government was to be short lived. It was defeated by the guerrilla force led by Museveni in 1986. After more than ten years (in which there was an extension of term in office) of Museveni rule, elections were held in 1996. The 1996 elections were seen as returning political power and decision making to the people (McKinley, Jr. 1996). Museveni won a second/third term in 2001.

He influenced parliament to change the constitution in 2005 and he stood for presidency and won by 62 % majority. The win was challenged in court by the opposition, however, the court ruled in favour of the government.

This case study analyzes the results from the 1996 presidential elections to understand the nature and implication of the recent elections. A central theoretical focus of this section is to provide a better understanding of the nature of current elections outside the core states of the world system (Taylor, 1990), as well as to critically examine the meaning and implication of elections within Uganda.

The case study offers some insights and sheds light on the geography of voting in Uganda. Since political parties were not allowed in the 1996 and 2001 elections, a comparison of the three elections may not seem practical, but, in reality parties and groupings were present (Taylor, 1997, BakamaNume, 1998). This case study is in part an extension of electoral geography beyond the core of democratic societies normally dealt with in the literature.

9.6. Case Study of Electoral Geography

9.6.1 The Presidential Election of 1996: Political History of Candidates

In 1996, the presidency of Uganda was contested by three candidates. The three were the incumbent President Yoweri Museveni, the seasoned politician Paul Ssemogerere and the unknown Mohammed Mayanja. Ssemogerere represented the alliance of the two dominant political parties -- the Democratic Party (DP) and the Uganda Peoples' Congress (UPC). Mayanja's backers and majority support group are still unknown. By design, or by mere coincidence, the three candidates belong to the three major religions in the country. Ssemogerere is Catholic, Museveni is Protestant and Mayanja is Moslem. Several scholars (Mudoola, 1977; Mujaju, 1976; Welbourn, 1965) have shown the relationship between religion and Ugandan politics. The Democratic Party has a majority Catholic following and the Uganda Peoples' Congress has a generally Protestant following (Mudoola, 1988).

The leader of the DP, Ssemogerere came to political prominence after the death of Ben Kiwanuka, the original party leader. During the early 1980s, he led the opposition against the Obote government. He later served as a cabinet minister in the Okello government that overthrew Obote and in the Museveni government. Mayanja the other opposition candidate was a complete unknown. He was an administrator at Makerere University before declaring his candidacy. After losing the elections he went back to his old job at the university.

Museveni is a product of the UPC political machinery. He worked for UPC as a youth winger and as an official in the first Obote government. He broke ranks with UPC in 1980 and formed a political party -- the Uganda Patriotic Movement (UPM). He later contested and lost the 1980 elections against the two major parties -- Obote's UPC and Ssemogerere's DP. Soon after, he went to the bush and started a guerrilla war claiming that the elections had been rigged.

In 1986, Museveni came to power after fighting a guerrilla war for six years. Ten years later, in May 1996, President Museveni was elected to power by a seventy-five percent majority. Was this a mandate for Museveni? Or, was there a manipulation of voters? There are several possible interpretations of the results. The results could be a product of the need for peace. After 30 years of political turmoil, the majority of Ugandans may have cast their votes in approval of the peacefulness they had enjoyed since Museveni took power. On the other hand, the opposition claims the lack of a level playing field during the elections. If true, the election results must be attributed to unfair machination rather than a search for peace. There was also a possibility of voting due to intimidation. This is because the role of the army in voting though questioned has always been very prominent. This section of the chapter examines the recent voters' behaviour, to shed some light on the complexities of politics in Uganda. A single critical election, as that of 1996, has the ability to provide useful insights.

9.6.2 Electoral Geography of the Periphery: Characteristics of Politics of Failure

Uganda belongs to the periphery in terms of economic development. Electoral geography beyond the core of the world-system, in what Wallerstein (1974) described as the periphery, has been described as the politics of failure (Taylor, 1990, Osei-Kwaku and Taylor, 1984). According to Taylor (1990), the politics of failure has four characteristics.

1. The governments may be continually voted out of office because of *unstable geography of support*. Every election produces a new government – musical chair elections.
2. There is an '*ethnic geography of support*.' Despite different economic principles, parties rely on ethnic support to win elections.
3. There is a '*geography of aggregation*'. Parties lack popular support, but win or rule through forging coalitions.
4. The fourth characteristic and an outcome of the politics of failure, is the *use of coups* to overthrow unpopular governments.

Uganda and the Four Characteristics

Uganda has not had a history of voting or electing governments. The first government was elected at independence. The next four were not elected. Then came an elected government in 1980 followed by two unelected ones. The first elections were consequent upon independence in 1962. Since independence, there have been four other general elections (1980, 1996 and 2001 and 2006). The elections in Uganda have been few and irregular. It is, therefore, reasonable to suggest that the first characteristic, *unstable geography of support*, does not fit the Ugandan example. The second characteristic, *ethnic geography of support*, seems to have been present in the elections of 1980, 1996, 2001 and 2006. The third characteristic, the *geography of aggregation*, does not directly apply to Uganda. The 1962 elections had some element of aggregation. The Uganda Peoples' Congress party and Kabaka Yekka (KY) joined forces to defeat the Democratic Party. The fourth characteristic is clearly applicable to the Ugandan case. There have been five coups (others scholars account the 1966 crisis). Two coups overthrew the elected governments of President Obote. Three coups overthrew military governments. The discussion addresses the second characteristics, the ethnic geography of support.

9.6.3 The Geography of Voting in Uganda: Analysis of Voting Patterns

Studies of voting patterns and voter turnout in the core (developed) countries have focused on three issues; demographics, spatial and socio-political characteristics. Demographics include education, occupation, income, age, marital status, sex and rates of mobility; spatial characteristics focus on regional differences; and socio-political characteristics look at political efficacy, information, or awareness. Socio-economic status (SES) variables (income, higher education, occupation) are associated with the tendency to vote (Kaplan and Venezky, 1994). The analysis here examines the voters' choice of candidate and participation in relation to demographics and spatial factors. The third characteristic (socio-political variables) is not examined because of lack of good measures of these variables in the Ugandan case.

9.6.4 Spatial Characteristics and Choice of Candidate in the 1996 and 2001 Elections

There were regional differences in the voting choice for president in 1996 and 2001. In 1996, Museveni took the south and Ssemogerere took the north (see Map 9.2). Mayanja, the third candidate did not win a single district and his share of the total vote showed no clear regional patterns. This pattern was repeated in 2001 when the opposition candidate was Kiiza Byesigye. Besides the general regional differences observed from the mapping of percent voters, the analysis examined the urban and rural differences. Simple correlations between number of votes for a candidate (choice of candidate) in each district and percent urban population in the district were computed.

The simple R squares were 0.455, 0.255 and -0.114 for Mayanja, Ssemogerere and Museveni respectively. Mayanja voting was associated with increasing percent urban population of the district. Museveni voting was least associated with percent urban population. Mayanja and Ssemogerere voting is positively correlated with percent urban population in the district. Why were urban voters more likely to vote for Mayanja than Museveni? Why were rural voters more likely to vote for Museveni than the other two candidates? These two questions are addressed in the examination of voting for individual candidates.

Map 9.2: Winner of District in 1996 Elections

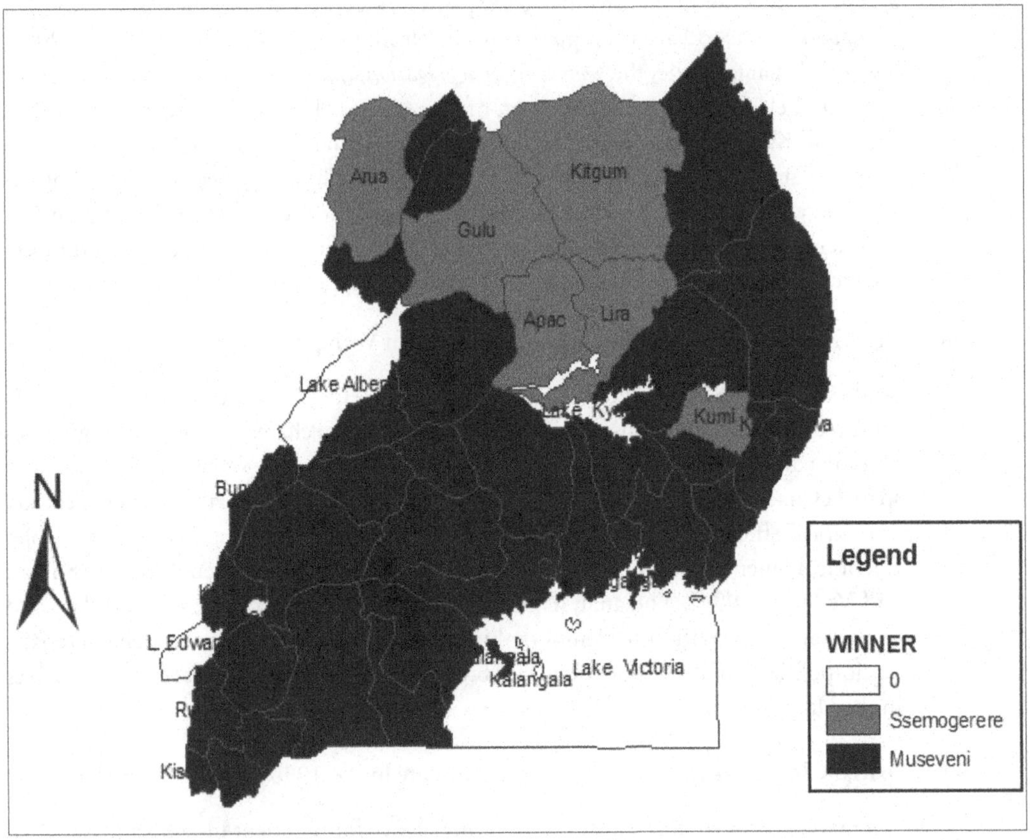

Source: Uganda Electoral Commission

Map 9.3: Mayanja'a Geography of Support

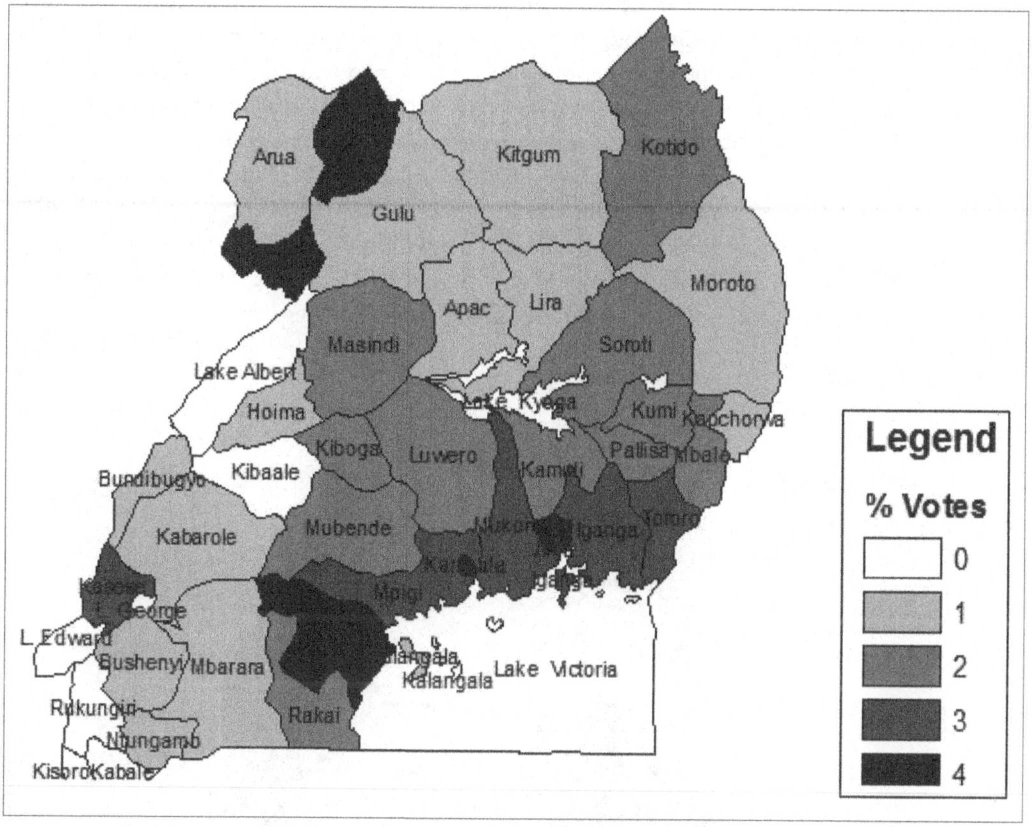

Source: Uganda Electoral Commission

Museveni won in most districts in the South, West and North East. Ssemogerere won the North Central and Northwest.

Mayanja got 4-6% of the votes in Kampala, Jinja, Masaka, Moyo and Nebbi districts. Kampala, Jinja and Masaka are the most urban districts in the country. The urban factor was important in Mayanja voting. Urban areas are easily accessible. Given the fact that Mayanja had very little time to campaign (39 days), he could only effectively reach the voters in accessible areas. His decent showing in most urban districts may be partly a function of accessibility. His performance could also be attributed to some of the population characteristics in these districts. Mayanja is part of the urban elite of Uganda, and was a university administrator. It is possible that his views were more appealing to the urban residents than the rural dwellers. The urban population generally has a higher literacy rate than the rural population and this was the group that provided most of the votes for Mayanja. It is possible that information awareness in the rural population could also have contributed to voting for the incumbent.

Mayanja voting was affected by several factors. He entered the presidential race late, he lacked a good campaign network and he was politically inexperienced and unknown. Despite these apparent weaknesses, he did fairly well in the two rural districts of Moyo and Nebbi! His respectable performance in these two remote, rural districts is hard to explain. The factors that seem to have affected his performance elsewhere were not important here. It is possible that other factors contributed to his popularity and acceptance in these two districts. Mayanja also managed to get at least 3% of the votes in the remaining Lake Victoria districts. Mayanja did fairly well in the core districts (most urban, most populated and most developed). He also did well in two outlying districts (Nebbi and Moyo). Map 9.3 shows the geography of support for Mayanja. Maps 9.4 and 9.5 show the geography of support for Museveni and Ssemogerere respectively.

Map 9.4: Museveni's Geography of support

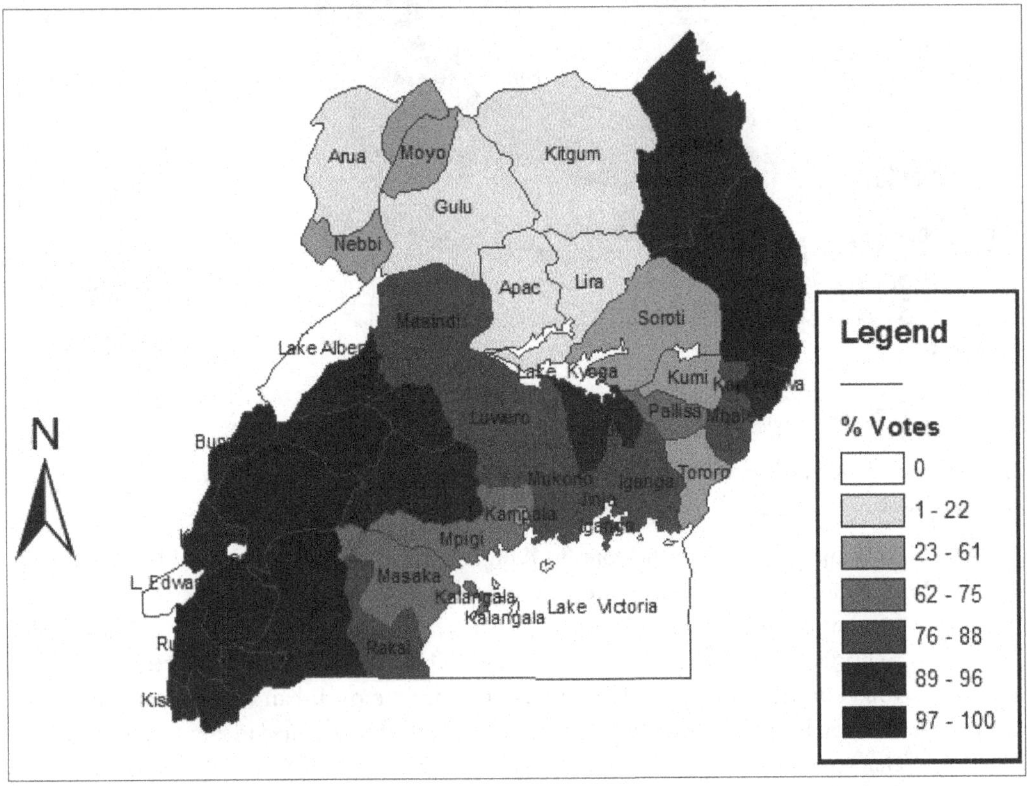

Source: Uganda Electoral Commission

Map 9.5: Ssemogerere's Geography of support

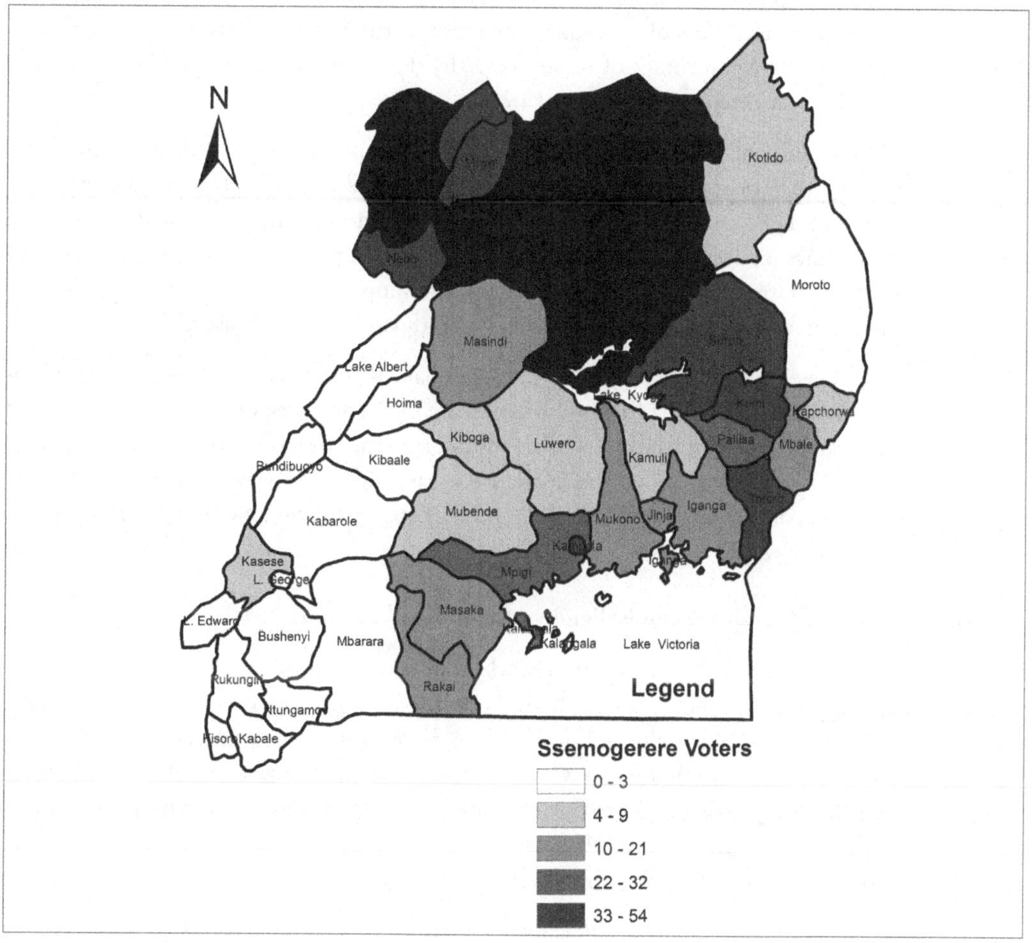

Source: Uganda Electoral Commission

In the 1996 and 2001 elections, the incumbent President Museveni carried the districts south of Lake Kyoga and the northeast districts of Moroto and Kotido. In his own area of western Uganda, he got 80-99% of the votes (Map 9.4). This may be evidence of ethnic voting patterns. The results clearly represent some of the advantages the incumbent enjoyed. He had well organized campaign machinery in place. He had more money to spend and more time to campaign than his challengers (The Economist, 1996). Unlike Mayanja, Museveni was well known everywhere. So why do we see an inverse relationship between percent urban population and choice for Museveni? This may be partly a function of population distribution, or where people voted.

Just before the elections, many urban dwellers in places like Kampala and Jinja returned home to the rural areas to vote (see Monitor, New Vision, 1996). Museveni supporters in urban areas who voted in their birth place areas reinforced the Museveni rural vote. Another important factor for the overwhelming choice of Museveni in rural areas could be the numbers. 89% of Uganda's population is rural (1991 Census). The rural areas have the largest percentage of voters. Thus, by sheer numbers, the incumbent president was better favoured to win the rural areas.

The seasoned politician Ssemogerere carried the north but lost in his birth area of Buganda. His win in the north can be attributed to the alliance between the Democratic Party (DP) and Uganda Peoples' Congress (UPC). The two parties formed an alliance to improve their chances of defeating Museveni. The Inter Party Cooperation was formed in December 1995. Its purpose was to unite and consolidate support in order to win the presidency. Ssemogerere was viewed as the most likely candidate to defeat Museveni.

Ssemogerere got 15-36% of the votes in Kampala, Rakai, Masaka, Mpigi, Tororo, Kumi and Mbale districts. The first four districts are in Buganda region. Since Ssemogerere is a Muganda (a person from Buganda), the choice could be a function of ethnic support. Tororo, Kumi and Mbale districts are in the east and have always been strongholds for the UPC party. Ssemogerere, the DP and UPC party candidate was expected to have a decent show here.

9.6.5 Choice of Candidate and Religious Affiliations

Ugandan politics has been influenced by religion (Mudoola, 1977; Mujaju, 1976; Welbourn, 1965). For example, Uganda Peoples' Congress (UPC) and the Democratic Party (DP) (the two major political parties), have strong ties to the two major religions the Protestant church and the Catholic church respectively. The political history in Buganda gives some insight to the relationship. Protestant chiefs had most of the powers during the colonial rule and DP was formed as a result of the protestant chiefs' monopoly (Twaddle, 1988). Was religion a factor in the 1996 voting?

In 1996 the three candidates represented the three major religions in the country. Ssemogerere is a Catholic, Museveni is a Protestant and Mayanja is a Moslem. The analysis of religious affiliation and choice of candidate involved computation of the simple correlation between the two variables. The religious data is published by the 1991 Census. For each of the 39 districts, data is given on percentage religious affiliations of the population. Correlations between votes obtained in district and religious affiliation in the district are shown in Table 9.7.

Table 9.7: Simple Correlations between Religion Affiliation and choice of Candidate

	Mayanja		Museveni		Ssemogerere	
Catholic	.204	.050*	-.411	.169*	.414	.179*
Moslem	.642	.412*	-.241	--	-.125	--
Protestants	-.367	.132*	.324	.106	-.308	.095

P=.05

Museveni seems to have appealed to voters across religious lines. Museveni's support has, however, an inverse relationship with Catholic and Moslem voters. It seems that Protestant Museveni did not do very well in districts with high Moslem and Catholic populations. The positive relationship between Museveni voters and Protestant population is probably a clear indication of the influence of religion on voting.

Mayanja did better in districts with a high Moslem population, while Ssemogerere had better results with Catholic dominated areas. Mayanja voting is inversely related to Protestant and Catholic populations in the districts. Similarly, Ssemogerere voting is inversely correlated with Moslem and Protestant populations. These are interesting findings given the polarity of the population. While we cannot conclude that Moslems voted for Mayanja, Catholics voted for Ssemogerere and Protestants voted for Museveni, finding the expected relationship along religious lines and the percentage of voters obtained, is a good starting point in this venture of understanding voting behaviour -geography of support.

9.6.6 Choice of Candidate and Education

Using the district education data on literacy rates, number of schools and number of colleges in the district, an attempt was made to understand choice of candidate on the basis of education (See Table 9.8). There is a significant positive relationship between Mayanja voting and the education level of the district. Districts with high measures of education had the highest number of voters for Mayanja.

Museveni and Ssemogerere voting and education has a negative relationship. Ssemogerere won in districts in the northern region. These are districts that lag slightly behind the southern region in education facilities. It may be true that education was not a factor where Ssemogerere choice prevailed. Museveni took the south (south of Lake Kyoga) and northeast. The south is relatively better off than the north in terms of education, but the northeast lags behind other regions. Museveni took both the most literate region and least literate region. In this respect regional education measures may not have a clear influence on Museveni support.

Table 9.8: Choice of Candidate and SES Variables

Variable	Mayanja		Museveni		Ssemogerere	
Employment	.204	.050*	-.411	.169*	.414	.179*
Education	.642	.412*	-.241	--	-.125	--

P=.05

9.6.7 Voting Participation

Africans as a group take voting seriously (The Economist, 1996). In the 1996 elections, the national average voter participation was 70%. Eleven districts out of 39 were below the national average. Six of the eleven districts were in the south, three in the north and two in the northeast. What factors influenced voter turnout? Studies have shown that several factors tend to influence whether or not an individual will vote (Kaplan and Venezky, 1994). Most studies of voter turnout have examined demographics and socio-political characteristics. Rarely are regional differences examined. The analysis here incorporates all three factors.

Using social economic status (SES) variables (education, income, occupation, age), the analysis examined regional variations in voting. The general view is that educated individuals are more likely to vote than the less educated (Kaplan and Venezky, 1994). Studies of voter turnout in the core (developed) countries have found that education leads to greater awareness of one's responsibility and that vote participation will vary with level of education. Using literacy data on a district level, simple correlations were computed between percent voters (turnout) and the literacy rate.

There was no significant difference between the most literate district and the least literate district. The number of voters was not associated with literacy rate, number of schools and number of colleges in the district. The other SES variables were equally irrelevant to voter turnout. Low voter turnout in the northeast can be attributed to remoteness and the semi-nomadic life styles of the Karamajong people (New Vision, 1996; Electoral Commission, 1996). The way of life of the Karamajong conflicts with structured activities such as political elections.

On the other hand, the below average turnout in the northern districts could have been partly a function of instability in the region, lack of faith in the electoral process, or apathy on the part of the voters. The presence of the Kony rebels in the region has been blamed for low turnout here (Vision, 1996; Monitor, 1996). While evidence for low turnout in the northeast and north is more grounded, there is lack of justification for the below average participation in the six southern districts. The six districts include the most urban districts of Kampala and Jinja.

There was no statistically significant variation in voting in relationship to income or occupation. As indicated in the discussion above, some of the most developed districts had below national average participation. The most urban districts have the highest literacy rates and have the most non-agricultural jobs. These are the areas where voter turnout should be highest. There is no evidence to support the view that higher education, income and employment are highly associated with voter turnout in Uganda. This is true despite the 70% turnout. Ugandans, like most Africans, take voting seriously irrespective of educational level and economic well being. It is a pilgrimage that comes once after a very long time – (16 years in 1996) for Ugandans. Therefore, it is an opportunity to participate.

9.6.8 Core and Periphery Model and Voting Patterns

This section focuses on the implications of Wallerstein's core periphery model as applied to political geography by Taylor (1990, 1991 and 1995). I argue that two cores influence elections in the periphery - the internal core, in this case inside Uganda and the external core. The internal factors which relate to the internal core of the periphery are what Taylor (1991) called elections beyond the world core. Taylor assumed minimum or no influence from outside the country and that only the internal factors would influence voting patterns. The external core discussion assumes influence from outside. This means that elections must be understood within the world system. The external core discussion assumes influence from outside.

The four characteristics of governments produced by internal factors were introduced earlier: (1) Unstable geography of support characterized by democratic musical chairs (Taylor, 1985, Dix, 1984). (2) Ethnic geography of support. (3) Geography of aggregation. (4) The use of coups to overthrow unpopular governments. Democratic musical chairs are where the government is routinely swept out of office. There is always a new government after every election. Uganda's politics do not fit this category. Only three of the nine governments have been elected.

In the politics of failure, support is dominated by ethnicity. There is a lack of a national movement, or rather, disconnected politics (Taylor, 1990). In the case of Uganda, two previous elections produced the same leader at different times. Obote was elected in 1962, with the support of Kabaka Yekka (KY) and the support of dominant ethnic group - the Baganda. He was elected again in 1980. It is debatable whether Obote's win in 1980 was a result of rigging, or his ability to form a coalition across the nation. President Obote belongs to a small ethnic group, the Langi. Therefore, ethnic support was not a determining factor. His victories were based on support from outside his ethnic group. Although Obote was elected with national support, both of his governments later faced strong opposition from the very people who helped elect him.

Obote's fallout with the Kabaka and the Baganda in 1966, caused disunity between and within the two camps (UPC and KY). This alliance between UPC and KY was instrumental in defeating Ben Kiwanuka and the DP in 1962. Twaddle (1994) argues that Obote's failure to achieve an effective national movement was the root of the first coup of 1971. Amin took advantage of this rivalry and exploited ethnic differences in his early years. Indeed, Amin's action of returning the remains of the Kabaka Mutesa, initially gave him the support of the Baganda.

Obote's second term in office did not have the support of the Baganda (Twaddle, 1994). Instead, he gained wide support both from different ethnic groups and from the army. Nonetheless, his second term in office failed to produce national unity. There was ethnic rivalry between the Langi and the Acholi. This conflict was of a similar nature to the conflict between the Kabaka Yekka and the Uganda Peoples' Congress. The very centripetal forces which helped Obote win election turned into centrifugal forces for his governments. Obote's second overthrow in 1985 made political history.

After Obote's removal by General Okello, Museveni took advantage of internal ethnic rivalry between the Langi and the Acholi and came to power in 1986 with the support of the Baganda. (He used Luwero Triangle in Buganda as a base.) Museveni's ascendancy to power was far from a nationally supported one. He attempted to honour his promises to the Baganda. For example, the return to traditional rulers - kings - was an important condition of support from the Baganda. On the other hand, the past fighting in the north, as well as past instability in the east and Buganda, point to some level of national disunity. So, how did he win the 1996 presidential elections by 75% margin?

The voting pattern and choice tend to be influenced by the level of economic development of the country. The South, the internal core region south of Lake Kyoga, which is more economically developed than the North, voted for Museveni. The peripheral region - the North, supported the opposition. Two districts in the periphery, Moroto and Kotido, voted for Museveni. This maybe evidence of Museveni's national base of support. The core region has endorsed the Museveni status quo.

Ssemogerere took most of the northern periphery but lost in his own birth district! This is clear evidence of lack of ethnic support for his candidacy. The Museveni candidacy was supported by the west, his area of birth, but he also won in other areas. For Museveni, there was some element of ethnic geography of support in the west. Museveni's win in 1996 and Obote's first win in 1962, have some common elements. Both wins had the support of the Baganda and both could be classified as national support elections. Could Museveni have won without the support of the Baganda? Probably, Obote did, in 1980. Could he win without the support of the West? Probably not. So we may have some ethnic support geography.

The geography of support for Museveni is sometimes seen as geography of fear. Voter behaviour and choice of presidential candidate is seen as partly a product of the uncertainty of the future. One of Museveni's campaign commercials showed a pile of skulls and bones at a mass grave. The message to the people was: "Don't forget the past. One million Ugandans ... lost their lives. Your vote could bring it back." Such a campaign advertisement can easily produce geography of fear leading to the necessary voting behaviour. The voters could have voted for Museveni because of such public threats. Cabinet members openly stated that a coup would be mounted against Ssemogerere, were he to win! Fear may be another characteristic of the politics of failure. Fear produces the wanted results. Another indirect characteristic is an external one. It is wrong to assume that election results in the periphery have no imprint of events in the core of the world system.

9.6.9 Elections and External Support Geography

While it is extremely difficult to measure direct influence, there is some evidence that supports the indirect external influence hypothesis. The core of the world system supported Museveni's candidacy. Museveni had been billed as an outstanding African leader, Uganda's man of vision. Articles in The Financial Times, The Economist and The New York Times gave him credit for substantial improvements in the Ugandan economy long before the elections. Lorch (1995), in an article in the New York Times, pointed out that the Ugandan leader was popular with world lenders. The World Bank and IMF have made and forgiven, loans to Uganda under his regime both before elections and after. Presently, Uganda is considered a model for economic development in the periphery.

Today, after more than twenty years in office, Museveni has changed. The lean guerrilla leader of 1986, is now stately. His Marxist orientation has been replaced by a western model of development, thus making him one of the popular African leaders to the donor countries of the world core, IMF and the World Bank. The donor countries financed part of the election despite the uneven playing field. Election observers agreed that some irregularities existed, but they endorsed the results. Could the core allow the defeat of one of its favourite supporters? Evidence of the external geography of support for Museveni is found in praise by the world newspapers and indirect backing of his incumbency.

9.6.10 Implications of the elections

If we assume that broad national support was achieved, would Museveni take advantage of this support to cultivate a far broader support base? Or, will he use his mandate to build more ethnic support and thus risk regional imbalances? Some interpretations may be found in one of the statements by President Museveni: "... those who supported us will be rewarded..." Regional imbalances exist in his assignment of cabinet positions. The majority of the ministers are from three districts of formerly, Ankole.

Museveni rewarded those who supported him. Unfortunately, the geography of rewards may create further regional economic imbalances and dissatisfaction. Now that a political mandate has been won, Museveni has the power to shape and design his geographies of support for the future. The African proverb of eating with one's enemy might be a good point to keep in mind.

9.7 Conclusion

The politics of success ought to be the opposite of the politics of failure. The 1996 presidential elections can be viewed as an event of the politics of success. Such politics reflects national unity, a geography of support that has a national base, voting choice that is not based on ethnicity and a possibility of defeating the incumbent without the problem of democratic musical chairs.

In examining the choice of the candidate by districts in Uganda, we find some evidence of national support but we also observe some ethnic support. Museveni's support in the West is close to the ethnic geography of support. On the other hand, Ssemogerere's win of the North is evidence of national support. Ssemogerere, clearly won in the North because of the political marriage between the UPC and the DP. Within this political marriage, we find some evidence of the third characteristic - the 'geography of aggregation.' The parties lack popular support but win or rule through forging coalitions. Although, Ssemogerere did not win, the coalition between the UPC and the DP was an element of the geography of aggregation. Such political marriages had worked in the past for the UPC.

The fact that the coalition did not win the elections, points to some characteristics of the 'politics of success' rather than the politics of failure. Can we have a politics of success without the democratic process? The electoral geography of core societies is characterized by a multiparty democratic system. In 1996, Museveni's movement rejected the multiparty system in favour of the movement. He transformed himself from a military-guerrilla ruler to an elected president.

Rather than political parties, he argued for a "Movement". The idea that individuals must stand on their own platform rather than a political party platform was not popular with the opposition. This idea advocates for "no-party democracy." But, were political parties absent in the 1996 elections? Was Ssemogerere representing DP and UPC? Was Museveni's Uganda Patriotic Movement (UPM) party not represented?

We may have political representation within the concept of "Movement". Parties may have been absent in name, but ideologies and interests predominate. Besides the absence of political labels, the groupings reflect the old politics. Within the idea of movement there may exist the geography of national support. Museveni had clusters of support, which he has fused together for his win. Would this support remain in place five years down the line? Or are we likely to see the unstable geography of support?

This question is in the minds of many today. In 1996, the movement democratic process worked in favour of the incumbent, with legitimization from the core of the world system. Some observers, at the time, saw Museveni's politics of the movement rather than of the political parties, as a step towards the politics of 'success' in Uganda. Hopefully, this represented the change from the gun to ballot but not a complete politics of success. To attain the politics of success, political leaders must keep their electoral promises. Let us hope the current leadership in Uganda can live up to their promises. Let us not forget that Museveni is also a product of situation that can be described as unfulfilled promises by the UPC.

Uganda may have experienced some politics of success during the 1996 presidential elections. The 2001 elections were viewed by some as rigged. The 2006 were also said to have been compromised. Uganda, like Ghana, was once hailed by aid donors as a recent political and economic success story (The Economist, 1996, The New York Times, 1996). The politics of such success ought to be the opposite of the politics of failure. Such a political story should reflect national unity, a geography of support that has a national base, voting choice that is not based on ethnicity and a possibility of defeating the incumbent without the problem of democratic musical chairs.

Discussion and Exercise Box 2

Compile the data on elections using the parish as the unit of measurement. Map the data using the different party affiliations as a classification. Interpret your results.

Questions

1. What is the significance of changes in administrative units?
2. Do more districts lead to better distribution/ accessibility of resources?
3. What is gerrymandering?
4. What does the term politics of failure mean?

5. What is electoral geography of the periphery?
6. What are the characteristics of the politics of failure?

References and further readings

Anonymous. 1996. Africa: Electing elites, *Economist*, 33:7951, 35-36.

BakamaNume, B. 1997. An Electoral Geography of Uganda: From the Gun to the Ballot, A Politics of Success of Legitimization? *The East African Geographical Review*, 44-56.

Chan, S. D. 1992. Democracy in Southern Africa, *The Round Table*, 322, 183-201.

Forrest J, C. Pattie and R. J Johnson 1995. Continuity or Change? The Geography of the Labor Vote at the New Zealand General Election of 1990, *Electoral Studies*, 14:1, 47-66.

Hartshore, R. 1954. Political Geography in Clarence F. Jones and P. James (eds.) *American Geography: Inventory and Prospect*. Syracuse University Press, 1954, 167-225.

Jackson, D. W.A (ed.) 1971. *Politics and Geographic Relations*, 2nd ed. Englewood Cliffs, NJ: Prentice Hall.

Johnson, R. J., F.M. Shelley and P.J. Taylor (eds.). 1990. *Developments in electoral Geography*, Routledge: London.

Kaplan, D. and R. Venezky. 1994. Literacy and Voting behavior: A Bivariate Probit Model with Sample Selection, *Social Science Research*, 23, 350-367.

Kibble, E. 1991. Political Change and Political Research in Africa: Agenda for the 1990s, *Issue*, 20:1, 50-53.

Lorch, D. 1995. Uganda Strongman a Favorite of the World Lenders, *The New York Times Sunday*, January 29, 1995.

McKinley Jr., J. 1996a. After a String of Dictators, Ugandans Get to Vote, *New York Times* May 10.

McKinley Jr., J. 1996b. Uganda Leader Looks Set for Election Victory, *New York Times* May 11.

Mudoola, D. 1983. Religion and politics in Uganda: the case of Busoga, 1900-1962, *African Affairs*, 77 306. *Monitor December* 15, 1995.

Navicki, M. 1993. Interview: President Yoweri Museveni *Africa Report*, 38: 23-25. *New Vision December* 16, 1995.

Osei-Kwame and P. Taylor. 1985. A politics of failure: The Political Geography of Ghanaian Elections, 1954-1979. *Annual of the Association of American Geographers*, 74: 574-89.

Taylor, P. 1997. Direct communication with the author.

Taylor, P. 1990. Extending the world electoral geography, in Johnson, R. J., F.M.

Shelley and P.J. Taylor (eds.). 1990. *Developments in electoral Geography*, Routledge: London, 257-271.

Vengroff, R. 1994. The Impact of the Electoral System on the Transition to democracy in Africa: The Case of Mali, *Electoral Studies*, 13:1, 29-37.

Welbourn, F. B. 1965. *Religion and Politics in Uganda 1952-1962*, Nairobi.

Wiseman, J. 1992. Early Post-redemocratization Elections in Africa, *Electoral Studies*, 11:4, 279-291.

Web Page CMW-Uganda.org/Uganda/administration.htm

Uganda Bureau of Statistics Web page.

Wikipedia. 2005. Politics of Uganda, http://en.wikipedia.org.wiki/Election in Uganda.

Chapter 10

ECONOMIC GEOGRAPHY OF UGANDA

BAKAMA BAKAMANUME

10.1 Introduction

Economic geography is a study of spatial variations in economic activities. Economic activities are classified into five main sectors – primary, secondary, tertiary, quaternary, and quinary. Primary activities involve extraction and harvesting of resources. These activities include agriculture, mining, forestry, hunting, and quarrying. Secondary activities involve adding value to a raw material. Manufacturing and construction are secondary activities. Tertiary activities are concerned with bringing the good or service to the customer – wholesale and retail activities belong to this class. Quaternary activities comprise economic activities rendered by white collar professionals. This sector includes the following types of workers: government, education, health, research, management, and information processing. Quinary is a sector of decision makers. The activities are linked by transport and communication.

Figure 10.1: Economic Activities

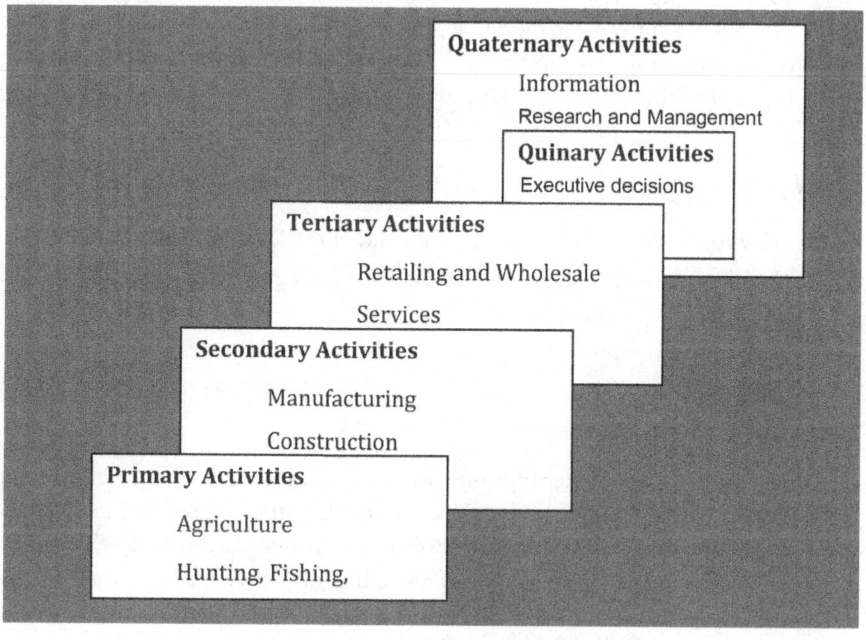

This chapter focuses on two sectors of the Ugandan economy. The chapter examines primary and secondary sectors. The emphasis is on agriculture, fisheries, and manufacturing.

Uganda's economy has been growing at a rate of 5-6% per annum, for the last 15 years (UBOS, 2004). The economy's most important sectors have been agriculture, manufacturing, and service.

10.2 Agriculture in Uganda

Historically, agriculture has been the leading primary economic activity in Uganda. Its importance is due to the fact that the country enjoys good climatic conditions and fertile soils. The country offers rich opportunities in the agricultural sector. Opportunities in this sector include agribusiness, forestry, food processing, and the manufacture of inputs. Recent trends show two sub-sectors of agricultural production – a traditional and non-traditional.

10.2.1 Non-traditional agriculture

This sub sector includes production of cereals, oilseeds, fresh and preserved fruit, vegetables and nuts, fresh, dried ground species, vegetable oils, essential oils, plants, orchids, flower seeds, sericulture (silk), and vanilla. These agricultural products are being marketed in larger volumes than before. They are the non-traditional cash crops.

10.2.2 Traditional export agriculture (Commercial/Large Scale Farming)

Traditional cash crops include coffee, tea and cotton growing, processing and packaging, growing of high quality tobacco, cocoa rehabilitation and expansion. The country's main opportunities are in the development of modern large scale farming, and processing for both the domestic and export markets. Uganda is currently the second largest producer of coffee in Africa. Uganda's cotton is of the best quality, and compares most favourably in the world market (Ministry of Agriculture, 2004).

10.2.3 Factors Influencing Location of Agriculture

1. Environmental factors – soil fertility and climatic conditions
2. Land availability – land rent (areas with high population density are intensively farmed)
3. Market for the produce
4. Cultural factors

1. Environmental factors

Ugandan agriculture depends on climatic conditions – rainfall amount, temperature of the area (Rugumayo, et al, 2003), and soil fertility. The country receives favourable rainfall and temperature ranges between 30 to37 degrees centigrade. These conditions allow for a wide range of crops to be grown. In regards to soil fertility, Uganda's agriculture largely depends on topsoil where most nutrients are concentrated; hence loss of this topsoil is synonymous with loss of soil productivity (Tenywa, 1999).

2. Land Availability

Population settlement patterns influence agricultural practices due to their effect on land rent. The most densely populated areas have limited agricultural land, thus, high land rents. The arable land in these areas has been put to use for more profitable activities than agriculture. It is important to note that there is land scarcity in some areas. Land scarcity is often measured by physiological density – the amount of arable land per person in a given region/ area.

The relationship between population and availability of land for agriculture is explained in theories of intensive agriculture. Ester Boserup (1965, 1981) argued that agricultural change was directly in proportion with population growth. In other words, the growth in population is associated with growth in intensity and production – the case in China and India. The theory is the opposite of the Malthusian theory. Malthusian theory argues that agricultural production declines as population increases, and it is the check to population growth.

3. Market

Distance to market and availability of market for an agricultural produce is another important factor influencing agriculture. There has been some shifting in terms of crops and their region of production. Cotton production has declined throughout the country. On the other hand, the area under cultivation for non-tradition crops has increased. Vanilla crop and rice have the highest increase. The area of production has also shifted. For example, Bugerere in Kayunga district used to be a dominant banana producing area for the Kampala market. Today, Ntungamo, Mbarara, and Bushenyi districts are the dominant producers/suppliers for Kampala.

4. Cultural Factors

There are some agricultural activities that are influenced by cultural factors. The Karamajong, Hima people of Ankole, and Iteso groups have always been known as cattle keepers. Livestock keeping is the main agricultural activity by culture.

10.2.4 Distribution of Agriculture

Access to market is a very important factor influencing the distribution of commercial agriculture. The distance from the farm to market may influence a farmer's choice of crop. Geographers use the Von Thunen model to explain the importance of access to market in determining what type of crops will be cultivated.

Von Thunen's Model and Agriculture Activities

The model was first proposed in 1862 by Jonathan Heinrich van Thunen, a farmer in northern Germany. According to the model, a commercial farmer's decision on what crops to cultivate and animals to raise is based on market proximity. The farmer looks at two factors: the cost of land (land rent), and the cost of transporting products to market.

Land values are highest in urban areas, and values decline with distance from the urban areas. Therefore, land use intensity is greater near the urban centres. Transport costs increase with distance from the market (urban centres). The figure below shows the relationship between distance from the market and agricultural type.

Figure 10.2: Distance from the market and agricultural type

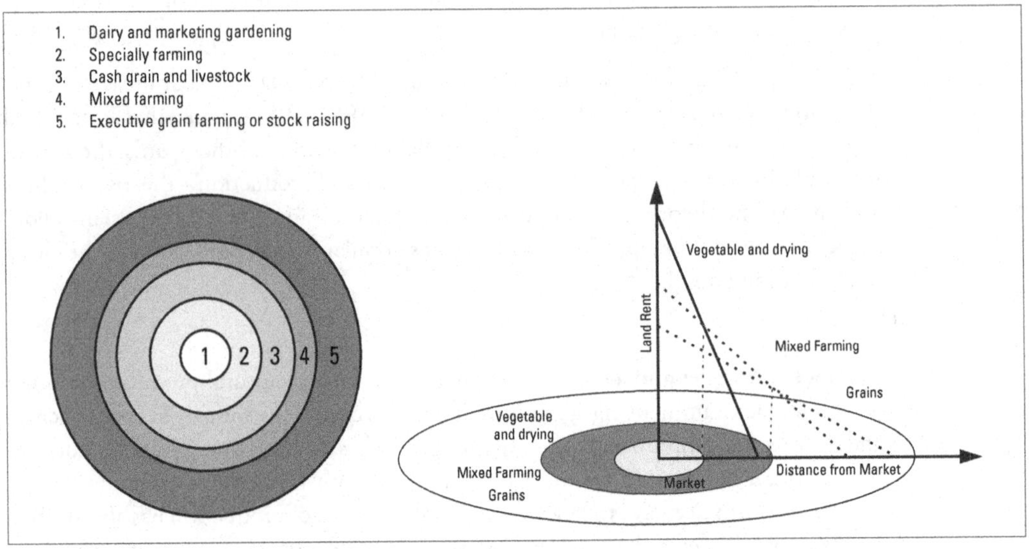

(a) Shows zones of agricultural practices. Zone 1 has the most intensive agriculture and zone 5 has most extensive agriculture type.

(b) Shows the land rent in relation to the location of type of agricultural activity. Land rent decreases with increase in distance from the urban areas.

The Ugandan agricultural practices do not exactly fit the Von Thunen Model of agricultural production. The land gradient aspect, however, is clearly applicable to Uganda. The price of land is highest in urban centres. It decreases with distance from urban areas. It is also true that the most extensive agricultural practices are located furthest from urban centres.

10.2.5 Characteristics of Uganda Agriculture

There are several characteristics of Ugandan agriculture. Agriculture is characterized by the following:

1. It is not well coordinated;
2. It is composed of scattered small scale farms;
3. It is based on low level of technology;
4. There is limited extension services;
5. Grows a variety of crops with little market orientation or commercialization.

There is no direct government intervention in marketing of crops. At one time, there were government agencies (cooperatives) that bought and marketed the major cash crops such as cotton (Uganda Cotton and Lint Marketing Board), coffee (Coffee Marketing Board), but they were abolished in 1980s. Since the liberalization of the economy, farmers have been faced with limited access to market information and credit (Ministry of Agriculture, 2004). The poor market conditions, limited access to financial services, and poor crop quality have led to unprofitable rural farming in Uganda.

Most farming is based on small scale cultivation, and close to 80% of the population is involved in agriculture. The farming in Uganda is labour intensive rather than capital intensive, so there is little technology involved. The farmers grow a range of crops, and intercropping is common. For example, in Bushenyi District, it is common to find a field with three crops – beans, maize, and ground nuts or another crop.

Despite these conditions, agriculture in Uganda has continued to grow but at a lower rate than other sectors. The growth rate for 2002/2003 was 2.2%, and for 2003/2004 was 2.3%. Compared to other economic sectors of the country, this sector has registered the lowest growth rate (UBOS, 2004). For example, in the financial year 2003/4, the economy registered a growth rate of 6% compared to 5.2% that was registered in 2002/3. The overall Gross Domestic Product (GDP) growth rate has been driven by better performance of all sectors. The agricultural sector has grown by 5.2% in 2003/4 compared to a lower growth rate of 2.3% in 2002/3 (Ministry of Finance and Economic Planning, 2005). Another notable fact is that the percent contribution of agriculture to the GDP has declined (See Table 10.1 below.)

Table 10.1: GDP contribution by sector

Financial Year	1999/00	2000/01	2001/02	2002/03	2003/04	2005/06
Agriculture	40.9	40.7	39.7	38.7	38.5	30.2
Industry	18.6	18.7	19	19.5	19.4	24.7
Services	40.5	40.6	41.2	41.8	42	45.2

Source: Uganda Bureau of Statistics, 2006

Table 10.1 and Figure 10.3 show Gross Domestic Product (GDP) by sector. GDP share of agriculture has declined from 40.9% to 30.2%. The share for the industrial sector has increased from 18.6% to 24.7%. The service sector has increased from 40.5% to 45.2%. A fast growth of the service sector has been reported in many countries. However, Uganda would like to see higher growth rate in the industrial sector. The government has policies to improve the industrial sector.

Figure 10.4 shows growth trends in agriculture in relationship to national growth rates. The two trends show some similarities in peaks and slumps. It is also clear from the graph that agriculture in Uganda has performed below the national growth rate. However, as shown elsewhere, cultivated areas for most crops have increased, but production tonnes or yields have not increased significantly.

Figure 10.3: GDP Contribution y Economic sector

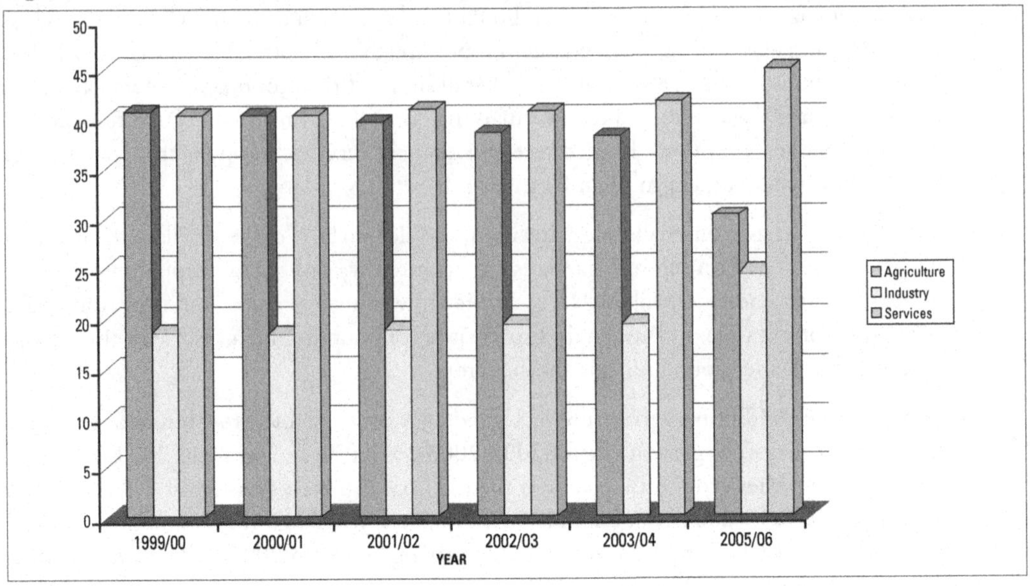

Source: Uganda Bureau of Statistics

Figure 10.4: Agricultural Growth Rates VS National Growth Rates for 1989/90 - 2002/03

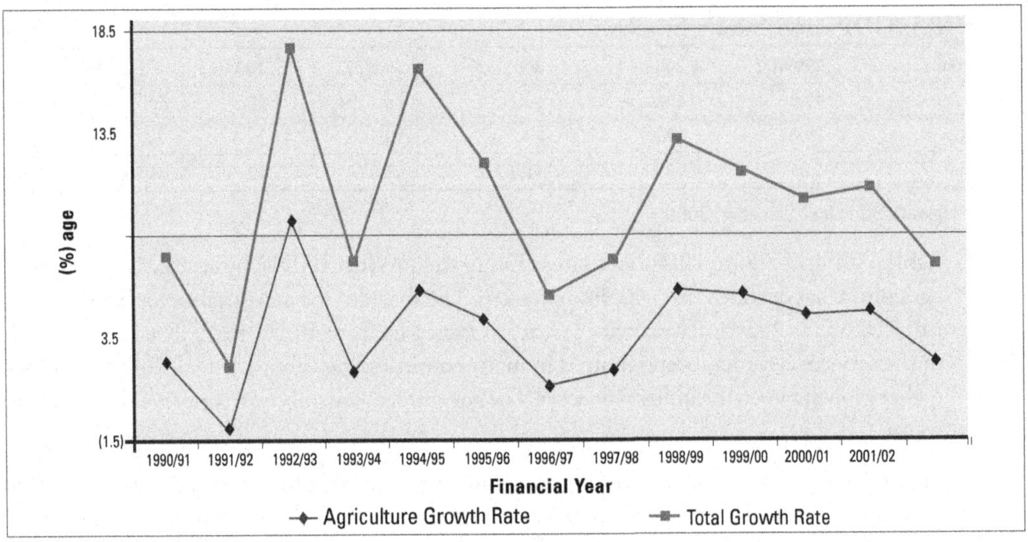

Source: Ministry of Agriculture, 2005.

10.2.6 Trends in Agricultural Production as part of GDP

Table 10.2 and Figure 10.5 below show the agricultural trends in Uganda. Food cropping remains the most important activity. Farmers grow food for their own consumption as well as for the market. This makes some food crops part of the cash cropping sector. The significance of food cropping is also shown in the graph below (Figure 10.5).

Table 10.2: Agricultural Contribution to GDP (000s)

Industry/ Group	1999	2000	2001	2002	2003
Agriculture					
Monetary					
Cash crops	327,489	236,510	208,797	274,716	351,512
Food crops	870,313	958,168	946,929	882,445	1,186,642
Livestock	278,271	282,394	296,144	322,626	355,392
Forestry	51,228	57,322	63,784	65,757	73,061
Fishing	163,661	168,069	209,852	228,996	248,282
Total monetary	1,690,962	1,702,463	1,725,506	1,774,540	2,214,889
Non – monetary					
Food crops	1,109,598	1,205,811	1,109,196	980,391	1,311,623
Livestock	151,691	155,795	167,207	184,071	213,535
Forestry	91,953	100,310	105,196	108,530	111,973
Fishing	20,642	21,198	26,468	28,882	31,315
Total Non monetary	1,373,884	1,483,114	1,408,067	1,301,874	1,668,446
Total Agricultural GDP	3,064,846	3,185,577	3,133,573	3,076,414	3,883,335
Total GDP	7,995,220	8,671,303	9,347,279	9,940,724	11,634,441

Source: Uganda Bureau of Statistics 2004

Figure 10.5: Agricultural trends 1999 - 2003

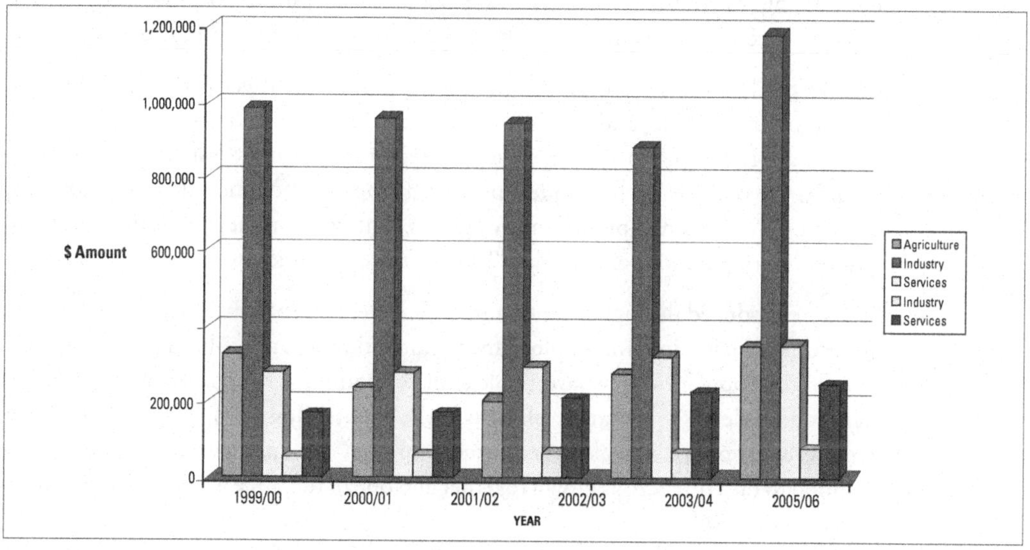

Source: Uganda Bureau of Statistics, 2004.

10.2.7 Crops

Food crops have continued to dominate crop husbandry. Except for a slight decline in 2002, food cropping output has increased every year. On the other hand, traditional cash cropping has declined in importance, partly because of non-traditional agriculture. The non-traditional sector is now regarded as a cash generating sector. Cash cropping declined in years 2000, and 2001. It has shown some slight gains in the last two years.

Other primary activities under the Ministry of Agriculture include livestock farming, forestry, and fishing. Livestock farming has increased since 1999 (MOA, 2004). This section of agriculture is almost equal to cash cropping in monetary terms. Forestry is another primary activity. The forestry sector is discussed in Chapter 4. Fishing is another section within the Ministry of Agriculture. Fishing is discussed under Water and Wetland Resources (Chapter 5).

Description of Maps

A diagonal strip of high agricultural land use is shown on the map of agricultural plots by districts. It extends from the east to northwest. The southwest is also an important area for agricultural production while the lake basin area is no longer a dominant agricultural region. The map shows that the east and central northern areas are important agricultural areas.

Banana production is concentrated in the southern part of the country. Bananas were the major staple food of the people in the south but that is changing. The leading banana producing districts are Rakai, Mbarara, Bushenyi, Ntungamo, and Kabarole. Mbale, Masaka, and Rukungiri are also important banana producers. Bugerere area – Mukono and Kayunga are no longer the most important banana production areas. The decline in banana production in this region has been attributed to fungus in the region.

Cassava production is important in the region west of the River Nile. The West Nile districts of Arua, Nebbi, and Yumbe are important cassava producers. Other important districts are Kalangala in the Lake Victoria region, and Busia District in the eastern Uganda. Cassava used to be a widely produced crop. At one time, all households were required to have a small plot of cassava plants. This was a policy to ensure that there was food during periods of droughts. Today, this requirement is no longer in place.

Beans are produced throughout the country. Kabale, Kisoro, Kibale, Apac, and Lira are the leading districts in terms of plots under cultivation. Beans are both a cash and food crop for the population. The favourable climate conditions in Uganda plus fertile soils make it possible to grow beans in all parts of the country. It is a crop that is often grown together with maize. This is called inter-cropping. In some parts of Bushenyi it is common to find three crops grown together and one of the three will be beans or maize.

Maps 10.1: Agricultural Plots by District in Uganda

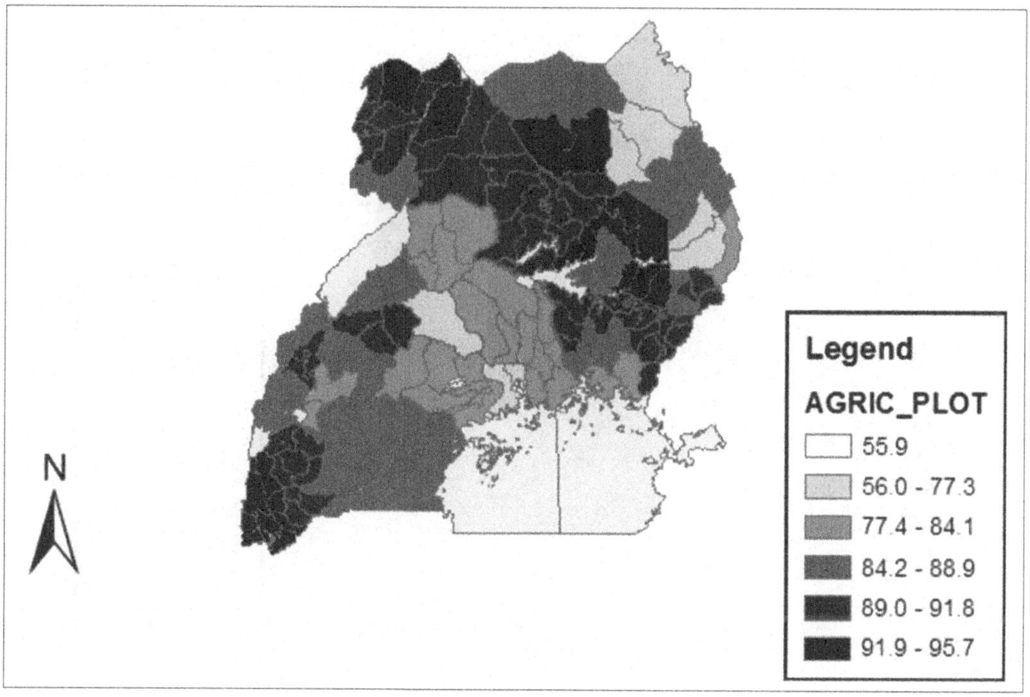

Maps 10.2: Banana Production in Uganda

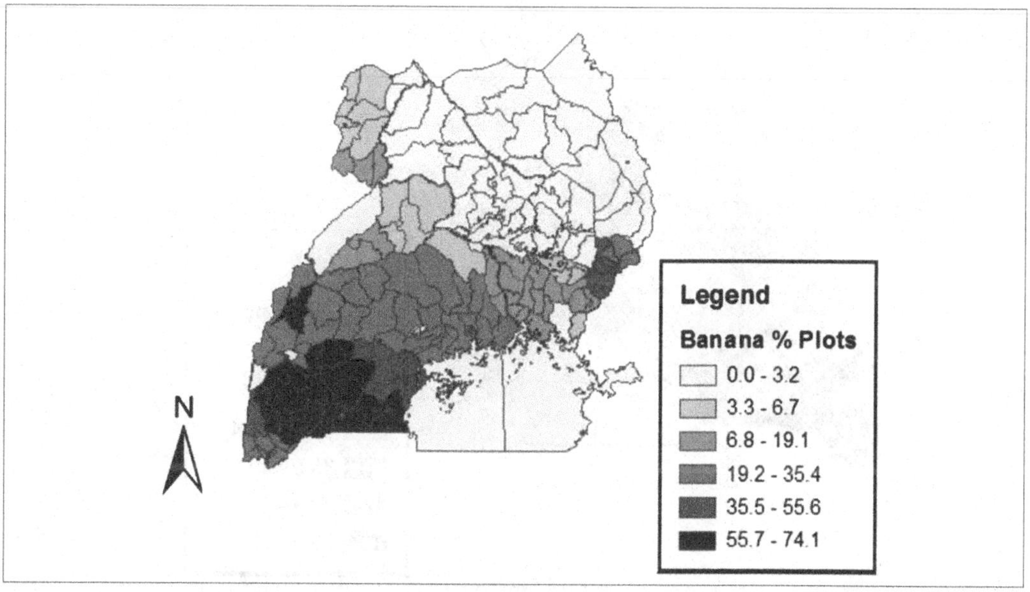

Source: Ministry of Agriculture 2003

Maps 10.3 Cassava Production in Uganda

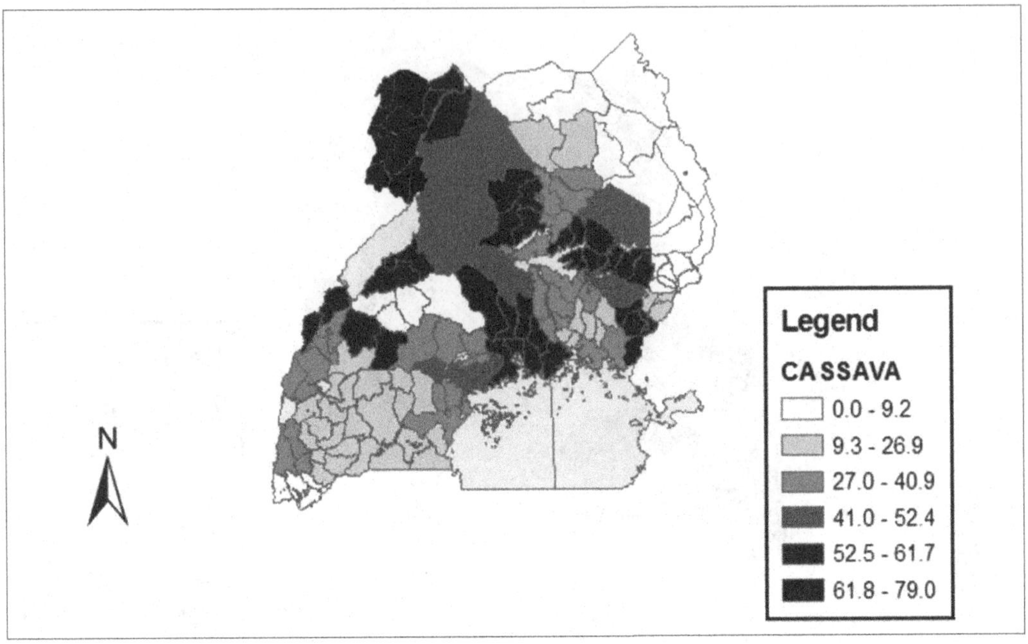

Source: Ministry of Agriculture 2003

Maps 10.4 Beans Production in Uganda

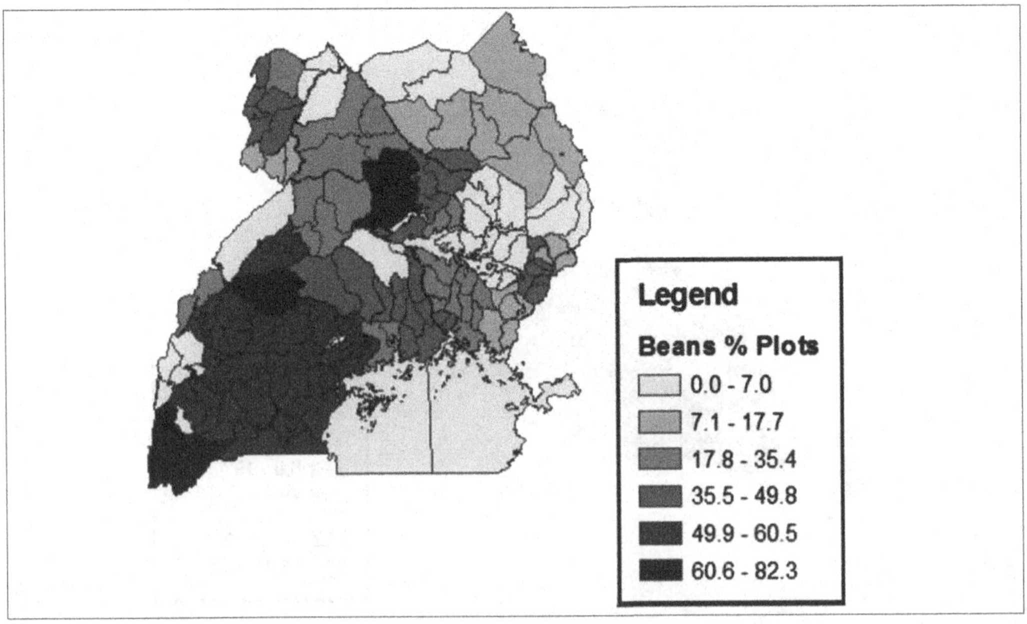

Source: Ministry of Agriculture 2003

Maps 10.5: Millet Production in Uganda

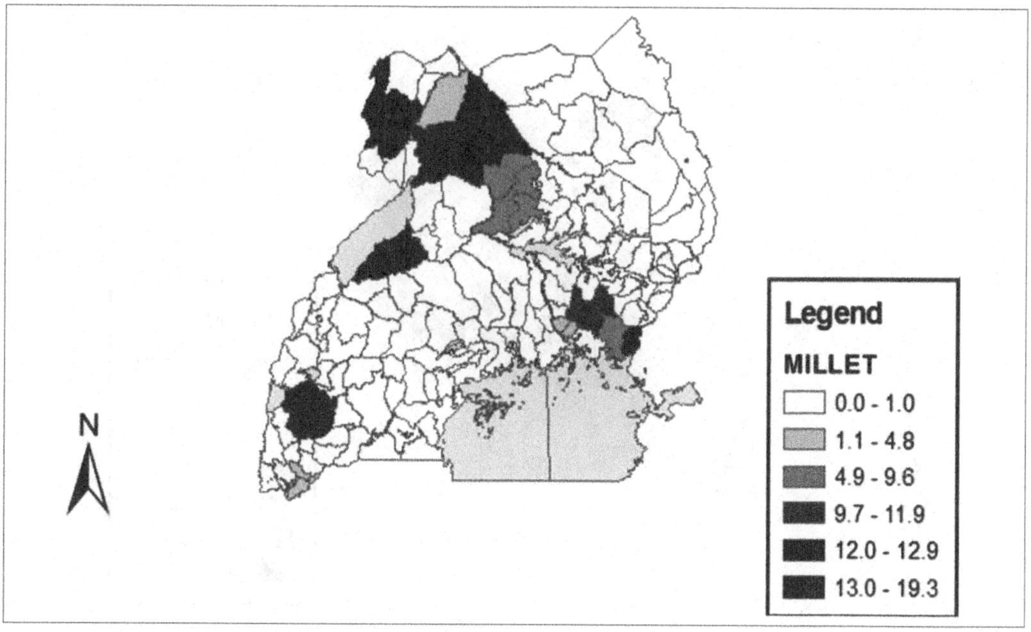

Source: Ministry of Agriculture 2003

Maps 10.6: Maize Production in Uganda

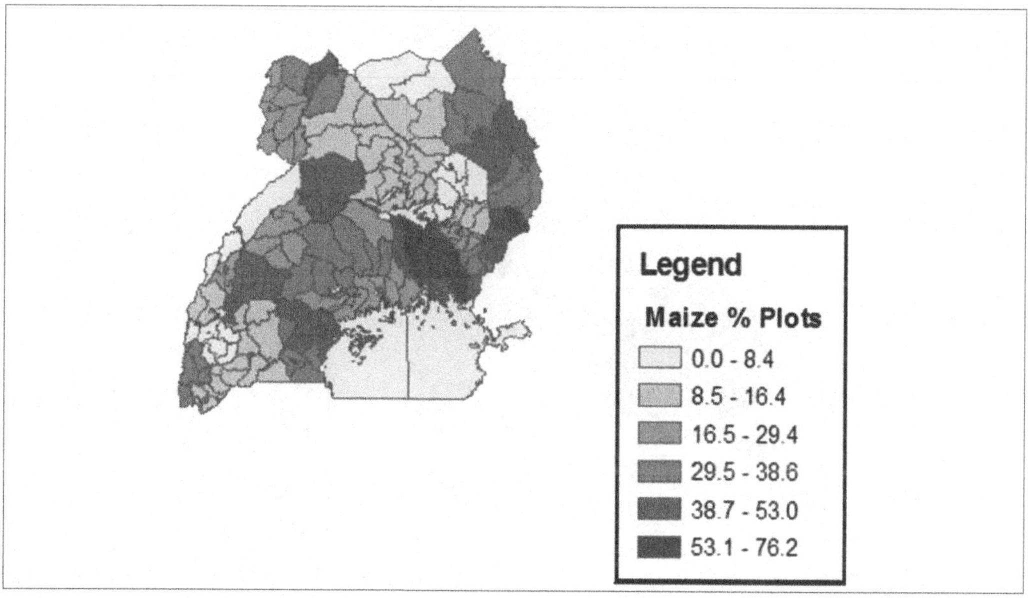

Source: Ministry of Agriculture 2003

Maps 10.7: Sweet Potatoes in Uganda

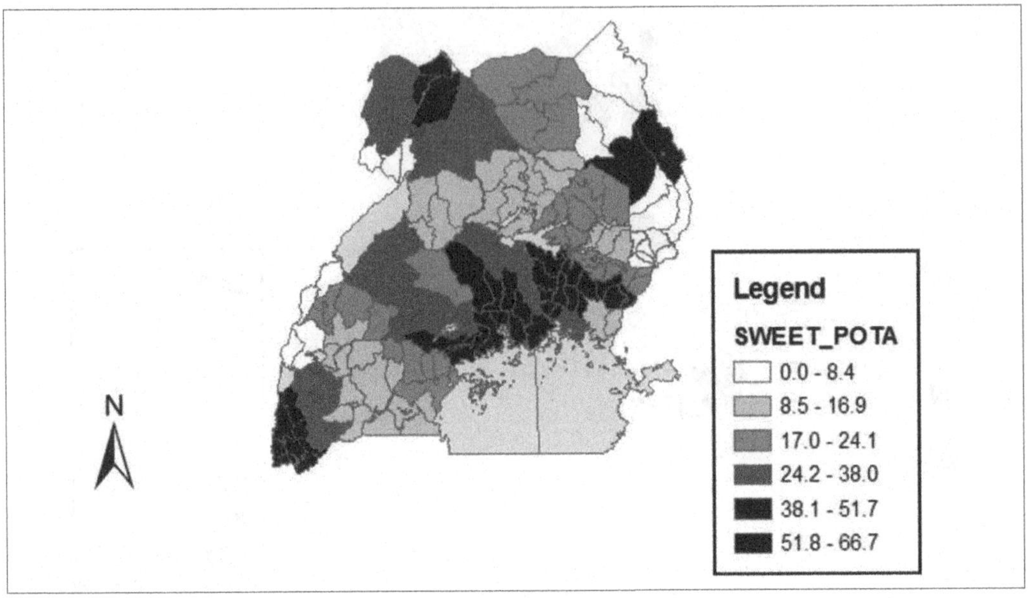

Source: Ministry of Agriculture 2003

Maps 10.8: Coffee Production in Uganda

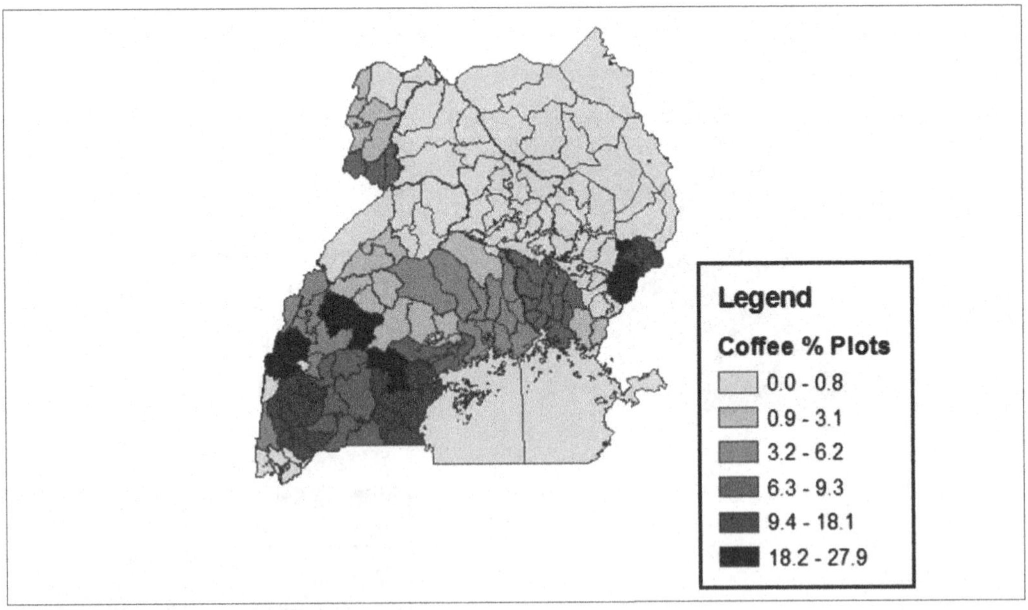

Source: Ministry of Agriculture 2003

Millet production is concentrated in the northwest, near northwest, and the west. There are some small pockets of production in the east, and west. In most of the production areas, millet used to be a staple food crop. It is also used to make local brew.

Maize production is important in eastern Uganda. The districts of Kamuli, Bugiri, Jinja, Kapchorwa, and Kalangala have the highest number of plots under maize production. The weather is very suitable for maize production. Most of the maize produced here finds a ready market in neighbouring Kenya. In many households, maize is now the "cash crop".

Sweet potatoes are one of the important food staples especially in the southern part of the country. The leading producing districts are Rukungiri, and Adjumani. Both the Lake Victoria and Lake Kyoga basins are important producers of sweet potatoes. Sweet potatoes like maize are both a food and cash crop. Sweet potato crop is less labour intensive than a banana plantation. Sweet potatoes grow better in regions with good rainfall. This explains why the Lake basin areas are important.

Coffee production has declined in Uganda. At one time coffee was the leading cash crop. The leading districts in coffee production are Mbale, Kasese, Lira, Rakai, and Sembabule. Coffee is mostly grown in the southern and eastern districts. There are two types of coffee produced in Uganda. Arabic coffee produced in high elevations such as the slopes of Mt. Elgon and Kasese, and Robusta coffee grown in low elevations.

Table 10.3 and Figure 10.6 show acreage and production of selected food crops. Bananas, maize, sweet potatoes, cassava, millet and sorghum are the leading food crops in terms of area planted. Wheat, Irish potatoes, and rice are the least important food crops in terms of area cultivated. Bananas and maize are both leading food and cash crops.

Photo 10.7: Picking Tea in Uganda

Source: Ministry of Agriculture 2003

Table 10.3: Area Planted and Production of Selected Food Crops

Year	Plantains	Cereals						Root crops			
Area planted	Bananas	Finger millet	Maize	Sorghum	Rice	Wheat	Total	Sweet potatoes	Irish potatoes	Cassava	Total
000 hectares											
2000	1,598	384	629	280	72	7	1,372	555	68	401	1,024
2001	1,622	389	652	282	76	8	1,407	572	73	390	1,035
2002	1,648	396	676	285	80	8	1,445	589	78	398	1,065
2003 Estimates	1,661	400	710	290	86	9	1,495	595	80	405	1,080
2004 Projections	1,670	412	750	285	93	9	1,549	602	83	407	1,092
Production											
000 tonnes											
2000	9,428	534	1,096	361	109	12	2,112	2,398	478	4,966	7,842
2001	9,732	584	1,174	423	114	14	2,309	2,515	508	5,265	8,288
2002	9,888	590	1,217	427	120	14	2,368	2,592	546	5,373	8,511
2003 Estimates	9,700	640	1,300	421	132	15	2,508	2,610	557	5,450	8,617
2004 Projections	9,686	659	1,080	399	121	15	2,274	2,650	573	5,500	8,723
Pulses				Others							
Beans	Field Peas	Cow Peas	Pigeon Peas	Total	Groundnuts	Soya Beans	Slim Slim	Sun Flower	Total		
699	29	64	78	870	199	106	194	79	578		
731	26	65	80	902	208	127	203	99	637		
765	27	66	82	940	211	151	211	124	697		
780	27	67	84	958	214	145	240	145	744		
812	25	70	84	991	221	144	255	145	765		
420	16	60	78	574	139	128	97	79	443		
511	15	59	80	665	146	144	102	99	491		
536	16	59	82	693	148	166	106	124	544		
525	14	67	84	690	150	160	120	160	590		
545	15	69	84	623	137	158	125	162	582		

Figure 10.6: Area Planted by Crop

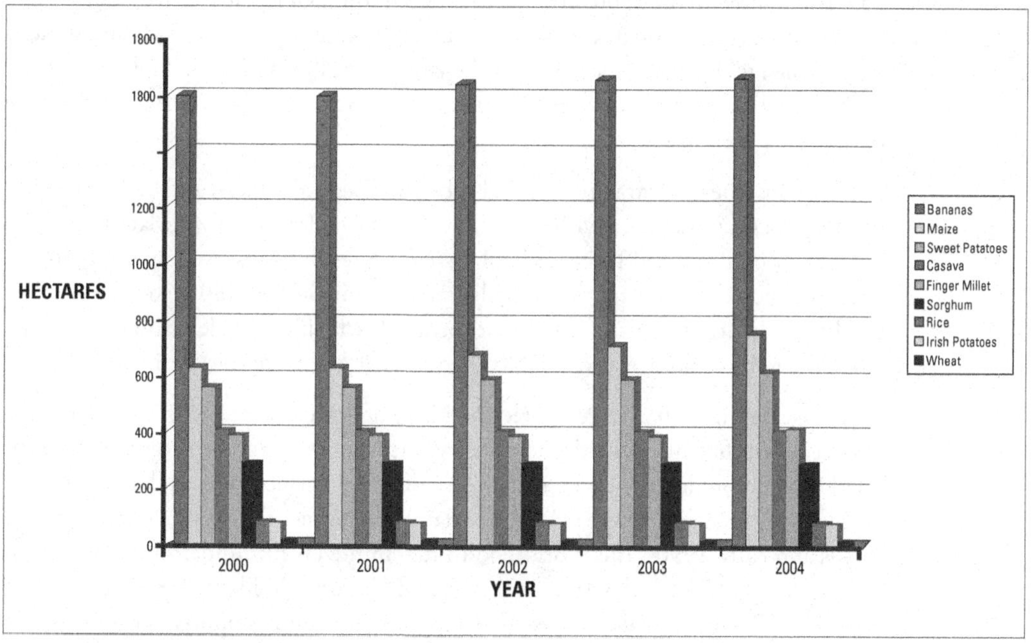

Source: Ministry of Agriculture 2003

10.2.8 Livestock

The livestock sector of agriculture has made tremendous recovery in the last few years (See Table 10.4). However, current output in this sector can only meet two thirds of the domestic demand. There is, therefore, great need for expansion and investment in the animal husbandry sector such as ranching (cattle, goats, sheep), poultry (chicken, pecking duck and ostrich), hides and skins, bee-keeping for honey products, leather processing and preservation and canning of the various livestock products such as meat and milk (Ministry of Agriculture, 2004).

Table 10.4 Number of Livestock, 1998-2002

Type	1998	1999	2000	2001	2002
Cattle	5,651	5,820	5,966	6,144	6,328
Sheep	1,014	1,044	1,081	1,108	1,141
Goats	5,999	6,180	6,396	6,620	6,852
Pigs	1,475	1,520	1,573	1,644	1,710
Poultry	22,293	24,622	26,622	29,671	32,639

Source: Uganda Bureau of Statistics

The country's good natural environment provides good grazing for cattle, sheep, and goats, with indigenous breeds dominating most livestock in Uganda. The livestock sector of agriculture has always been dominated by small scale farmers. They own about 95% of all cattle. In the 1960s and 1970s modern commercial ranches were established by the Ministry of Agriculture, in areas that had been cleared of tsetse-fly infestation. Large scale ranching was successful in the late 1960s, but during the upheaval of the 1970s many ranches were looted, and most farmers sold off their animals at low prices to minimize their losses.

In the 1980s, the government provided substantial aid to farmers, and by 1983 eighty ranches had been restocked with cattle (MOA, 2004). However, by the late 1980s, the livestock sector suffered heavy animal losses as a result of disease, especially in the northern and north-eastern regions. Political instability in the 1980s, led to a complete breakdown in disease control, and to the spread of tsetse flies. Cattle rustling, especially along the Kenyan border, also depleted herds in some areas of the northeast.

Uganda's dairy industry has recovered (Ministry of Agriculture, 2004). The country is almost self-sufficient in the dairy industry but has been hampered by a number of problems. Low producer prices for milk, high costs of animal medicines, and transportation problems are especially severe obstacles to dairy development (MOA, 2004). In the late 1980s and early 1990s, the World Food Programme (WFP) undertook an effort to rehabilitate the dairy industry, and the United Nations Children's Fund (UNICEF) and other UN agencies also helped subsidize powdered milk imports, most of it from the United States and Denmark.

Uganda's beekeeping industry has also suffered throughout the years of civil unrest. In the 1980s, the CARE Apiary Development Project assisted in rehabilitating the industry, and by 1987 more than fifty cooperatives and privately owned enterprises had become dealers in apiary products. Today, the beekeeping industry boasts of over 6,000 hives in the field, and an estimated 800 tonnes of honey and 650 kilograms of beeswax were produced.

The poultry sector is the fastest growing among livestock. The growth between 1998 and 2002 is 46%. Cattle registered the lowest growth rate of 11.9%; pigs were 15.9%, goats 14%, and sheep 12.5% (MOA, 2004).

10.3 Fisheries

The fishery sector has been discussed under water resource and wetlands (Chapter 5). It is, however, part of the Ministry of Agriculture, and for government accounting purposes it is included in agriculture. Uganda's fishery sector has grown rapidly in recent years and current export earnings are in excess of US$100 million per year (UBOS, 2004). Uganda is blessed with large fresh water lakes that are home to a wide variety of fish products. Opportunities are also available for fish farming.

10.4 Prospects for Agriculture in Uganda

There are four reasons for agricultural prospects in Uganda.

1. Increasing Demand at Home

As the population of the country increases, the demand for agricultural goods increases. Uganda's population is growing at a rather alarming rate (see Chapter 6). That may be regarded as a sure domestics market, but it can also reduce agricultural production potentials. In the neo-Malthusian tradition, increase in population is inversely proportion to agricultural production. However, the Boserup argument is of the view that population pressure (increase in demand) will lead to agricultural development.

2. Natural Conditions

The country is blessed with excellent natural endowments, fertile soils, and adequate rainfall. It is a low cost producer of tea, coffee, cotton, tobacco and a number of high value agricultural export crops such as cut-flowers and asparagus. With improved efficiency, this natural advantage can be exploited to make Uganda a major exporter of agricultural produce and agricultural products. The country has a very sizeable arable area. Uganda has over 18 million hectares of arable land, of which less than 30% is currently under cultivation (MOA, 2004).

3. Export market

The market prospects in Europe and the Middle East for Uganda's non-traditional agricultural export products such as flowers and horticultural products are excellent. The country's agricultural exports to the European Union are tariff free under the provisions of the Lomé Agreement. The country has food surpluses and is already exporting food to the neighbouring countries – a ready market already exists.

4. Foreign Investments in Agriculture

Outside donors have put in place a number of schemes to support the development of Uganda's agricultural potential. These schemes include credits to private farmers, export incentives as well as marketing assistance.

Despite the good prospects for agriculture in Uganda, there are some barriers to increasing agricultural production, namely:

1. Land tenure systems
2. Civil war in the Northern Uganda
3. Disease prevalence – such as malaria and HIV/AIDS
4. Access to market
5. Poor farming methods
6. Small scale/ subsistence farming practices (limit access to technology and financing).

Prospects of Agriculture in Uganda

Source: Uganda Tourist Board

10.5 Manufacturing in Uganda

The country's manufacturing sector is one of the fastest growing sectors. Manufacturing has grown by more than 10% annually, over the last five years (Ministry of Industry Tourism and Trade, 2004). Several opportunities exist in all areas ranging from beverages, leather; tobacco based processing, paper, textiles and garments to pharmaceuticals, fabrication, ceramics, glass, fertilizers, plastic/PVC, and assembly of electronic goods, hi-tech and medical products (MoITT).

There are three reasons for the existing wide range of potentials in manufacturing.

Firstly, the potentials are a result of the devastation of the sector in the 1970s and 1980s. In 1972, President Idi Amin ordered non citizens especially Asians, who controlled the manufacturing sector to leave the country. Their departure created a void in manufacturing expertise. The sector declined until its revival in the late 1980s.

In the late 1980s, most manufacturing industries relied on agricultural products for raw materials and machinery, and as a result, the problems plaguing the agriculture sector hampered both production and marketing in manufacturing. Processing cotton, coffee, sugar, and food crops were major industries, but Uganda also produced textiles, tobacco, beverages, wood and paper products, construction materials, and chemicals. During the 1980s, most industries ceased to produce. There is, therefore, a vacuum (opportunities) in manufacturing at the moment.

The second reason is market availability. The opportunities to market manufactured products from Uganda are not limited to the domestic market. An export market exists in the neighbouring countries within the Common Market of Eastern and Southern African countries (COMESA) region of which Uganda is a member. The geographical location of Uganda makes it accessible to Southern Sudan, Eastern Congo (DR), and Rwanda markets.

Thirdly, Uganda has a large, cheap labour force and abundant resources. The country has attractive fiscal (tax), and physical export incentives that make it a favourable location for targeting the rest of the Sub-Saharan and Southern African region.

Until the late 1980s, most manufacturing industries in Uganda depended on agricultural products for raw materials and machinery. The consequent dependence was that problems of the agriculture sector became problems of production and marketing in manufacturing sector. The major industries were processing cotton, coffee, sugar, and food crops. The country also produced textiles, tobacco, beverages, wood and paper products, construction materials, and chemicals.

During the same period of 1980s, the government began to return some nationalized manufacturing firms to the private sector in order to encourage private investment. The goal was to attain self-sufficiency in consumer goods and strengthen linkages between agriculture and industry. Manufacturing production has increased since the 1990s. More industries have been built, and efficiency at individual plants has improved.

Uganda's industries can be classified into two types:

1. Construction Industries

   Production of Steel Sheets

   Hoe Production

   Scrap Iron

   Cement manufacturing

   Brick making

   Cables

   Metal

   Some assembling plants

2. Consumer Goods

   Production of Beverages – alcoholic and soft drinks

   Sugar Production

   Grain milling – wheat and corn

   Cigarette manufacturing

   Edible oil

   Soap manufacturing

Leather and tanning

Bakeries

Furniture making (includes the coffin making sector)

Table 10.5: Types of Industries and Index Industrial Production

Industry	Number of Establishments	1998	1999	2000	2001	2002
Food Processing	50	110	123.6	118.2	131.9	126.2
Tobacco and Beverage	11	104	112.6	116	119	126.7
Textiles, etc	8	128.4	185.4	178.9	166.9	165.9
Paper and Printing	14	115.3	134.1	163.5	183.8	186.6
Chemicals	19	209.4	125.3	124.8	138.2	133.8
Bricks and Cement	11	109	118.6	136.2	148.6	167.9
Metal	16	111.6	126.6	155.9	204.9	199.6
Miscellaneous	15	101.8	98.198	98	103.7	154
Other		109.7	123.4	127.5	141.4	144.5

Source: Uganda Bureau of Statistics (2004)

The index of industrial production measures trends in manufacturing. All indexes except for the chemicals have improved since 1998. The 2002 indexes are higher than those of 1998 for all the categories (See Table 10.5). The trends shown indicate some ups and downs for some of the manufacturing categories, but the 2002 index is higher than that of 1998. The indexes reflect the fact that the manufacturing sector has grown. The tobacco and beverage index has steadily risen. Food processing declined in 2002 but picked up in 2003.

10.5.1 Factors influencing Manufacturing activity in Uganda

There are several factors that influence manufacturing activity in Uganda: market, government fiscal and physical policy, transport, and resource availability. These factors have helped change the manufacturing landscape in the country. Uganda's manufacturing sector has an aspect of the classical industrial locational theory, and the current factors. In the classical industrial locational theory, Alfred Weber considered *raw material, transport costs, labour cost*, and *market* as the important factors for industrial location. Today's industrial location is influenced by capital, political stability, technology, government regulations, existing infrastructure, and labour costs.

Market oriented industries locate at or near the market. The raw materials are transported to the point of production, and the finished good is then distributed to consumers. The largest concentration of consumers in Uganda is found in Kampala city and other urban centres. It is also true that most of the manufacturing activities in the country are located in urban centres. In this respect, Uganda's manufacturing sector may be seen as market oriented. The urban centres with most manufacturing activities are Kampala, Mbarara, Jinja, Tororo, and Soroti.

Resource/Material oriented industries are influenced by availability of the raw materials. Some of the manufacturing activities in Uganda are closely associated with sources of raw materials. The manufacturing of cement at Tororo and Hima, sugar processing in Kakiira and Lugazi, tea processing in Fort Portal and Lugazi, and fertilizer industries are some of the good examples of raw material oriented industries in Uganda.

Policy oriented industries are those that locate at a point as a result of government incentives. The Uganda Investment Authority, a government agency, has gazetted over 1,000 hectares of prime industrial land to be developed into fully serviced industrial estates and export processing zones. The Uganda Investment Authority is also actively courting private sector participation in the development of industrial estates and export processing zones. Most of these industrial parks are located near Kampala or in Mbarara town. Unfortunately, the former industrial town of Jinja has lost most of its industries to these two urban centres. Manufacturing activities have moved to the market centres of Kampala and Mbarara.

Political stability has played an important role in the revival of the manufacturing sector in Uganda. This point is discussed at the beginning of this section. The opportunities that exist today are partly a product of the political instability of the past. During the 1970s and 1980s, instability caused a decline in industrial activity. This vacuum is being filled today.

Inertia of infrastructure (facilities and suppliers) existing also influences location of manufacturing. Uganda abandoned the import substitution policy in the 1980s. This is a policy of protecting infant industries from outside manufacturers. According to recent Global Competitiveness Report 2005-2006, the country dropped eight positions in ranking. It ranks 87th among 117 countries. Tanzania ranks 71st on the list. Uganda has the 11th best macroeconomic environment in Africa (World Economic Forum, 2005).

Discussion Questions

1. What are the advantages and disadvantages of import substitution?
2. Would such a policy benefit the Uganda manufacturers?
3. What would be the implication of the East African Common Market and COMESA?

10.6 Export Trade

The agricultural sector continues to dominate in terms of exports. There have been some changes in the importance of individual export crops, for example, coffee is down from 60% in 1999 to 19% in 2003 (UBOS, 2004). This is an indication of a diversifying economy. The diversification reduces over dependence on one or two exports. Coffee, cotton, tea and tobacco have been the traditional exports. But other non-traditional exports such as fish, flowers, fruits, vanilla, and bee honey are now important exports.

Table 10.6: Exports of Goods

	2000/01	2001/02	2002/03	2003/04 (est)
Coffee	109.64	85.25	105.47	107.53
Volume (60 – kg bags)	2.84	3.16	2.99	2.5
Unit value (US$/kg)	0.64	0.45	0.59	0.7
Cotton	14.08	18	16.88	28.87
Volume ('000 mtons)	12.14	22.5	16.36	20.45
Unit value (US$/kg)	1.16	0.8	1.03	1.41
Tea	35.93	26.85	29.45	38.26
Volume ('000 mtons)	28.09	30.3	31.14	35.27
Unit value (US$/kg)	1.28	0.89	0.95	1.08
Tobacco	27.64	32.27	39.89	38.94
Volume ('000 mtons)	12.77	17.62	23.48	26.48
Unit value (US$/kg)	2.16	1.83	1.7	1.47
Fish & its products	50.11	80.85	83.77	98.44
Volume ('000 mtons)	22.31	27.37	24.13	31
Unit value (US$/kg)	2.25	2.95	3.47	3.18
Maize	6.13	13.07	8.15	18.03
Volume ('000 mtons)	29.59	89.97	33.82	91.66
Unit value (US$/kg)	0.21	0.15	0.24	0.2
Flowers	13.22	15.91	17.04	26.9
Volume ('000 mtons)	3.47	4.29	4.74	6.57
Unit value (US$/kg)	3.81	3.71	3.6	4.1

Source: Bank of Uganda

When exports are grouped into traditional and non-traditional categories the advantages of the traditional individual export crops are lost (Ministry of Planning and Economic Development, 2004). Current trends show an advantage of the non-traditional exports (See Figure 10.5). The growing importance of non-traditional exports (NTE) is also shown in Figure 10.7 below. Since 2000, the non-traditional exports have performed better than the traditional exports (TE). In 1998, traditional exports accounted for 66% of the export value, and non-traditional exports contributed 34%. The 2004, statistics show the reverse, NTE account for 63% and TE provide 37% of the export income. The non-traditional exports have superseded traditional exports for the last four years.

Figure 10.7: Comparisons between Traditional and Non-Traditional Exports(Export Performance 1998-2003)

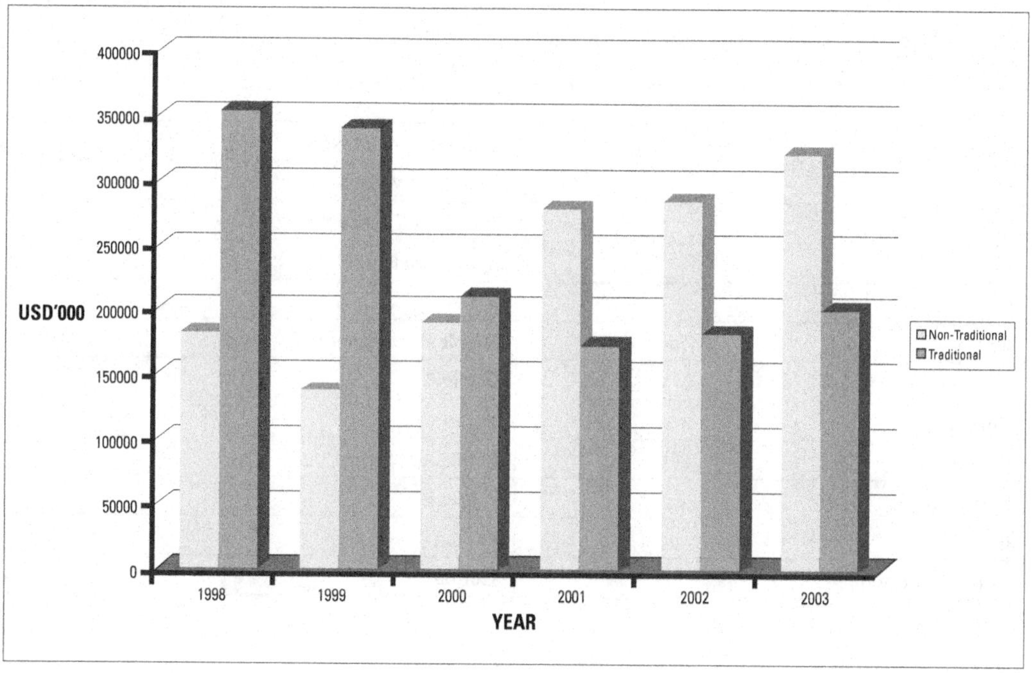

Source: UBOS, 2004

10.6.1 Export Opportunities

The country has the potential to improve its export trade. Several reasons have been presented to support this position. Firstly, there is a structure to help in exporting Ugandan goods – Uganda Export Promotion Board (UEPB) and Uganda Services Exporters Association (USEA). The UEPB was established in 1983 to promote exports. It has been very instrumental in developing export trade with other countries. Secondly, markets are available for agricultural products. This point has already been discussed under the agricultural prospects. The market for the Ugandan exports exists in Africa and outside the continent. Table 10.8 shows the opportunities for exports from Uganda.

Table 10.8: Export Opportunities

Denmark	Beans
Egypt	Soybeans
India	Red Kidney Beans, Green Mung, White Sesame seeds, Chich peas,
Turkey	Pulses
Egypt	Fish

Singapore	Fish
Spain	Fish Products
Saudi Arabia	Foodstuffs
United Arab Emirates	Foodstuffs
Yemen	Foodstuffs
France	Dried Fruits and Nuts
Jordan	Fruits and Vegetables
Turkey	Poultry and Game
Egypt	Sesame seeds and Maize
Germany	Spices
Jordan	Whole Black Pepper
Morocco	Garlic
Panama	Spices
Barbados	Honey and Syrups
Hong Kong	Apparels
Iran	Cotton
Russian Federation	Cotton

Source: UBOS

10.7 Remittance of Money from Outside

Remittance of money from outside the country is another important source of foreign exchange. Remittances account for 9.3% of the Gross National Product (World Bank, 2006). Remittance is now the second most important foreign exchange earner in Uganda (Bank of Uganda Report, 2006). Money remitted has increased every year (See Figure 10.10). The remitted amount must be higher than the reported figure. The figure does not include unrecorded flows through the informal channels. The World Bank report shows that less than 25% of the remittances to Uganda use formal transfer system. The remittance activity should be researched well in order evaluate the impact of this important economic activity.

10.8 External Debt

Uganda like most developing countries faces a range of economic problems. It is an agricultural country, prices of agricultural products fluctuate and are generally far below those of finished goods, and there are production problems. Uganda's economic problems are compounded by a large external debt ($1.456 billion, 2006 estimate). The debt service burden level is 80% of the export revenues (World Bank Report, 2006). This means that most real inflow of foreign exchange comes in the form of remittances, external aid and or grants. Most of the earnings from exports are used to service the external debt.

Discussion

Given the fact that the majority of the export markets shown in Table 10.8 are for agricultural products, should Uganda concentrate on agriculture to maximize its production advantages? Or, should the country focus on manufacturing for the local and neighbouring market?

10.9 Conclusion

Uganda is underdeveloped but has some potential in most economic activities. We classify economic activities according to the stage of production – primary activities (food and raw material production), secondary activities (manufacturing and processing), tertiary activities (services), and quaternary activities (specialized services).

Agriculture is the most practised economic activity in Uganda (in terms of percent of the labour force). However, its share of the GDP has declined over the years. The service sector has the best share of the GDP. The agricultural sector is characterized by small farms, limited technology, cultivation of a variety of crops, and lack of a well coordinated progamme. As discussed earlier, the Von Thunen Model has limited application to location of agriculture in Uganda. Uganda's agricultural sector has the potential to increase because of good natural conditions, availability of an export market, increasing demand, and raising foreign investment. The leading crops are bananas, maize, sweet potatoes, cassava, millet, sorghum, and coffee.

Manufacturing is a growing sector. Uganda's products will find a ready market in Southern Sudan, Rwanda, Burundi, and probably Eastern Congo. The availability of cheap labour and raw materials are other advantages. The service sector is now the largest contributor to GDP. Its share of the labour force is still low compared to agriculture, but most of the service sector can be described as informal. This may explain the low (recorded) employment in the sector.

Key Terms

Gross Domestic Product	Import-Substitution
Market-oriented manufacturing	Raw material-oriented manufacturing
Malthusian theory	

Questions

1. What are the major classifications of economic activities?
2. What factors influence the location of agricultural activities in Uganda? What are the characteristics of agriculture in Uganda?
3. How have agricultural production areas changed?
4. Explain the factors influencing industrial activity in Uganda. What are the characteristics of the industrial sector?

5. What are the most important economic sectors of the country? Why? How have the sectors changed over time?
6. What are the export opportunities?
7. What is the impact of remittances and external debt on Uganda's economy?
8. What are some of the problems facing the Ugandan economy?

References and further readings

Berry B., E. Conkling, and M. Ray. 1997. *The Global Economy in Transition*, Upper Saddle River, NJ. Prentice Hall.

Bird, J. (ed.). 2004. Uganda, Land of Opportunity, *Africa Travel Magazine*

Grigg, D. 1995. *An Introduction to Agricultural Geography*, New York: Routledge.

Mugyenyi, O. and R. Naluwairo 2003. *Uganda's Access to the European Union Agricultural Market: Challenges and Opportunities*, ACODE Policy Research Series No. 6.

Naude W. and T. Gries. 2004. *The Economic Geography and Determinants of Manufacturing Exports from South Africa*. A Conference paper.

Rugumayo, A., Kiiza, N., and Shima, J. 2003. *Rainfall Reliability for Crop Production: A Case Study in Uganda*, Diffuse Pollution Conference, Dublin.

Sachs, J.D. and Warner, A.M. 1997. "Sources of slow growth in African Economies", *Journal of African Economies*, 6(3): 335-376

Stutz, F and de Souza, A. 1997. *The World Economy: Resources, Location, Trade, and Development*, Upper Saddle River, NJ, Prentice Hall.

Warner, A.M. 2002. *Institutions, Geography, Regions, Countries and the Mobility Bias*. CID Working Paper no. 91, Harvard University.

Ministry of Planning and Economic Development (MOPED)

The Ministry of Agriculture, Animal Industry and Fisheries. Uganda Bureau of Statistics

World Economic Forum. 2005. *Global Competitiveness Report 2005-2006*

Chapter 11

GEOGRAPHY AND DEVELOPMENT

JOCKEY B. NYAKAANA

11.1 Introduction

This chapter provides interpretation and elements of sustainable development and the relationship between geography and sustainable development in Uganda. Equally, it gives a conceptual framework for linking the two. It also shows the linkage between the Millennium Development Goal 7 (MDG) for ensuring environmental sustainability and other MDGs. Finally, policy implications for future sustainable development are derived.

11.2 Human beings at the centre of Geography

When the world leaders converged in Rio de Janeiro for the United Nations Conference on Environment and Development (UNCED) in 1992, it was after the realization that the linkage between Geography (environment) and economic development had been polarized at the risk of threatening human survival. They observed thus:

"Humanity stands at a defining moment in history. We are confronted with a perpetuation of disparities between and within nations, a worsening of poverty, hunger, ill health and literacy, and the continuing deterioration of the ecosystems on which we depend for our well-being." (UN, 1992: 199)

To restore the relationship between the environment and development, they set themselves to following 27 principles under Agenda 21, a programme of action for development worldwide. The principle, which puts human beings at the centre of geography states:

"Human beings are at the centre of concern for sustainable development. They are entitled to a healthy and productive life in harmony with nature." [UN, 1992. p9]

Agenda 21 acknowledged the close relationship between geography and development as well as the need to improve health in order to achieve sustainable development. Poverty eradication and economic development cannot be achieved where there is a high prevalence of debilitating illness. Likewise, the health of the population cannot be sustained without responsive health systems, a healthy environment and an intact life-supporting system. In short, sustainable development has its antecedents in the "Limits to Growth" (IUCN's advocacy in 1980). Economic development cannot be sustained unless it is founded on sound analytical geography.

11.3 Understanding sustainable development

According to the Brundtland Commission, sustainable development is *"development that meets the needs of the present without compromising the ability of future generations to meet their own needs"*. (Middleton, 1995: 286 from WCED 1987: 431)

Since 1987, literature has exhibited a multiplicity of interpretations of, and prescriptions for, sustainable development based on nine operational objectives as given by WCED (1987-49). (*Box1)*

Box 1: Operational objectives of sustainable development

- Reviving growth
- Changing the quality of growth
- Meeting essential needs for jobs, food, energy, water and sanitation
- Ensuring a sustainable level of population
- Conserving and enhancing the resource base
- Reorienting technology and managing risk
- Merging environmental and economics in decision making
- Reorienting international economic relations
- Making development more participatory

Source: WCED (1987-49)

Figure 11.1 presents the dimensions of sustainable development in an interactive and dynamic manner among social, economic, institutional and environmental dimensions. The implication is that the impact of those interactions must be monitored on a continuous basis and the negative ones addressed on time to avoid irreversible damage. The other implication is that balancing the various development objectives, or even making well-informed trade-offs are central to the decision-making processes.

Figure 11.1: The dimensions of sustainable development

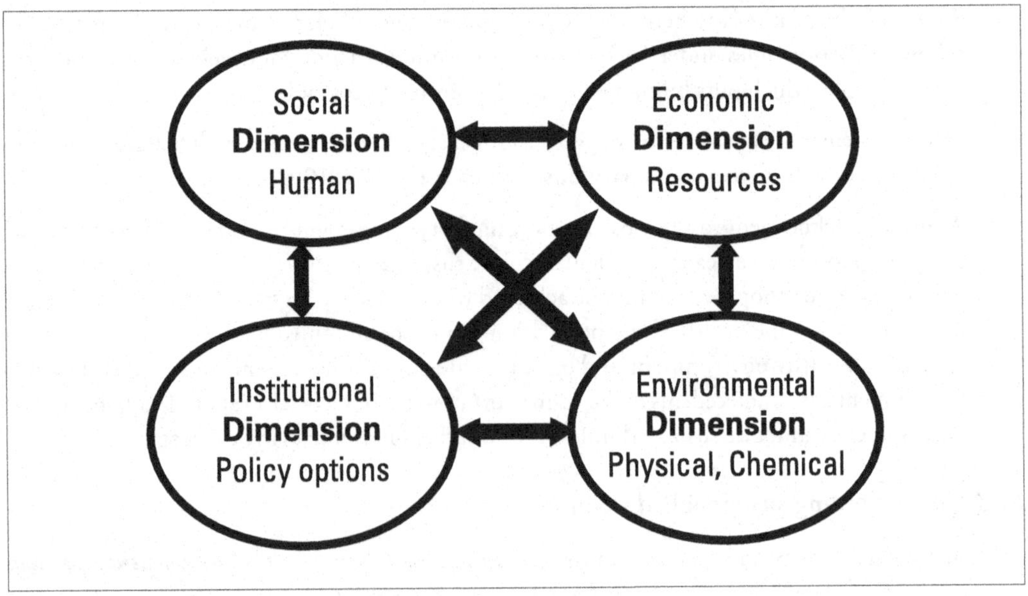

Owing to the close linkage, dynamism, and human development, UNDP too has greatly evolved the concept of Human Development (HD) since it first defined it in the first Global Human Development Report (GHDR) as 'a process of enlarging people's choices' (UNDP: 1990). At the time, it listed the three essentials as: people leading a long and healthy life, acquisition of knowledge and access to resources needed for a decent standard of living. It asserted 'that if these essential choices are not available, many other opportunities remain inaccessible' (UNDP: 1990).

In 1994, UNDP, in the GHDR went a step further and introduced the concept of Sustainable Human Development (SHD). According to the Report, SHD means that we have a moral obligation to do at least as well for our successor generations as our predecessors did for us. It also means that current consumption cannot be financed for long by incurring economic debt that others must repay. It equally implies that sufficient investment must be made in the education and health of today's population so as not to create a social debt for future generations. Finally, it means that natural resources must be used in ways that do not create ecological debt by over exploiting the carrying and productive capacity of the earth.

All postponed debts mortgage sustainability (economic, social or ecological) as they borrow from the future. They rob coming generations of their legitimate options. That's why the strategy for SHD is to replenish our capital (physical, human, and natural) so that it maintains the capacity of future generations to meet their needs, at least at the same level as that of the present generations (GHDR, 1994). "Human development and sustainability are thus essential components of the same ethic of universalism of life claims. There is no tension between the two concepts for they are a part of the same overall design" (UNDP: 1994).

11.4 Linking Geography to Development

Whereas it appears obvious to talk about a close link between geography and human development, in practice, that relationship is sometimes marginalized, or forgotten, with the consequence that environmental management and human development are taken in parallel rather than in an integrated manner. The consequence is that the country continues to fall short of its targets and aspirations.

In order to bring ourselves back to build a synergistic and symbiotic relationship between geography and development, we must adopt a conceptual framework. (Figure 11.2) The framework strongly builds on other frameworks, namely those by DFID (2002), UNDP/World Bank/EU/DFID (2002) and UNEP (2002).

The framework is founded on the Brundtland Report, and takes sustainable development as both an outcome ("development that meets the needs of the present without compromising the ability of future generations to meet their own needs" (WCED: 1987) and a process ("a process of change in which the exploitation of resources, the direction of investment, the orientation of technological development and institutional change are in harmony and enhance both current and future potential human needs and aspirations").

As a process, sustainable development uses three basic capital inputs, human, physical, and natural. The three forms of capital make up the total stock of a nation. According to estimates, the natural capital in Uganda makes up 15% of total capital stock, while human and physical capital contribute 38% and 37% respectively (Kunte, et. al. 1998).

In addition, there are social and financial capitals. Social capital is related to human wellbeing, but on a societal rather than individual level. It consists of the social networks that support an efficient cohesive society, and facilitate social and intellectual interactions among its members.

As defined by the Civic Practices Network, "social capital refers to those stocks of social trust, norms and networks that people can draw upon to solve common problems." Examples of social capital are neighbourhood associations, and civil organizations. In the framework, social and financial capital are shown as some of the transformation factors. Further, the framework recognizes a reciprocal and cyclical relationship between geography and human development. The Physical Geography (natural capital) is transformed into development outcomes using social and financial capital, technology, information, research, etc. The transformation is also influenced by external factors like globalization and processes of regional and international cooperation.

The natural capital is entitlements people have as either individuals, or collectively as a community, the risk they face from vulnerability factors, and the transformation factors at their disposal influence their choices of survival strategies. They may solely depend on the natural capital, or non-natural capital, or both. The development outcomes from those strategies are many and include reduced poverty and hunger, improved health and education as a result of investing natural resource based incomes, reduced burden from environmental diseases, reduced vulnerability, and enhanced environmental values.

Finally, Figure 11.2 shows, the relationship between geography and development. From a sustainable development perspective, the cardinal rule is that the present generation must leave geographical/capital/resources in an equal or better position than they found them for the future generation. That is to say, it must not leave an ecological debt, hence the need to assess the significance of geography to the development process.

Figure 11.2: Conceptual Framework linking Geography and Development

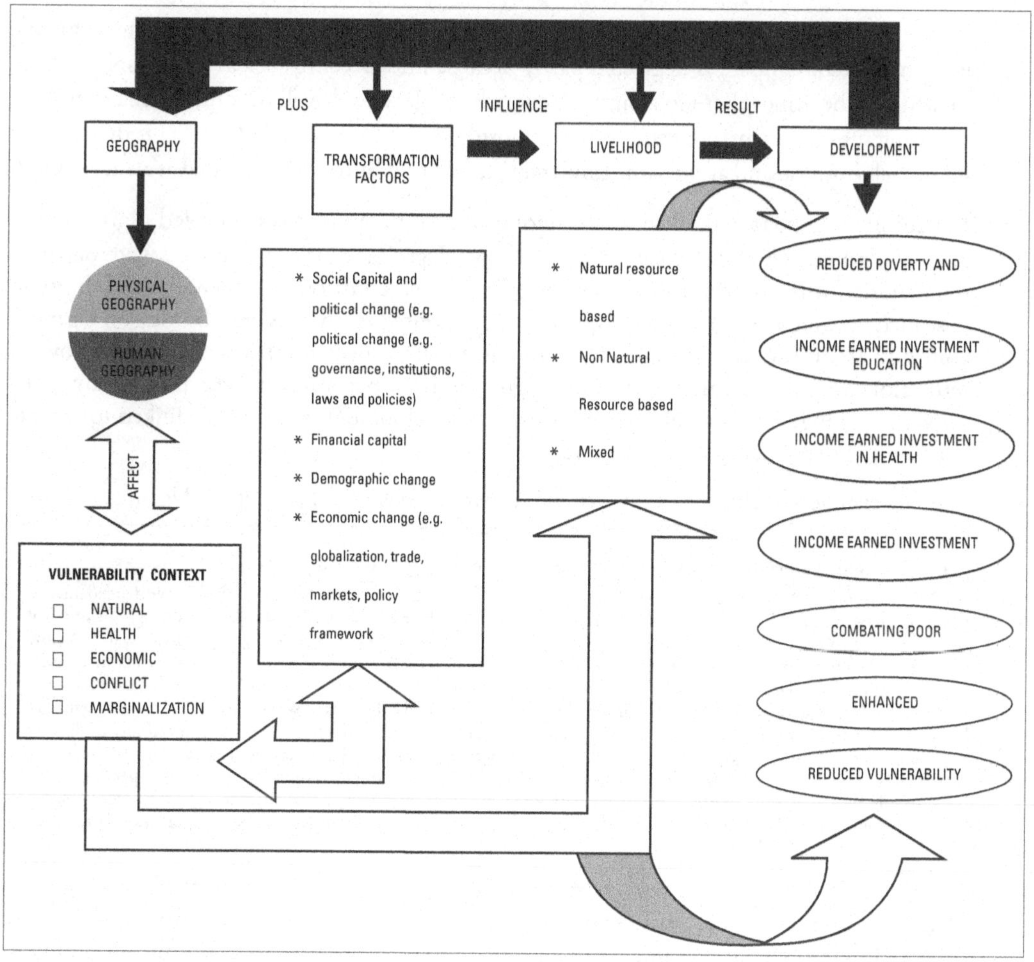

Modified from: DFID (Sustainable Livelihoods Framework (UNDP/World Bank/EU/DFID (2002) (linking Poverty and Environment Management: Policy, Challenges and Opportunities), UNEP (2002) (Poverty and Ecosystems: Conceptual Framework), Kazoora (2003), Linking sustainable livelihoods and ecosystem management

Box 2: Wrath of El Nino in Uganda

- 525 people died and an additional 11,000 and more were hospitalized and treated for cholera which was triggered by the El Nino induced floods and landslides
- 1,000 people were reported to have died in floods-related accidents
- Damage to trunk and rural roads infrastructure was estimated at about $400 million
- Infiltration of water resources and the flooding (sub-merging of some pumping stations)

11.5 Climate and Atmosphere

Climate, perhaps Uganda's most valuable geographical resource, is also, at the same time the most neglected. Climate is neglected partly because until recently, it was predictable and the capacity of the atmosphere to assimilate degrading effects was well beyond the magnitude of polluting substances introduced into the atmosphere. A second cause of neglect is that climate is a global common and at the same time transboundary. Thirdly, it is regarded as an act of God.

Agriculture, Uganda's main engine of economic development is mostly rain-fed. Agriculture is profoundly influenced by and dependent on climate. Uganda's subsistence farmers who constitute the majority of the poor are dependent on climate, which directly influences their ability to raise incomes and improve upon the quality of life. Consequently, this has a bearing on Uganda's ability to attain the Millennium Development Goals, especially the one related to poverty reduction, the Government's ambitious goal of 10% per annum mass poverty by 2017 notwithstanding. For the majority of Ugandans, development is intricately linked to climate as a result of dependency on agriculture.

Climate change and climate variability are the two main issues of concern to Uganda. (Climate change refers to the long-term change of one or more climatic elements from a previously accepted long-term mean value, which must be statistically proven as significant. Climate variability is the sharp, short-term variations of meteorological elements as compared to the long-term value of the elements)

Box 3: Highly vulnerable to sudden shocks and changes in physical conditions

- During the Second Participatory Poverty Assessment, the poor explained that due to their heavy dependence on environmental resources, their livelihoods are highly vulnerable to sudden shocks and changes in physical conditions. In 9 out of the 12 districts assessed, people reported that unpredictable weather patterns and climatic conditions, characterized by usually heavy and erratic/unreliable rains, lead to crop and infrastructure damage causing food insecurity. Strong winds and hailstorms, which destroy crops or disrupt fishing activities, were noted to be causes of poverty by people in five districts. In 10 of the 12 districts, it was reported that flooding damages crops and physical infrastructure and contaminates domestic water sources, increasing vulnerability to diseases such as cholera. Drought was also said to be common in 11 of the 12 districts studied, causing declining crop yields and food shortages. In this case, households cope by reducing the number of meals eaten per day, which in turn affects the health of women and children. Water problems associated with drought were mentioned in four districts by mainly cattle keepers, forcing them to move their livestock, leading to degradation of watering points (MoFPED, 2002).

Climate change is caused by the release of greenhouse gases. (Green house gases are: carbon dioxide (Co2), carbon monoxide (Co), methane (CH4), sulphur dioxide (SO2), nitrous oxides (Nox), volatile organic compounds (VOCS) and water vapour released during the combustion of woodfuel, charcoal and petroleum products.)

Biomass for energy, non-metallic mineral production and bushfires are the main sources of greenhouse gases in the country. Fortunately, the country's extensive plush green vegetation is able to absorb all the greenhouse gasses it generates plus those from other countries, since these gases know no national boundaries. The accumulation of greenhouse gases in the atmosphere is causing a rise in the world's surface temperature with dire consequences for countries in the tropics, including Uganda. Climate change and environmental degradation have led to food shortages and increased pressure on available land and water resources in the Horn of Africa

(Mkutu, 2004). Sub-Saharan Africa (SSA), including Uganda, is the most vulnerable to climate change because widespread poverty limits its capabilities to adapt to a continuing changing climate (TAC, 1999).

Climate variability is currently one of the most pressing environmental problems in Uganda (NEMA, 1997). Persistent droughts as a result of prolonged dry seasons, and flooding due to flash storms and hailstorms, including shifts in seasons are of great concern, and impact directly on agricultural production, and sustainable development. The last few decades have seen an increase in the frequency and intensity of these extreme weather events, with adverse socio-economic consequences, including loss of human life as illustrated in Box 2.

Of great concern in particular, is the monthly variation of rainfall within a year that has made it impossible for farmers to accurately predict when to plant annual crops. As a result, the failure of these crops is now a common occurrence triggering famines and reliance on relief food aid, thus having a negative effect on the country's food security situation and sustainable development.

To make matters worse, traditional crop storage in granaries at household level as a food security measure has long been forgotten. Government also has six-grain silos, but these are empty! Bye-laws requiring each household to have a minimum acreage devoted to the growing of cassava, a drought-resistant crop, are not enforced. The northern and eastern regions are particularly vulnerable since they depend mostly on one rainy season per year and grow annual crops, unlike their central and western counterparts who depend on perennial crops.

The poor consider themselves highly vulnerable to sudden climate-related shocks and changes in physical conditions (Box 3). Except for a few areas, rainfall in Uganda is fairly generous. It is the uncertain start and cessation of the rainy season, which is of concern.

Seasonal changes also have significant influence on intra-household and community level poverty in rural areas. The government has responded to this challenge by being a signatory to ratifying the United Nations Framework Convention on Climate Change (UNFCCC) and other regional agreements. The Government with the assistance of development partners is going further to domesticate such instruments.

Unfortunately, as of now, there are no significant investments being made to improve the country's ability to monitor and forecast climate change and climate variability for the establishment of early warning systems. Furthermore, the capability of the Ministry of Disaster Preparedness and Management in the Office of the Prime Minister (OPM) to prepare for, and handle disasters as a result of adverse climatic changes is also limited. Endorsing the principle that 'prevention is better (cheaper) than cure', would make it relatively easy to advocate for a higher level of resource allocation for the Department of Meteorology and the Ministry responsible for disaster preparedness and management.

11.6 Land and Soils

Land forms the basis for all plant and animal life, a store for waste materials, and a basis for human settlement and transport corridors (NEMA, 2003). Land is, therefore, central to sustainable development since it is the basis for virtually all activities in Uganda (NEMA, 2003). Yet, land resources are finite, fragile and non-renewable (UNEP, 2002). The total area of Uganda is 241,551 km2, made up of farmland, grasslands, bush lands, forests and woodlands, built up areas, wetlands, and water bodies.

Table 11.1 shows that, of the total area, 84,694 km2 is farmland. In turn, 84,010 km2 of this farmland area is under subsistence agriculture, while a mere 684 km2 consists of commercial farms, illustrating the importance of land in supporting rural livelihoods and perhaps the major constituent of household assets, an important component of the measure of sustainable development.

Table 11.1: Area and distribution of land use/cover at national level

Stratum	Area (ha)
Plantations hardwoods	18,682
Plantations softwoods	16,384
Tropical high forests (normal)	650,150
Tropical high forests (degraded)	274,058
Woodlands	3,974,102
Bush lands	1,422,395
Grasslands	5,115,266
Wetlands	484,037
Subsistence Farmlands	8,400,999
Commercial Farmlands	68,446
Built up areas	36,571
Water	3,690,254
Impediments such as rocks, etc	3,713
Total	24,155,058

Source: NEMA (2003); NBS (2002)

Article 237 of the Constitution of the Republic of Uganda provides that land belongs to the citizens of the country under four tenurial arrangements, namely: customary, freehold, *mailo* and leasehold (RoU, 1995). These tenurial regimes are further provided for in the Land Act, 1998. (The Land Act is in the process of being revised.) The different land tenures and their features are presented in Table 11.2.

Table 11.2: Land Tenure Categories, Key Features and Geographical Incidence

Tenure/Issues	Key Features	Geographical Incidence
Customary	"Traditional" land tenure, varying in different areas. More individualized in south and west, more communal in north and east. Can be issued a customary certificate of ownership and this is an incentive to the customary tenant to invest in proper land management practices which are long-term	Countrywide
Leasehold	49 or 99 year leases, with development conditions. The conditions can be used to promote conservation or increase agricultural productivity. Ground rent and premium payable. Leasehold title issues	Countrywide especially in urban areas

Freehold	Registered ownership in perpetuity, with full powers of ownership including development and disposing the land at will. Can encourage land fragmentation, which is not conducive to proper soil management and conservation and this undermines production. Freehold title issues	Predominantly in southern and western Uganda
Mailo	Limited form of freehold, which recognizes tenants' right. This ownership refers to the holding of registered land in perpetuity. Squatters have subjected some tracts of land to degradation for a long time. Mailo title issues. This should motivate tenants to invest in land improvement technology and increase agricultural productivity	Central region of Uganda
Occupancy	Right to occupy land under specific conditions based on occupation prior to 1983	Countrywide on any registered land
Renting	Use right to land for a defined period subject to payment of rent	Varies countrywide
Borrowing	Use rights to land for a defined period subject to payment of part of harvest	Varies countrywide

Source: MWLE (2001)

Land degradation, principally caused by soil erosion, itself a result of different ownership types, utilization and management systems, is on the increase. Slade and Weitz (1991) estimated that soil erosion accounted for 4-10% loss of the Gross National Product (GNP) and represented as much as 85% of the total annual cost of environmental degradation. Yaron and Moyini (2003) suggested that the economic cost of soil erosion was about 11% of the GDP, which meant that the total annual cost of environmental degradation in 2002 was about 13% of GDP. Drawing on the 2002 IFPRI soil nutrient loss studies and the 2002 Census Data, the value of soil nutrient loss is about $625 million per annum! Capitalized at a time preference rate of 12% (the social cost of capital for public projects), the annual cost of soil erosion translates into a debt of $5,200 million.

Unfortunately, this debt is invisible and unrecorded in the official system of national accounts. Nonetheless the burden is equivalent to per capita soil erosion debt of about $210. Who is to pay this debt? Will it be the present generation or future generations' responsibility? Clearly choices have to be made. In the meantime, what is true is that when the loss of soil nutrients is taken into account, it becomes obvious that Uganda's annual net savings is negative, meaning current practices and the country's development are not sustainable. The formation of physical and human capital is too slow to offset the loss of natural capital. Practical steps need to be taken, therefore, to reduce soil degradation, if it is not to undermine sustainable development, and the strategy for doing so needs to be explicitly defined in the Plan for Modernization of Agriculture (PMA), the framework for transforming subsistence agriculture into commercial farming.

In general, where population density is high, so is the proportion of land affected by erosion. However, even in some of the districts with low population densities such as Kotido, the severity of soil erosion is quite high due to the fragile nature of the rangeland ecosystem exacerbated by over-grazing. The rangelands are typically found in what is referred to as the 'cattle corridor' which covers 16 districts, representing about 37% of the total area of the country, close to 28% of its rural population, and an average population density of 65 persons per square kilometre.

While land is a key resource of production and the main capital available to the majority of the people (MoFPED, 2003) and supports agriculture on which the country depends, access to land is the basis for rural livelihoods (MoFPED, 2003) and hence sustainable development. During the Second Uganda Participatory Poverty Assessment Project (UPPAP 2), it was evident that access to land was increasingly becoming a problem for poor people (MoFPED, 2003). In Uganda, households are not accumulating land; rather, this asset is diminishing in size (MoFPED, 2003). Households owning most land in 2002 found themselves owning the same acreage as in 1993 having recovered from a dip in 1996. The middle and poorest households on the other hand saw their landownership decreasing significantly (MoFPED, 2003). The challenge, therefore, is to identify opportunities in which poorer households are enabled to take advantage of alternative sources of income while making the best use of the little land they are left with (MoFPED, 2003). The root causes of land scarcity are large families, particularly polygamous ones; distress sales by poor people; insurgency in the north; inadequate land use planning; commercial farming; and rural-urban migration (MoFPED, 2003).

One of the manifestations of land scarcity is fragmentation. Land fragmentation and sub-division of holdings in parts of the country constrain individual investment in erosion control and, therefore, reduce agricultural productivity. The highest incidence of land fragmentation is found in Mbale District in eastern Uganda (MoFPED, 2003). Here, the median household held five plots; and the median distance between the home and land plots ranged from 0.25 km to 1.0 km in a hilly terrain. Therefore, a significant amount of time is lost daily just to reach these plots. The main causes of sub-division and fragmentation in Mbale District were reported to have been families selling off parts of their land to raise money and traditional inheritance practices where land is divided between family members (MoFPED, 2003). The major areas of concern regarding land are small and unviable holdings, rights and management of some gazzetted areas, rights to and benefits from certain natural resources, declining land productivity, women's land rights, urbanization, and corruption and lack of transparency in the land sector (MoFPED, 2003).

Government policy regarding land is to provide security of tenure for all, particularly the poor and create an enabling environment for participation of all stakeholders in effective planning, management and use of Uganda's land resources (MoFPED, 2003). The main challenges to achieving the land policy objective include unequal gender relations which constrain the ability of the poor to raise their incomes and are likely to reduce the effectiveness of efforts to stimulate new export-oriented production (Eilor and Giovarelli, 2003; MoFPED, 2003), legal changes under the Land Act aimed at strengthening land rights of married women, dependent children, and orphans, through the requirement of consent to transactions on family land, have little effect on the ground; and there is a need to 'revisit with caution' the issue of co-ownership of family land (MoFPED, 2003).

11.7 Fisheries

Uganda has substantial resources comprising of capture fisheries and aquaculture with the latter currently contributing only a small fraction of the country's annual catch. The country is estimated to have the potential to sustainably produce about 330,000 metric tonnes of fish annually through capture fisheries. Artisanal fisher folks, estimated at over 136,000, dominate the fishing industry and over 700,000 people are directly involved in fisheries related activities such as processing, transporting, trading and boat-building. Others are involved in industrial fish processing, fishnet making, fish equipment trade, research, extension and administration (RoU, 2004).

Annually, Ugandans consume about 13 kilograms of fish per capita, a high quality protein. Fish is the most important source of protein for the people in central, eastern, and northern Uganda (particularly West Nile) and less so in the west (Atukunda, 2003). Fish is also now one of the country's major sources of income and an export commodity. Thus, fisheries are an important resource for the development of the country.

Unfortunately, over-fishing, especially closer to the shoreline, is reported on all Ugandan lakes. The number of boats per lake far exceed the recommended numbers. In addition, there are excess nets per boat and a significant number of under-sized nets used to harvest immature fish. Another threat to the fisheries resources is the use of poor fishing methods. In total, fishing effort has increased, but the catch per unit effort is on the decline. It is also reported that species diversity, an indication of ecology resilience, is on the decline.

The trends within the fishing industry towards booming fish exports have development implications. On the positive side, some local people have benefited from the boom in terms of increased incomes, better employment opportunities, and improved standards of living. On the negative side, many small-scale fisher folks are being 'edged out' by increased competition from commercial operations. Furthermore, because the fisher folks are poorly organized, they have no capacity to determine their returns, and the middle people get a disproportionately larger share of the economic rent accruing from fisheries. The women among the fishing community who are mainly involved in fish processing and trading may also be slowly rendered less employed as commercial processing gains significance.

As investments in industrial processing increase, more people are getting involved as middle agents. Fish prices have increased sevenfold between 1990 and 2002 (FIRRI, 2002). The fish boom has also led to increased competition and more fishing effort, including the use of illegal and unconventional fishing methods, subsequently degrading the resource. Catches are sporadic and fish sizes smaller. Immature fish is illegally harvested with obvious adverse impact on the fish population.

There are two broad approaches that provide opportunities for economic growth and improved development within fisheries. The first is to raise the productivity of existing exploited fish stocks through some form of management intervention such as co-management with local communities. The second is to develop new fisheries based on under-exploited fish species. Since Uganda's fish harvesting is still largely artisanal, rural communities are likely to benefit from such approaches.

Aquaculture currently accounts for less than 1% of all fish production (Banks, 2003). Nonetheless, it is the main option available to meet the estimated 90,000 metric tonnes/year of extra fish that will be required to meet the needs of a growing population as well as maintaining exports in 10 years' time. At least 28 reservoirs and small lakes have been stocked with three million fish fingerlings (DFR, 2003). Fish stocking of reservoirs is a national initiative actively promoted by the Department of Fisheries Resources (DFR) with the aim of increasing livelihood options in rural areas, increasing incomes of resource users, and improving food security and nutrition. Out of the additional 90,000 mt required annually, at least 10,000 mt/year is expected to come from small-scale or artisanal aquacultural operations while the remaining 80,000 mt/year would be from commercial aquaculture.

An emerging problem in the demand-supply dynamics of fisheries resources is the potential danger of satisfying international demand for fish at the expense of local needs. As fish exports and the fish processing capacity have increased, prices have risen and this affects local demand. Traditionally, fish has been the most affordable form of solid animal protein for local people. Increased fish exports and prices may thus lead to reduced local fish consumption, which could in turn contribute to malnutrition and food insecurity. There is the already decreased consumption of Nile Perch as a result of high prices, possibly due to increased demand created by an expanding export market and the increased export of tilapia is jeopardizing local fish consumption. For example, in the central region, people buy fish skeletons, leftovers after fillet for export has been removed instead of buying whole fish, which has now become too expensive and out of reach, especially for the poor.

In as much as the fishing industry promises to contribute more to economic development and structural transformation in the future, it faces some critical challenges, including the inadequate legal and regulatory framework for fisheries and aquaculture; high levels of post-harvest losses and multiple taxation at all levels; and women and the poor (such as boat crews) have less access to capture fisheries due to problems with licensing and the aforementioned multiple taxation systems (MoFPED, 2003). Some of these challenges are being tackled. For example, one unique opportunity of international significance is the granting of decentralized legal powers, through the Beach Management Units (BMUs), to local people for sustainable utilization and management of aquatic resources (MoFPED, 2003). There is also a move to manage Lake Victoria on a regional basis through the efforts of the East Africa Community (EAC) and its Victoria Fisheries Organization (VFO).

11.8 Wildlife and Tourism

Uganda is rich in wildlife diversity both within and outside protected areas with 338 mammals, 1,010 birds, 5,406 plants, 149 reptiles, 98 amphibians and 1,245 butterflies Bakamwesiga 1999, (UTB, 1996). Most important is the mountains gorilla found in Mgahinga and Bwindi National Parks. The current spatial size of wildlife-protected areas is about 20% of the country's total surface area. The wildlife estate includes 10 national parks, 29 wildlife reserves, sanctuaries and controlled hunting areas. The total land coverage is 56,000 km2 and another 14,900 km2 of gazetted forests. There are fourteen community wildlife management areas representing different levels of conservation effort. Uganda is home to more than half of the world's population of Mountain Gorillas, which places a great responsibility on the country to conserve this world heritage. Notwithstanding issues related to adequacy, the international community is supporting some conservation efforts. Case in point is the GEF-funded Mgahinga Bwindi Impenetrable Forest Conservative Trust (MBIFCT), which provides funding for community development projects, park management and research.

Early in the 20th Century, Uganda's human population was small and scattered. It was during this period that herds of elephants, buffaloes and other 'plain wildlife' ranged over the areas. However, during the 1970s and early 1980s, many of Uganda's wildlife protected areas were severely encroached upon, and their animal populations drastically reduced by poaching and habitat alteration. The decline is still continuing especially outside protected areas. Contributing factors include alteration of wildlife habitats, competition with livestock for water and pasture, increased human settlements around protected areas and within wildlife migratory routes. The latter leading to increased human-wildlife conflicts.

Wildlife resources provide subsistence and employment through tourism, which is fast regaining its previous economic importance of being the third largest foreign exchange earner in the 1960s. Over the period 1994 to 2003, tourists visiting Uganda increased from 31,259 to 102,567 respectively (Figure 11.3).

Figure 11.3: Tourists Visiting Uganda's Wildlife Areas, 2000-2004

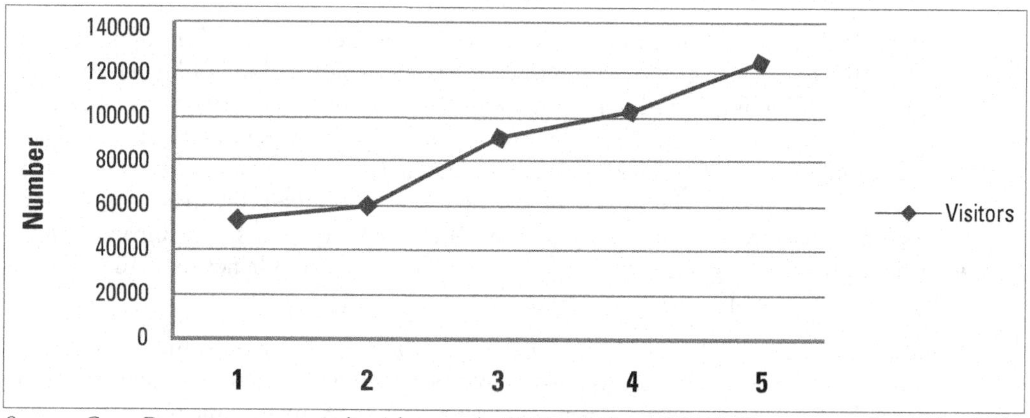

Source: Game Department reports and aerial surveys by UWA and (then MTWA). These are cases for which reliable previous estimates are available, from which to determine trends. Numbers are approximate.

Tourism contributed about $ 163.1 million, or 24.7% of total export earnings in the year 2001 and $449 million or 64.1% of service exports in 2006 (UBOS 2007). In addition to this, there are more benefits derived from tourism (in terms of direct and indirect employment, foreign exchange earnings, revenue generation for government, and development opportunities for rural communities). Where people get involved as tour guides, sellers of handicrafts and workers in the accommodation sector, the multiplier effect of tourism in rural areas is significant. This has important implications for sustainable development conditions in rural areas.

The government has demonstrated its support to the revitalization of the tourism industry, through the empowerment of local governments and communities to manage and benefit from sustainable use of natural resources through a number of policy actions as well as institutional reforms. As a result, the Uganda National Parks (UNP) and the Game Department (GD) were merged in 1996 to form the Uganda Wildlife Authority (UWA) so as to increase the efficiency and effectiveness of conservation efforts, and develop tourism products. Currently, UWA has the overall responsibility for wildlife management in the country. The government also established a number of parastatal and mixed public/private sector entities responsible for different tourism sub-sectors, many of which were previously under the direct responsibility of government departments. These include the Uganda Wildlife Education Centre (UWEC), the Uganda Wildlife Training Institute (UWTI), the Uganda Tourist Board (UTB), the Hotel and Tourism Training Institute (HTTI), Uganda Community Tourism Association (UCOTA), Uganda

Tourism Association (UTA), Community Based Tourism Initiative (COBATI) and the Department of Antiquities and Museums.

The move to transfer what were traditionally government functions to semi-autonomous entities was to enhance their sustainability through the generation and retention of revenues, mainly from tourism-related activities. They are also intended to provide a synergistic linkage between the public and private sectors, and to facilitate the development and growth of the tourism industry. Thus, they represent an important model for public/private cooperation for private sector development. Tourism training is increasingly becoming important, both in public and private institutions, where certificate, diploma and degree courses are offered. This has created employment opportunities, which is an important ingredient to sustainable development.

Poverty reduction through tourism can be achieved as the industry expands its opportunities. Special emphasis has been given to community tourism as a means of addressing poverty at the grass roots. Tourism is more labour-intensive in low-labour-cost countries such as Uganda. Besides, the direct jobs created within the tourism sector, both at the investment stage (construction, manufacturing, etc.) as well as jobs for the labour required to run tourism establishments, other jobs are created in outsourcing of certain services, small-scale enterprises in agriculture, food processing, transport, distribution and light manufacturing industries. The linkage with the informal sector is strong (e.g. handicrafts being sold at roadsides) hence the significance of tourism in achieving sustainable development.

Women have access to jobs and it is estimated that where tourism is a mature sector women usually account for 50% of the tourism workforce. For example, 50.1% of the accommodation sector workforce in Uganda are women. The fact that tourists move to the production point may lead to enhanced interaction between the tourists and the local community. There could be some negatives to this, but on the positive side it enables exchange of knowledge that can lead to a reduction in poverty. The multiplier effect of tourism spending has a catalytic effect across the economy in terms of production and employment because of the links that tourism activities have with local small producers and the informal sector. Tourism development ensures conservation of the environment and biodiversity hence ecological sustainability, which is a cornerstone to sustainable development.

11.9 Biodiversity

Biodiversity or biological diversity is defined as the variability among living organisms from all sources including, inter alia, terrestrial and aquatic ecosystems and ecological complexes of which they are a part. This includes diversity within species, and of ecosystems. Uganda is rich in biodiversity (Pomeroy et al, 2002).

Due to its unique bio geographical location, Uganda has seven of Africa's 18 plant kingdoms, more than any other African country, and one of the highest on the Continent (Box 4). There are more than 18,783 species that are known or have been recorded. They include resources that can nurture or are nurtured by humans. It is for this reason that Millennium Development Goal 7 has special significance for Uganda.

Box 4: Biodiversity endowment

- Ugandans have inherited a very rich flora and fauna, but the country is rapidly losing its biodiversity: a preliminary estimate (Arinaitwe, et al 2000) suggests an overall rate of loss of about 1% per year. Planned agricultural development, urgently needed to improve peoples' lives, will further reduce the habitats of many species, whilst a wide range of human activities continues to degrade non-farmland areas, especially (but by no means only) outside protected areas.

Source: Pomeroy et al (2002)

The contribution of Uganda's biodiversity resources including genetic resources, organisms or parts thereof, populations or other biotic components of ecosystems is about $1,000 million per year and probably greater when valued more comprehensively (Emerton and Muramira, 1999). Of the total amount, direct benefits were worth $411.5 million, while indirect benefits constituted the remainder at $588.5 million. However, these benefits accrue at a cost. Biodiversity economic costs were estimated at a minimum of $253 million/year; opportunity cost at $202 million per annum; and losses to other economic activities at $48.5 million/year (Emerton and Muramira, 1999). Other estimates are presented in *Box 5.*

Box 5: Suggested values of environmental goods and services

- Emmerton & Muramira (1999) estimated the total economic benefits of natural resources at about $ 1,000 million annually, and the total costs to be about $504 million (mainly production foregone). On a finer scale, Howard (1995) estimated the benefits of national parks, wildlife reserves and forest reserves to be $123 million annually, while the annual cost of conserving them was estimated at $200 million, more than half of which is attributable to the opportunity cost of land. Mason (1995) estimated that by the time the land becomes the most limiting constraint to agricultural production in Uganda as a whole, perhaps by the 2020s, the opportunity cost of conservation of Uganda's national parks and wildlife reserves will be between $450 million and $1,100 million per year (in 1995 prices), whereas the revenue flowing to Uganda from tourism is likely to be only around $200 million per year.

Source: IFPRI (2003)

Although Uganda is recognized as a biodiversity rich country, this biological wealth is experiencing stress despite national efforts aimed at conservation. Biodiversity is being lost through encroachment, over-exploitation and depletion, pollution and ecosystem degradation, poaching and illegal trade, and the introduction of alien invasive species into ecosystems where they have not been before. However, the principal loss of habitats and some species is mainly due to conversion of natural ecosystems into agricultural land needed to feed an ever-growing population (Box 16) and the expansion of urban and industrial centres.

Box 6: Opportunity cost of conservation

- The potential cost of conserving natural resources to a nation such as Uganda, which has both a disproportionate share of conservation treasures and highly fertile and well-watered land, is the farmland foregone by setting aside land for conservation. Looking ahead 25 years to a largely rural population more than twice that of today, these empty tracts of land will appear to be a massive under-utilized resource.
- It is likely that tourism will go some way towards offsetting the opportunity costs of agriculture foregone, but is unlikely to be anywhere near valuable enough. The problem is compounded by the fact that the highest value assets for tourism are the open, relatively less fertile plains and their mega fauna, whereas the areas of greatest conservation interest are the fertile mountain forests and wetlands which, it has been said, offer tourists little but bugs, rain and difficult access. Others, however, take a kinder view and the real position at present is not as dismal as it may appear, certainly for the next 20 or so years.

Source: Pomeroy, et al (2002); emphasis added

In the context of agricultural production, there is a fear that the rate of adopting exotic livestock species, coupled with cross-breeding between exotics and indigenous breeds might accelerate the rate of displacement of indigenous species by those introduced. This may make livestock populations more vulnerable as a result of the much narrower genetic base. This worry is based on the fact that the country has already lost 12 indigenous breeds of cattle, three breeds of goats and one of sheep (NEMA, 2003). The second worry is a global one. There has been an imbalance in the exploitation of the world's food plant genetic resources. For example, 45% of the world's food supply is from three species (wheat, rice and maize) out of an estimated 7,000 plants, which have been utilized in one way or another previously (NARO, 2000). This over-reliance on a narrow genetic base has serious implications for food security and hence sustainable development.

Local communities in biodiversity rich areas, on the other hand, bear a disproportionately higher responsibility for, and costs of, conservation. They also have the highest potential to impact on biodiversity through their day-to-day economic activities.

At the upper end, 40% of annual biodiversity values represent direct benefits; but local communities receive only a fraction of these direct benefits, the rest being appropriated by governments and intermediaries. It is partly for this reason that local communities, especially those neighbouring protected areas, feel 'cheated', or alienated in the face of the multiple livelihood challenges they face. This calls for a range of economic measures, or instruments to rectify the situation. First, the establishment of clear property rights so that local communities are fully involved in the management of land and biological resources. Second, is market creation to target the residents, and to increase their economic gains and control over biodiversity. Third, the introduction of financial instruments so that communities can invest in alternatives to biodiversity.

11.10 Energy

There are a variety of energy sources, ranging from biomass, petroleum, and electricity. Biomass represents a consumption level of 93%, and petroleum is at 6%. The hydroelectric power potential is 3,200 MW of which only 10% is developed. The low electricity consumption can partly be explained by the fact that only 3% of the population in rural areas and 8% in urban areas have access to grid electricity, while the rest of the population relies on biomass (MEMD, 2003). Although electricity coverage is likely to expand as a result of the Rural Electrification Programme, the ability of communities to afford electricity, is yet, another issue. Hence, in the foreseeable future, biomass is likely to remain the dominant source of energy for rural people.

Oil and gas exploration is going on. Five basins with hydrocarbon potential have been identified, however, no production has begun as yet. At the moment, all petroleum products consumed in the country are imported and their consumption is sector specific: gasoline, aviation gas and diesel consumed in the transport sector; fuel oil consumed in the industrial sector; and kerosene and liquid petroleum gas (LPG) consumed by the residential and commercial sectors. A significant share of Uganda's foreign exchange earnings goes to satisfy oil imports, leaving little else to development.

The national geothermal energy potential of 450 MW is yet to be developed; while currently wind energy is used mainly for pumping water as opposed to electricity generation. Biogas technology which provides a cheap and clean source of energy was introduced in Uganda 20 years ago, but has not been widely adopted. The startup costs are significantly high for the majority of Ugandans and this has discouraged widespread adoption. There is a noticeable increase in the use of solar energy, but the cost of this energy source is also high. Therefore, the principal and cheaper source of energy available will continue to be biomass, up to 2015 at least (MEMD, 2003). However, the availability and access to biomass resources is itself becoming a problem in many parts of Uganda.

There is evidence that links woodfuel types and people's nutritional behaviours, hence the contribution of anecdotal woodfuel to people's livelihoods through improved nutritional values. Ddungu et al (1998) observed that women in Molo Sub-county in Tororo District were unable to regularly prepare foods such as dry beans, which require sizeable amounts of energy because of woodfuel scarcity. The use of crop residues (which in turn robs the soil of nutrients), tree roots and grass for cooking was observed to be a common phenomenon in the district.

A solution involves the encouragement of the population to establish woodlots or practise agro forestry systems as a source of biomass energy to improve their quality of living and also relieve pressure on the country's natural forests.

Improved access to energy has a direct impact on poverty and overall development since access to electricity supply raises incomes by permitting the introduction of new technologies and services, which in turn expand the range of productive opportunities in rural areas (MoFPED, 2003). The quality, reliability and access to power have been identified as major impediments to sustained investment and growth during the 1998 Private Investment Survey (MoFPED, 2003). The Government's overall energy policy objective is to 'improve the quality of life in rural areas and facilitate rural non-farm income by accelerating access to rural electricity from about 1% in 2000 to 10% by 2010, through public and private participation and other forms of energy (MofPED, 2003).

11.11 Minerals

The mineral sector in the country (as of 2006) is largely unknown. It has not been significantly described in detail. Only 50% of the country has been mapped. However, from the little data available, indications are that the country possesses significant mineral resources. Modern mining in the country started during the 1920s, principally in the south-western part concentrating mainly on gold and tin. From that date onwards, other minerals attracted interest and were mined. These included beryllium, asbestos, copper, cobalt, limestone and phosphates. In the 1960s, Kilembe Mines was the largest mining operation in Uganda and produced copper ore and employed thousands of people. Other mining activities that have gone on for a long time and are still robust are: sand, clay, gravel and aggregate extraction.

Generally speaking, the mineral sector in Uganda is still largely under-developed, contributing a paltry 1% to GDP. On a base case scenario, the value of mineral production is expected to rise from the current $12 million to over $100 million; while on a best case scenario the value is expected to increase to over $200 million (MEMD, 2003).

The formal mining sector employs about 15,000 people. Due to the labour-intensive nature of mining, the sector has the potential to offer significant non-farm employment opportunities in rural areas. Private operators currently hold a total of about 140 to 200 exploration and development licences. One of the major constraints they face is lack of capital to initiate production.

Gypsum and copper production ceased; while that of wolfram declined as of 1997 due to lack of market and low prices. Katwe Salt Works need to be modernised. On the other hand, marble and limestone production have increased significantly.

Mining could become a key engine for enhancing development. However, for this potential to be realized, proper management procedures need to be adhered to. Otherwise the activities could generate negative environmental impacts such as pollution and other social ills (Box 7). For example, various aggregate quarries, sand clay and gravel excavation pits have been abandoned all over the country, leaving scars on the natural landscape.

The Government is responding to this challenge by encouraging community development and small-scale mining, and putting in place environmental and social management safeguard policies and frameworks.

11.12 Difficulties in implementing sustainable development

In attempting to implement sustainable development, Uganda is faced with some difficulties which include:

Inadequate funding as development funds generated through trade and savings are inadequate. There is high competition for the inadequate funds and environment is not a prioritized sector and yet it is very vital for the success of sustainable development. Donor funding is still substantial, it makes up 55% of the National Budget and 85% of the development budget (Uganda 2003). This makes the country vulnerable and sustainable development difficult.

There are problems associated with governance and politic will. The government is not enforcing the public trust doctrine entrusted to it by the constitution, corruption is rife, there is limited political tolerance and the Leadership Code is abused with impunity. These undermine investment, donor confidence and subsequently sustainable development.

In Uganda, environmental enforcement is still weak due to limited financial, human and physical resources. Planning is not integrated and some laws are irrelevant while others are conflicting; for example the Land Act and Wetland Act. These need to be updated, harmonized and coordinated if sustainable development is to be achieved.

Sustainable development is also beset by socio-economic barriers especially inadequate public awareness on environmental issues, government policies and programmes. This has led to limited public participation in planning and holding leaders accountable for their actions. Insecurity, high population growth, poverty, high mortality rates all hinder sustainable development. It is therefore important for the government to re-assess its approaches to sustainable development and address the difficulties highlighted in this section.

Box 7: Mining and Environmental Decline

During UPPAP2, environmental decline resulting from mining was reported in Buungama in Bundibugyo District, and Lorukumo, in Moroto District. The local people involved in mining barely earn a living from selling the ore because it fetches very little. They have no control over the price of the ore, which is determined by the buyers. Those involved in the mining expose themselves to several dangers. Huge craters are formed as the ore is removed from the ground by the mining process; these craters, in due course, become a health hazard.

11.13 Policy Implications

Since geography is central to MDGs, there is justification to raise sufficient resources for investment in environment.

- The first strategy should involve mainstreaming environmental management in all spheres of development.
- Additional funds need to be raised to fund new or expanded investments, e.g. for water, sanitation and energy.
- Government should provide policy framework to enable institutions and people to forego certain opportunities e.g. complying with the requirement to reduce ozone-depleting substances in the spirit of Kyoto Protocol.
- Government must continue to address the high population growth rate, HIV/AIDS, resettling and rehabilitating Internally Displaced People (IDPs), unemployment, low revenue base and restoring public accountability.
- Uganda should try to restore resource productivity as there is an ecological debt to the future generation.
- Environmental awareness campaigns must be expanded to enable communities to tap opportunities within existing policies and programmes and to avoid risks of pollution.

11.14 Conclusion

Since human beings are at the centre of concern for sustainable development, they are entitled to a healthy and productive life in harmony with nature. In Uganda, there is evidence that the linkage between geography and sustainable development is both practically and theoretically relevant. Under the Uganda Participatory Poverty Assessment Process (UPPAP), the main causes of poverty are development and geographical perspectives. Poor health and diseases continue to be the most important causes of poverty, followed by limited access to land, and land shortage. The location of communities in relation to natural resources is likely to be a source of vulnerability e.g., location of gardens close to forests can lead to vermin destroying crops hence a source of conflicts.

It is important to develop the human resources for the protection of geographical resources if sustainable development is to be achieved in Uganda. Government should support environmental agencies to develop methodologies, tools, and capacities to cautiously integrate sustainable development in geographical resources management.

Key terms

Sustainable development

Environmental sustainability

Questions

1. How is geography central to Millennium Development Goal 7?
2. How does geography contribute to sustainable development?
3. How does sustainable development help to improve the social welfare of the community?

References and further readings

Atukunda, G. 2003. *Contribution of small-scale aquaculture to farmers' livelihoods in Uganda.* Uganda Journal of Agricultural Sciences, Vol. 8, No 10, Kampala, Uganda.

AWF (African Wildlife Foundation). 2001.

Bakamwesiga, H. 1999. *The Distribution, Diversity and Status of Species in Sango Bay Area.* Unpublished MSC. Thesis in Environment Management, Makerere University, Kampala, Uganda.

Banks. 2003. *The Uganda Fisheries Authority Draft Business Plan.* Department of Fisheries Resources. Ministry of Agriculture, Animal Industry and Fisheries Entebbe, Uganda.

Ddungu, M S, Abrua, H and Rusoke, C. 1998. *Land Management Practices and Communities' Priorities in Southwestern and Eastern Uganda.* A Participatory Rural Appraisal Report.

DFID (Department for International Development). 2002. *Integrating Land Issues in the Broader Development Agenda: Uganda Country Case Study.* DFID – Uganda Kampala, Uganda.

DFR (Department of Fisheries Resources). 2003. *Fisheries PEAP Revision Process Submission.* Ministry of Agriculture, Animal Industry and Fisheries, Kampala, Uganda.

Eilor, E and Giovarelli, R. 2002. *Land Sector Analysis: Gender/Family Issues and Land Rights Component.* Final Report. Government of Uganda, Kampala.

EMA (Ema Consult Limited). 2003. *The Evaluation of the Contribution of Forestry Resources to Poor People's Income and Livelihoods.* Reports to the ENR Sector Working Group and Poverty Eradication Action Plan (PEAP) Review, Kampala, Uganda.

Emerton, L et. al. 1999. *Economic Valuation of Nakivubo Wetlands.* IUCN, EARO, Nairobi, Kenya.

Emerton, L and Muramira, E T. 1999. *Uganda Biodiversity: An Economic Assessment.* IUCN EARO Biodiversity Programme, Nairobi, Kenya.

FD (Forest Department). 2000. *National Biomass Study Technical Report, 2000.* Ministry of Water, Lands and Environment, Kampala, Uganda.

FIRRI (Fisheries Resources Research Institute). 2002. *Technical Guidelines for the Management of Fisheries Resources, Biodiversity and Environment of Victoria Basin Lakes.* FIRRI Technical Document No 1 Jinja, Uganda.

GoU (Government of Uganda). 1995. *The Constitution of the Republic of Uganda, 1995.* Government Printer, Entebbe, Uganda.

Howard, P C. 1995. *The Economics of Protected Areas in Uganda: Costs, Benefits and Policy Issues.* M. Sc. Thesis University of Edinburgh.

IFPRI (International Food Policy Research Institute). 2003. *Strategic Criteria for Rural Investments in Productivity (SCRP). Phase II Completion Report.* Main Report submitted to USAID.

IUCN. 1980. *World Conservation Strategy: Living Resource Conservation for Sustainable Development* (Gland,

Switzerland, IUCN, United Nations Environment Programme and World Wildlife Fund, 1980).

Kazoora C IUCN [2002]: *Poverty Alleviation and Conservation: Linking Sustainable Livelihoods and Ecosystem Management.* A Case Study of Uganda.

Kunte A, Hamilton K, Dixon J, Clemens M 1998. *Estimating National Health: Methodology and Results.* World Bank Environment Department papers, paper No. 57.

Mason, P M. 1995. *Wildlife Conservation in the Long-Term – Uganda as a Case Study.* M. Sc. Thesis, University of Oxford, U.K.

MEMD (Ministry of Energy and Mineral Development. 2003. *Annual Report 2002.* Kampala, Uganda.

MFPED/UBOS. 2002. *Uganda Participatory Poverty Assessment Process 2* (UPPAP2).

Mkutu, K. 2004. *Pastoralism and Conflict in the Horn of Africa.* African Peace Forum/Safeworld, University of Bradford.

MoFPED (Ministry of Finance Planning and Economic Development). 1994. *Background to the Budget 1994/95.* Kampala, Uganda.

MoFPED (Ministry of Finance Planning and Economic Development) 2002. *Second Participatory Poverty Assessment.* Kampala, Uganda.

MoFPED (Ministry of Finance Planning and Economic Development) *Uganda Poverty Status Report, 2003 (Achievements and Pointers for the PEAP Revision),* Kampala, Uganda.

MoFPRD (Ministry of Finance Planning and Economic Development) .2001.

Moyini, Y. 2004. *The Potential for Increased Contribution by Environment and Natural Resources to Pro-poor Economic Growth in Uganda.* Paper prepared for OECD Working Group and funded by DFID.

MWLE (Ministry of Water, Lands and Environment). 2001. *Land Sector Strategic Plan 2000-2011.* Kampala, Uganda.

MWLE (Ministry of Water, Lands and Environment). 2001. *National Rural Water Coverage Atlas.* Kampala, Uganda.

MWLE (Ministry of Water, Lands and Environment). 2001. N*ational Forest Plan.* Kampala, Uganda.

NARO (National Agricultural Research Organization). 2000.

NBS (National Biomass Study). 2000. *Technical Report 2000.* Forest Department. Kampala, Uganda.

NEMA (National Environment Management Authority). 1997. *The State of the Environment Report, 1996.* Kampala, Uganda.

NEMA (National Environment Management Authority). 2001. *State of Environment for Uganda, 2000.* Kampala, Uganda.

NEMA (National Environment Management Authority). 2003. *State of the Environment Report for Uganda, 2002.* Kampala, Uganda.

NEMA (National Environment Management Authority). 2004. D*raft Guidelines for Solid Waste Management in Uganda.* Kampala, Uganda.

Ngategize, P, Moyini, Y and others. 2001. *Solid Waste Management Strategic Plan for Mpigi District.* Prepared for Mpigi District Local Administration.

Nuwagaba, A and Mwesigwa, D. 1997. *Environmental Crisis in Peri-Urban Settlements in Uganda: Implications of Environmental Management and Sustainable Development.* IDRC/Makerere Institute of Social Research (MISR) Kampala, Uganda.

Orone, P and Angura, T O. 1996. *The Role of Community Participation in Urban Environmental Management: A Case of Kampala City, Uganda.* Makerere Institute of Social Research (MISR). Kampala, Uganda.

Pomeroy, D, Tushabe, H and Mwima, P. 2002. *Uganda Ecosystem and Protected Area Characterisation* SCRIP Phase II Collaborator Report submitted by Makerere University Institute of Environment and Natural Resources (MUIENR) to the International Food Policy Research Institute (IFPRI), Washington, D C.

PPEA (Participatory Poverty Environment Assessment). 2002. *Participatory Poverty Environment Assessment. Brief 1: Why and how the Environment is Important: key messages from poor communities.* UPPAP/ MoFPED. Kampala, Uganda.

RoU (Republic of Uganda). 2004. *The National Fisheries Policy.* Ministry of Agriculture, Animal Industry and Fisheries. Entebbe, Uganda.

Slade, G and Weitz, K. 1991. *Uganda Environment Issues and Options.* A Masters Dissertations. Unpublished. Duke University, North Carolina, USA.

TAC (Technical Advisory Committee). *Climate Change and the CGIAR. TAC's Progress Report to the CGIAR.* The Consultative Group on Agricultural Research. Washington, D C.

UBOS (Uganda Bureau of Statistics). 2001. *Uganda Demographic and Household Survey, 2000-2001.* Entebbe, Uganda.

UBOS (Uganda Bureau of Statistics). 2002. P*rovisional Population Census Results.* Entebbe, Uganda.

UBOS (Uganda Bureau of Statistics). 2007. Ministry of Finance, Planning and Economic Development, Kampala, Uganda.

UNDP/World Bank/EU/DFID. 2002. *Linking Poverty Reduction and Environment Management:* Policy Challenges and Opportunities.

UNDP. 1990.Global Human Development Report.

UNDP. 1994.Global Human Development Report.

UNDP. 1995.Human Development Report.

UNDP. 1995.Governance Policy Paper.

UNEP. 2002. *Poverty and Ecosystems: A Conceptual Framework.*

UNEP (United Nations Environment Programme). 2002. *Africa Environment Outlook, 2002.* Earthscan

Publications, London, United Kingdom.

UN. 1992. Agenda 21.

UN (United Nations). 2002. *World Summit on Sustainable Development.* Johannesburg, South Africa.

UN (United Nations). 1992. *Agenda 21: The United Nations Programme of Action from Rio.* New York.

Sansa A. 2005. www.ictp.triesti.it.

Yaron, G, Moyini, Y. 2003. *The Contribution of Environment to Economic Growth and Structural Transportation.* Report prepared for the ENR Sector Working Group, Poverty Eradication Action Plan (PEAP) Review, supported by DFID. Kampala, Uganda.

Chapter 12

GEOGRAPHIC INFORMATION SYSTEM: APPLICATION TO URBAN GEOGRAPHY OF UGANDA

Shuab Lwasa

12.1 Introduction

The world is constantly changing although not all changes are for the better. Some changes seem to have natural causes (volcanic eruptions, dust storms) while others are caused by man (land use changes, land reclamation). Besides global changes, societies today are grappling with issues of human development including hunger, poverty eradication, improving incomes, public health management, agriculture development, city development, and infrastructure development. For both the global changes and the pursuance of transformation of our societies, humans endeavor to characterize an understanding of their environment. Environment means the geographic space in which pursuance of the human transformation desires take place. This can be a village, city, region/district, country, or even the global earth.

For an understanding of our environment, data are gathered, processed, analyzed to process information on the environment. Information about the environment has been produced by several techniques and methods for a long, time but the scale (implying amount of information and coverage) at which we can collect, process data and produce information has drastically changed in the last three and half decades. The digital era has had a tremendous effect on the means available for collection and processing of data on the environment. For example, several satellites are orbiting the earth for different purposes including recording of land cover on the earth's surface, while several satellites are also orbiting the earth daily and can be used to determine the geographic position, at any time, on the earth's surface.

Collecting data is a step in the production of information, that which can both serve a particular user for specific needs (for example land use change information), and that which can be for general use (for example a topographic map). Analysis of the data collected is pertinent in determining the quality, accuracy and up to datedness of the information produced. A Geographic Information System (GIS) is one such technique, tool and method for collection and analyzing of data to produce information. GIS has changed the way humans handle information regarding the environment, the way information is managed and the way humans think about their environment. The fundamental problem humans face in understanding the environment while using GIS is, understanding the phenomena that have a geographic dimension as well as a temporal dimension. In other words, we have 'spatial-temporal' problems. Due to vast improvements in hardware and software over the years, GIS has transformed the understanding of geographical phenomena, even complex phenomena that is caused by a mix of natural and human factors. But GIS has also transformed the way services and infrastructure are delivered, development planning and the way we deal with surprises of global change.

This chapter focuses on GIS as a new tool in Geography and geographic studies with some demonstrations of its capabilities in handling geographic data and information. The chapter is organized in three parts: the first part is an introduction to GIS in which definitions and theoretical underpinnings of GIS are elaborated; the second part gives a discussion on GIS as a tool and method with some review of literature on methodology in geography as well as the different fields where GIS is applied and used in Uganda; and the third part presents some case studies analyzed using GIS, focusing on the procedures, or algorithms and the uses of the outputs. The three case studies attest to the analysis of population dynamics and visualization and urban development. The chapter concludes with a brief on opportunities and limitations of GIS use in Uganda.

12.1.1 Definition of GIS

To define GIS implies providing a universal and definitive meaning of the concept. But this is a rather difficult task since there are many users and applications of GIS but also a lot of changes and transformations in the field occur regularly. It is also a difficult task due to the fact that the concept has evolved out of an amalgam of three distinctive fields each of which has elaborate procedures for interpretation of the environment. The three fields as derived from the acronym GIS are, first, the spatial science field for which geography could be taken as one of the branches, secondly, the information field, and thirdly, the systems analysis field. Before attempting to define GIS, each of these concepts of the acronym is defined.

Geographic, is a characteristic about any phenomenon which has a location on the earth's surface (such is not limited to area above the earth but also underground, as well as the universe). Geographic phenomena relate to space. Space is the three dimensional extensions around the earth. Location is defined by some reference system in terms of X and Y coordinates. The simplest phenomenon can be represented by a point which is defined the X and Y coordinates. Geographic phenomena have three characteristics; that which can be named or described; that which can be geo-referenced and that which can be assigned a time (interval) at which it is/was present.

Information on the other hand refers to the processed data. It is common to use the terms data and information almost interchangeably but there is need to be precise because their distinction matters. Data is the representation that can be operated upon either by computer or manually to produce information. More specifically, here, is spatial data which means data that contains locational values. Occasionally, one can find in literature a more precise concept, geospatial data, as a further refinement. This means spatial data that is georeferenced. But spatial data which is not georeferenced can have positional data that is not related to the earth. Therefore, information is data which has been interpreted and to which meaning (user dependent) is attached. Processing of data leads to information, understanding and knowledge by humans, for example it is not possible for a machine like a computer to 'understand' or even have knowledge.

Spatial information is therefore a specific type of information that involves the interpretation of spatial data.

System means an abstraction of objects that can be identified by some variable attributes. There is a relationship between the attributes of the object and relationships between the attributes of the objects and the environment. In the context of systems analysis and thinking, environment means any external object, or phenomenon outside the object of study or that considered relevant in any analysis. A system is characterized by four main issues; the structure, the behaviour, the relationships and the environment. The discussion of these in detail is beyond the scope of this book but mention of these system characteristics will be referred to occasionally in this chapter. Thus, a system is defined here as components that work, more or less, independently but related and each contributing to the achievement of a pre-determined overall goal.

Borrough (1986) defines GIS as a set of tools for collecting, retrieving, at will, transforming and displaying spatial data from the real world for a particular set of purposes. While Aronoff (1989) defined GIS as any manual or computer based set of procedures used to store and manipulate geographically referenced data. Borrough (1998) attempted to redefine GIS as a computer system capable of handling and using data describing places on the earth surface. Aronoff (1998) refined the definition as a computer based system that provides four sets of capabilities to handle georeferenced data: data input, data management (storage and retrieval), data manipulation and analysis and data output. Borrough (1998) contends that GIS is an organized collection of computer hardware, software, geographical data, and personnel to design, to efficiently capture, store, update, manipulate, analyse and display all forms of geographically referenced data. These are some of the definitions given but there are several other definitions by different scholars which is an indicator of the difficulty of defining GIS. This is partly because of the evolving nature of the 'field' of GIS as a science, profession and field of study. The difficulty also emerges as a result of the differences between GIS as an evolving science and as an application tool. It is therefore important for any reader of this text to consider GIS as a science and as an application support tool (taken broadly to include technique, methods in a problem solving context).

Thus, for purposes and scope of this chapter, GIS can be defined as a computerized system organized with a collection of computer hardware, software, geographic data, and personnel designed to efficiently capture, store, update, manipulate, analyse, and display all forms of geographically referenced information. This definition is derived largely from the application perspective. But it is possible to find in literature a definition which considers GIS as a generic software package that can be applied to different applications. For instance, determining suitability of land for urban development is an application but the software used therein can be used for analysing forest cover change and biomass loss.

12.1.2 Components of GIS

Aronoff (1998) identified what he considers the four main components of GIS. These include data input, data management, data manipulation and analysis and data presentation. These components are briefly discussed below;

Data input comprises procedures and methods for entering geospatial data into the computer-hardware system. Spatial data is collected from various sources and entered in the computer system through manual digitizing, keyboard entry, scanning and use of existing digital files. Data is entered for both the locational values and attributes of the objects.

Data management comprises of the procedures for storing and retrieval of spatial data. This component determines the model of representation of the spatial data as well as the structure for data storage.

Data manipulation and analysis, on the other hand, are the set of procedures available for processing the data to produce information. Manipulation of data can be done by re-classification procedures, overlays, statistical analysis and geospatial modeling.

Data output or presentation organizes all the analysed information into a format that best communicates the results of analysis to the user. Output can be in soft or hard copy form and it is mostly maps and tables accompanied by charts.

It is these four components which make a system of GIS but of late since the field is evolving, there has been an addition of the personnel. Though it is not considered a component, personnel are critical to a GIS especially to application.

12.1.3 Basic theoretical underpinnings of GIS

Being an evolving field of study and science, it is difficult to raise theories of GIS, and indeed there may not be such a thing as GIS theory. What is attempted here is a discussion of the basic fundamental concepts embedded in some of the main GIS components. These fundamental concepts include spatial data representation, spatial data entry and spatial data analysis. Each of these major concepts can be decomposed into various other concepts which will be discussed in the following section.

12.2 Geographic phenomena

In an earlier subsection, geographic phenomena, as a concept, was briefly discussed in the attempt to understand GIS. There is definitely a wide range of geographic phenomena around us but the relevant geographic phenomena for GIS use are dependent on the objectives of an application. Geographic phenomena exist in the real world but the real world is completely a different domain than the GIS/computer world in which simulations of the real world are done. This implies that whatever is represented in the computer can never be perfect as the real world. Therefore crossing the barrier between the real world and a computer representation of it is a domain of expertise in itself.

Geographic phenomena can be represented in various ways, the choice of which representation is dependent on mostly two issues,

- What original, raw data is available?
- What sort of data manipulation does the application want to perform?

12.2.1 Spatial data types

There are four distinctive spatial data types commonly used in representation of geographic phenomena. These include the point, line, area data types, and a special data type referred to as three dimensional data type. These data types are discussed below.

12.2.2 Point representations

Point data types also referred to as 0 (zero) dimension data, are defined by single coordinate pairs (x,y). Points are used to represent objects that are best described as shapeless and size-less, single locality features as shown in Figure 12.1 below. Whether this is the case, depends on the purposes of the spatial application and on the spatial extent of the objects compared to the scale applied in the application. For example for a city tourist map, telephone booths could be represented as point features. Besides the georeference, extra data is usually stored for each point object in which all thematic data relevant about the object can be captured. Taking our example of telephone booths, thematic data could be, owning company, phone number and date last serviced.

Figure 12.1: Single locality features

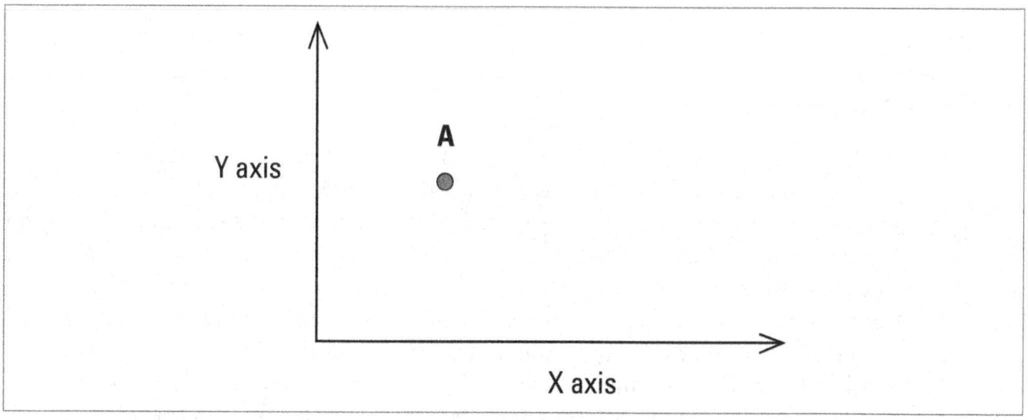

12.2.3 Line representations

Line data are used to represent one-dimensional objects for which length can be measured. Such features have a minimum of two zero-dimensional objects therefore the simplest one-dimensional object is a line. Features include roads, rail roads, canals, river and power lines. It is important to note again that it is dependent on relevance for the application and the scale that the application requires. For example application of mapping tourist information for bus routes is likely to be line features while for a cadastral system (land information) may consider roads to have width (two-dimensional objects).

In GIS literature you will find such terms as polyline, arc or line segments, these are used to represent the real world as much as possible. Associated with line features are nodes and vertices which are points but they only serve to define the line especially curvilinear features and have specific characteristics in relation to the GIS software. GIS will store a line as a simple sequence of coordinate pairs of its nodes and vertices assuming that all segments are straight as illustrated below. A collection of connected lines may represent phenomena that are best viewed as networks having connectivity and network capacity such as traffic routes.

Figure 12.2: Line representations

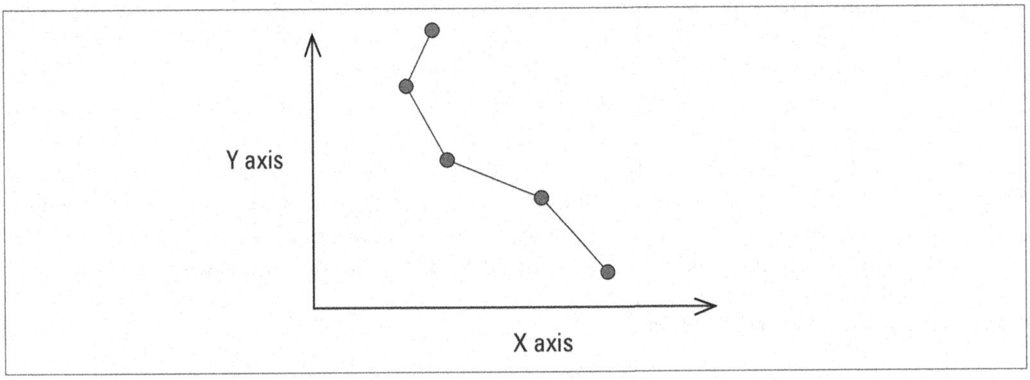

12.2.4 Area representations

When area objects are stored using a vector a model (model which uses lines, points and areas), the usual technique is to apply the boundary approach. This means that each area feature is represented by some arc/node structure that determines a polygon as the area's boundary. Such a feature is also called a polygon which approximates the extent of the feature. Additional data is also stored which may include area, perimeter and other thematic data as determined by the application. A simple but inappropriate representation of area features would be a list of lines that describe the boundary of each polygon. Each line in the list would be a sequence that starts with a node and ends with one possibly with vertices in between as shown in Figure 12.3 below. Under this representation area features that are adjacent to each other would have the boundary between them represented twice for each polygon. This creates data duplication or data redundancy. The boundary model is an improved representation that deals with the disadvantages of the data redundancy.

Figure 12.3: Area representations

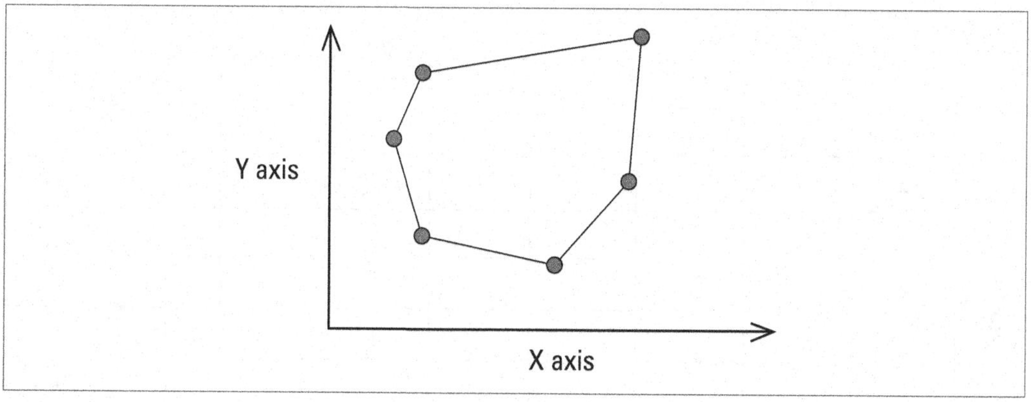

12.2.5 Computer representation of geographic phenomena

Geographic phenomena which can be described as fields and objects are represented in the computer in various ways. But geographic phenomena can be both continuous and infinite since there are many locations. Phenomena with intrinsic continuous and/or infinite characteristics have to be represented with finite means for computer manipulation. GIS represents fields using the *tessellation* approach while objects are represented using the *topological* (vector) approach.

A *tessellation* (or tiling) is a partition of space into mutually exclusive cells that together make up the complete space of interest. With each cell, some (thematic) value is associated to characterize that part of space. There are regular tessellations and irregular tessellations. Regular tessellations include *square cell* tessellations (as illustrated in Figure 12.4 below), *hexagon cell* tessellations and *triangle cell* tessellations while irregular tessellations can be irregular triangular networks (TINs), which are sometimes used to represent elevation field. All regular tessellations have cells that are of the same shape and size and the field attribute value assigned to a cell is associated with the entire area occupied by the cell. For example square regular tessellations, also known as raster, have the same size which is a raster's resolution.

Figure 12.4: Heading

COLUMN

ROWS

	1	2	3	4	5	6	7	8
1	1	1	1	1	2	2	2	2
2	1		1	1	2	2	2	2
3	7	7	3	3	3	3	2	2
4	7	7	7	3	3	3	1	1
5	7	7	7	7	3	3	1	1
6	7	7	7	7	7	7	1	1
7	7	7	7	7	7	1	1	1
8	7	7	7	7	7	1	1	1

Irregular tessellations, on the other hand, are partitions of space into mutually disjoint cells but cells vary in size and shape, allowing them to adapt to the spatial phenomena that they represent. Irregular tessellations are more complex than regular tessellations but they are also more adaptive, which leads to reduction in the amount of memory used to store the data. One of the well-known data structures of irregular tessellations is the region-quadtree, based on regular tessellations of square cells but takes advantage of cases where neighbouring cells have the same field value so that together they can be represented as one bigger cell.

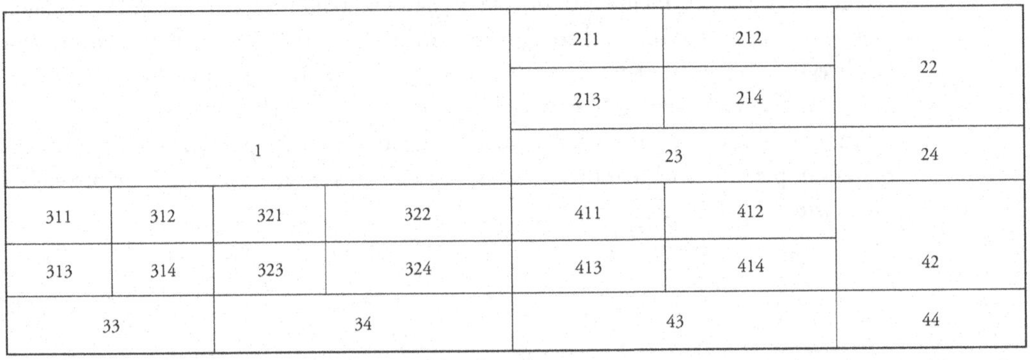

This type of tessellation has it that when the cells are in a quadrant, then they have the same field value. The procedure produces an upside-down, tree-like structure known as a quadtree. Quadtrees are adaptive because they apply the spatial autocorrelation principle; locations that are near in space are likely to have similar field values. When a conglomerate of cells has the same value, they are represented together in the quadtree, provided boundaries coincide with predefined quadrant boundaries. Thus, quadtrees provide a nested tessellation.

12.2.6 Topology and Spatial Relationships

Topology deals with spatial relationships between geographic objects and fields. Such spatial properties of relations do not change under certain transformations and are invariant under a continuous transformation. Topological mapping deals with topological space in which every point in space can have a neighbourhood around it that fully belongs to that space as well. We can use topological properties of interior and boundary to define relationships between spatial features. For example, if we take a spatial region A which has a boundary and an interior both denoted by boundary A and interior A. There are possible combinations of intersections between the boundary and the interior of A with those of another region B, to test whether they are the empty set or not. From these intersection patterns, we can derive eight (mutually exclusive) spatial relationships between two regions. If for instance the interiors of A and B do not intersect, but their boundaries do, yet a boundary of one does not intersect the interior of the other, we say that A and B meet.

12.2.7 Scale and Resolution

The term scale has many meanings and because of the transition from analogue to digital representations of information, it may have little meaning for such digital information (Goodchild, 2001). Scale has several meanings to different independent roots in which meaning is strongly dependent on context. For example, "the scales of justice", or scale of a hazard, have little connection to each other, or to GIS. Yet scale to the cartographer mostly relates to the representative fraction or the scaling ratio between the real world and a map representation on a two-dimensional surface. Scale to an environmental scientist is likely to relate either to spatial resolution (the representation level of spatial detail), or to spatial extent (the representation of spatial coverage). This implies that a simple phrase like "large scale" can send different messages when communities and disciplines interact- to a cartographer it implies fine detail, whereas to an environmental scientist it implies coarse detail. From these differing definitions of scale, four broad meanings of scale can be derived as observed by Goodchild (2001)

Level of spatial detail: to most scientists, scale implies level of spatial details, or spatial resolution which is often defined as the shortest distance over which change is recorded and having units of length. Level of spatial detail relates to the different representations of reality. A square raster provides the simplest instance because in this representation of the spatial variation of a field, the spatial resolution is clearly the length of a cell side. To the vector representations of a field, spatial variation is represented by capturing the value of the field at irregularly spaced points.

Representative Fraction: on a paper map all features on the earth's surface are scaled using an approximately uniform ration known as the representative fraction. The representative fraction in association with the paper maps, impose an effective limit on positional accuracy. Therefore the representative fraction serves as a surrogate for the features depicted on maps.

Spatial Extent: from the perspective of extent, scale is often used to refer to the extent, or scope of a study, or project in which a metric connotation can be derived. Spatial extent defines the total amount of information relevant to a project which rises with the square of a length measure. In digital databases, it is common to have process scales which refer to computational models of representation of a landscape-modifying process such as soil erosion. A process in this context is a computational model that takes a landscape from its existing state, to some new state. And in this sense, processes are a subset of the entire range of transformations that can be applied to spatial data.

Ratio scale: in the digital databases of spatial reality, ratio scales are best explained by the spatial extent and level of spatial detail. The comparison of these two meanings of scale to a geo-database yields the ration scale. But, both the spatial extent and spatial detail have dimensions of length thus their ratio is dimensionless. This differentiates the meaning of scale from the representative fraction since being dimensionless implies seamless. For example, remote sensing instruments as a range of spatial resolution from 1 m to 1 km. because a complete coverage of the earth's surface at 1 m requires on the order of 1015 pixels, data are commonly handled in more manageable tiles or approximately rectangular arrays of cells.

From the foregoing, it is clear that scale has many meanings and the context in which data are scaled is important for its definition. Digital data will have differing meaning to scale as compared to analogue map data. But in the context of this chapter definitions of scale in the digital sense are more appropriate for the understanding of how spatial data are represented.

12.2.8 Temporal dimension

Geographic phenomena change over time besides having geometric, thematic and topological properties. This characteristic of changing over time denotes temporal characteristic. As de By (2001) notes temporal changes address four main questions:

1. Where and when did change take place?
2. What kind of change occurred?
3. With what speed did change occur?
4. What else can be understood about the pattern of change?

Spatial temporal data structures are representations of geographic phenomena changing over time. Because change is time dependent, it has various properties including; time density (measured along discrete or continuous scale), dimensions of time (valid time or world time and transaction time), time order (considered to be linear extending from past to present), measures of time and time reference (time represented as absolute or relative.

Thus spatial data can be represented to denote phenomena using various time-dependent models. These models include; the snapshot model (in which data represents affairs for the valid time for which it was stamped), the event based model (in which we start with an initial state and record events along the time line) and the space-time composite model (which starts from the two-dimensional view of the study space at a given time.

12.2.9 GIS as an Integrative System

GIS is a system with capabilities of integrating information from different sources. Earth related phenomena has always been represented and modeled by maps, but such phenomena only used to show limited information which is socio-economic in nature. GIS has provided a means bringing together spatial information and social economic information in a similar database. The relationships between the spatial features and social processes that can be represented in the database are therefore easily captured. With this capability, GIS is an integrative system which can provide information hither to.

12.3 GIS and Methods of Geography

12.3.1 A brief on methodology in Geography

Methodology in geography has always been open to influence by the methodological attitudes developed in disciplines to which geography is related. Harvey (1967) identifies three sources of methodological influence including the fields of physics, social sciences and history. As a result, geographers have often considered themselves as either more related to the natural science or the social sciences. But at times geographers have appeared to reject any relation with other disciplines and instead sought refuge entirely within boundaries of their own discipline. It would therefore be wrong to assume that there is complete agreement as to the appropriate methodology in geography. But it has been generally understood that the 'standard model' which is derived from analysis of explanation in the natural sciences in general provides the equipment for discovering empirical statement about the world.

Geographical inquiry consists of a set of prescribed rules of behaviour in the pursuit of the aim. It is also important to regard statements which geography makes about the real world as being ordered in a consistent hierarchy. From the lowest order of the hierarchy are the factual statements, intermediate statements of generalizations or empirical laws and the highest order statements of theoretical laws. The achievement of such an inclusive system of explanation involves linking statements of extreme generality to statements of lesser generality. The different geographical inquiry techniques such as spatial statistics, descriptive techniques notwithstanding, the technique of mapping in geography has always provided the unique approach in explanation of the central concern of geographical inquiry; that is spatial phenomena. The technique of map use in geography has profound potential from providing descriptive models, integrating statistical analysis to modeling complex spatial relationships and their explanation. Thus any alternative means of handling spatial information and modeling such as GIS is by no means an ultimately valuable tool in geographical inquiry.

12.3.2 Maps and map making in Geography

The art, science and technology of making maps is concerned with compilation, design and drafting a map that can communicate all the desired information in such a way that the information can easily be understood and interpreted. This has traditionally been done by using conventional cartographic techniques that include the quill pen, drawing board, squares and tracing paper. But the challenge of meeting the demand for spatial information in an information society of today is almost difficult to be fulfilled by these conventional cartographic techniques. Spatial phenomena are changing very fast which requires up to date information to support decision making, administration, planning and management.

There are various sources of information on spatial phenomena. Conventionally, ground topo-survey and aerial photography processing have always provided the cartographers reliable information for map making. But the process of photographic interpretation is a time consuming and labour intensive process that by the time final information is published, it would be out dated. Thus the time at which information is captured about reality and the time it is produced on the map is an important characteristic of a system with usable spatial information.

The central idea of cartographic communication is that provision of information in a map for, and its subsequent interpretation by map users should be regarded as a single system, in which the graphic form of the representation strongly affects the efficiency of map use. In order to facilitate communication of information to the map user, the system use interacting components which symbolizes quantitative data that demonstrates the relationship between the interacting components for example population density, or ratio data for particular areas are represented by CHOROPLETH maps (from the Greek: 'choros' which means place and 'plethos' which means magnitude). In this case, shading of various intensities could be used to represent a feature of particular magnitude. For instance, areas of highest values, or greatest densities, should be assigned the darkest tones while the lowest are assigned the lightest tones. These conventional cartographic methods of map making were mostly used for representation of phenomena in a two dimensional nature (in which an object is represented by its outline along which length and width dimensions can be measured). Even when the geo-object, or phenomenon is in reality of three dimensional nature (in which an object is represented by its two dimensions of length and width in addition to the height dimension), maps have always represented such phenomena in a two dimensional way. Geographers have for a long time represented topographical profiles using graphical techniques by working and analysing maps.

Geographers have always used and relied on maps in their description or explanation of spatial phenomena. Thus, spatial analysis on maps and map making in geography are fundamental techniques for both analysis and presentation of information from geographic studies. As noted earlier, the process of data processing for final map production dictates those alternative techniques that are underway in current information technology developments have the potential for speeding up on the update of spatial information. There is a high potential for Remote Sensing, Global Positioning Systems and GIS in the line for production of information related to the earth. For example, whereas topographical profiles have 'traditionally' been derived from analysis of maps using scale-rulers, interpretation of terrain using contours and graphical techniques, with GIS all these can now be generated with limited human intervention within high degrees of accuracy. The image below illustrates a computer generated profile of a part of Kampala City indicating heights and draped with high resolution imagery.**Figure 12.5: Surface Profile of Makerere and Kololo in Kampala**

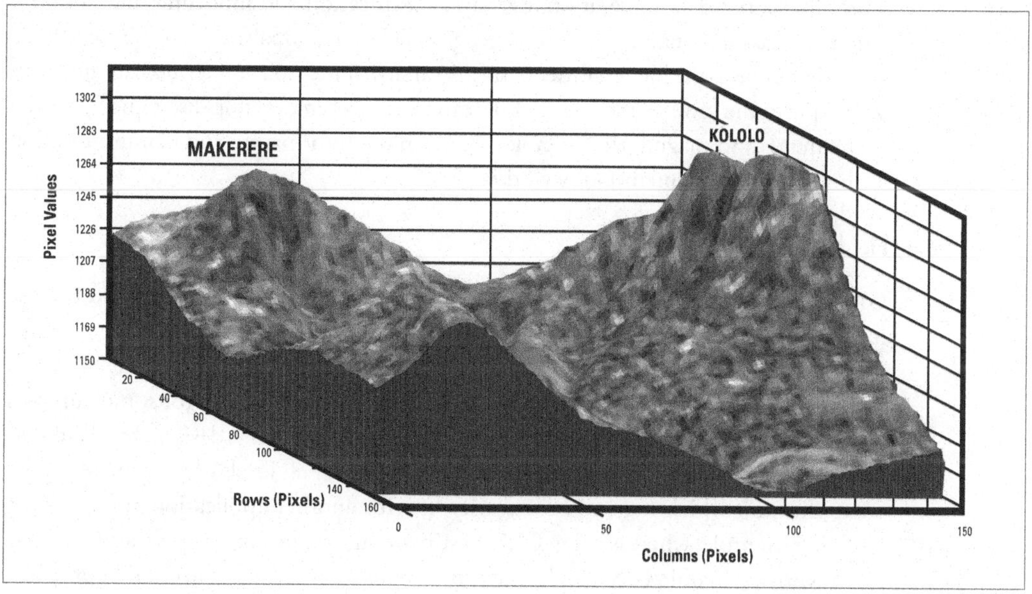

Source: Mapping and Surveys department

12.3.3 Is GIS a new discipline or method/tool in geography?

The analysis of spatial phenomena is definitely not a new field of inquiry. Spatial sciences including but not limited to geography, physics and geodesy have always employed methods and techniques in spatial analysis. Throughout the 19th and 20th Century, the search for knowledge in spatial sciences about the real world was fueled by developments in system thinking, and developments in the technology for data collection in respect to the earth. In the process, systems thinking, and increased use of information for decision making in various fields has led to the evolution of GIS, which is considered sometimes as a tool or method and currently as a science in some communities. Because the use of GIS is not limited to mapping but extends to knowledge systems, expert systems, decision support systems and modeling; it is starting to emerge in literature for reference to GIS as Geographic Information Systems Science (GISS).

Geographers, similarly, moved from analogue ways of handling and management of spatial information to what is being called 'modern' spatial information management, or GIS. So, it is also possible to consider GIS as having rejuvenated geography by enhancing its 'traditional' technique of analogue mapping with advanced tools of information analysis and management. There are various fields of GIS application and numerous techniques of analysis. GIS has created a hierarchy of applications from basic descriptive modeling (by use of overlays) to an explanatory method (by spatial integration and modeling), to decision making systems that take on different dimensions of operation (including databases, expert systems, plan support systems and systems modeling). Through spatial data integration, overlays, classification and modeling that does not require intensive human intervention, geography has nurtured what may be considered a sub-discipline at the highest order of the hierarchy, or method at intermediate and tool/ technique to the lowest order. Thus, GIS can be seen as an evolving discipline by nature of the trends and magnitude of developments in the Information and Communication Technology in the contemporary world.

12.3.4 Fields of GIS application

GIS application is a term which has different meanings depending on the level of use for a GIS as described by the hierarchy briefly presented in the preceding section. Basically, application could be taken as 'making use of'. But in GIS, application transcends this definition to the level of designing and implementing a system for some purpose. Such an application is usually software and data dependent that is designed and tailored to a specific operation, routine or activity. Therefore, at all levels of GIS, there can be an application which may not necessarily fit with another application at a different level. For example, whereas using GIS to estimate run-off can be an application, it may not fit with a dynamic GIS based database for management and collection of property rates. It is because of this diversity in application that the fields of GIS application are many. It is also important to note that in all applications, the central question or any of the questions/problem needs to be spatial in nature. Where?

Thus, many fields of study can, and have applied GIS in addressing the various problems associated with the earth. These fields include; land use planning and management, environmental studies, natural resource mapping and management, land tenure, forest resource management, mineral exploration, natural hazard prediction and management, geological studies, engineering, design and planning, utility management, marketing and retailing, urban management and planning, marine sciences and water resource management.

12.3.5 Organizations using GIS in Uganda

In Uganda GIS use has not penetrated many institutions as a tool for decision support. The only area in which GIS is being appreciated is mapping (which in essence is among the first steps in the development of GIS). Mapping is still an important activity since for a long time we have had our spatial information managed using analogue systems (including film-based map storage, aerial photo interpretation and stereo potting). The major mapping agencies in Uganda are the Mapping and Surveys department (responsible for surveys and production of topographical maps for the entire country), National Forestry Authority (NFA GIS centre, which has been responsible for the production of the National Land use/cover map) both of which are under the Ministry of Water, Lands and Environment. There is also a GIS laboratory at Makerere Institute of Environment and Natural Resources, Department of Geography at Makerere University as well and of late Kampala City Council which established a GIS unit in 2003 under the department of Urban Planning and Land Management.

Figure 12.5: GIS Unit at Kampala City Council

Source: Courtesy of Shuaib Lwasa

12.4 GIS applications in Geography

12.4.1 GIS and analysis of population

The focus of population geography is the characterization of population in space. One most commonly used measure of characterization is density which relates population variables with ground area units. But the population variables of size, distribution, structure and change continuously vary in space depending on various factors. Among the dynamics of urban populations is the mobility of residents. Usually urban dwellers continuously change residence within the city as, and when, their aspirations change driven by location and affordability (Wadhva 1988, 1989b, Lwasa 2004) but urban development policies are also influential in driving mobility.

The following case study demonstrates the ability and capability of GIS in the analysis of population and therefore provides powerful tools for the study of population geography. Traditionally, population geographers have always employed maps to illustrate population densities, migration and distribution in space. The various factors of urban development including housing markets, change of use policies and intensified activity at the core of cities fuel a process of population dynamics in the city. Such processes are manifest in the mobility of the low and middle income populations which occurs more frequently than the high income. In the case of Kampala, a comparison of population densities for the years 1991 and 2002 shows significant differences in densities. Neighbourhoods near and adjacent to the civic area, have higher densities which decrease away from the civic area. This is illustrated in Map 1 and2.

Map 12.1 and 12.2: Population Density in Kampala Neighborhoods

Source: UBOS, 2004

From the two maps, population densities are higher just outside the low density central administrative parishes -- the Central Business District (CBD). Obviously, density increases with the intensity of dots in the population map for 2002 compared to 1991. This is due to the changes in population. But the striking feature of the population distribution is the clustering of high densities to the north, west and south of the CBD and not as much to the east of the CBD. The variations explained by nature of developments to the east which mostly has the hitherto planned and properly planned settlements that have attracted less population.

On the other hand, population changes and distribution in Kampala city indicates that while some places in Kampala are gaining in terms of densities, the core part of the city is losing as conversion and change of use to civic or commercial takes over previously residential neighbourhoods within the CBD. For greater Kampala, there is an increasingly observed phenomenon in which the periphery areas are gaining population as suburbanization spreads to the rural areas at the periphery of the city. Mobility to periphery areas is mostly by middle and high income population but observations also indicate a sizeable proportion of mobile population being the low-income. Capitalizing on the improved transportation system with relatively low commuter charges, the low-income slowly establish outside the core inner city to the periphery zones.

Map 12.2: Percentile Map of Proportion of Population Change between 1991 and 2002

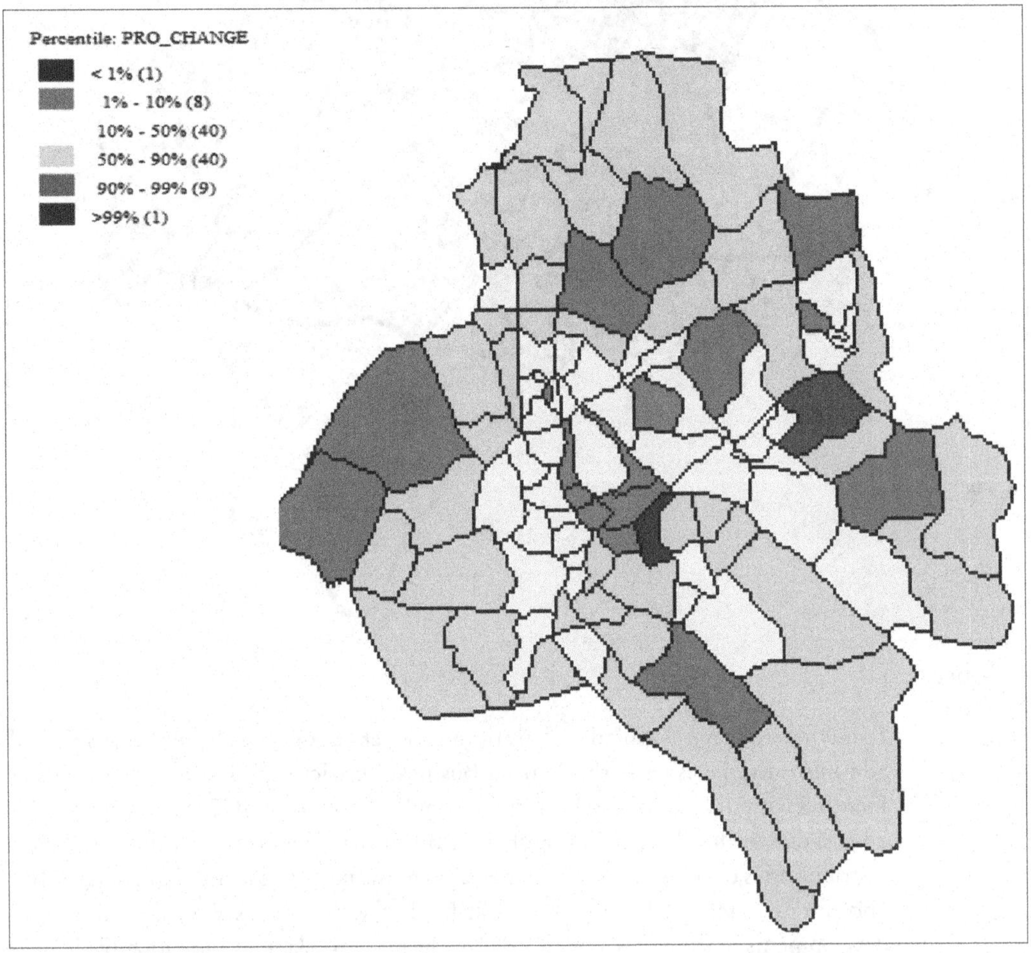

Source: UBOS

As shown in the above map, the parishes in the CBD had significant change in population with their percentile falling in the low ranges of less than 1% to below 50%. The parishes which gained in population are mostly the periphery parishes and the reason being mobility as driven by the land use change policies in the city, the housing enabling approach by government and promotion of owner-occupier type of housing. The middle and high income population have moved to the periphery parishes for more spacious land, aesthetics and utility derived from the settlements. Thus the capability of GIS in spatial population analysis is undisputed and replaces the traditional cartographic techniques of map production. This implies that geography and geographic analyses can be enhanced by the geo-information tools in understanding the spatial phenomena and how they relate to each other.

12.4.2 GIS applications in Urban studies and Development

In the earlier discussion, the applicability of GIS in activities from different fields identified several of the fields that have transformed method of inquiry and information production. The case study presented here illustrates the application of GIS in the preliminary stages of preparation of a spatial plan for urban development. The case study is based on Kampala city applying the data available but with the model having potential to be extended for other data. The case study proceeds with identification of criteria which in general were categorized as physical, social and economic those are decomposed to lower level for identification of the factors for analysis. The table below illustrates the criteria derived using a deductive approach.

The design of the suitability model is based on the selected criteria of soil type, slopes, drainage conditions and accessibility to transport network. The land use/cover map of 1996 and the proposed land use plan form the basis of evaluation of suitability level of each area for residential purposes

Table 12.1: Identified general and specific criteria

Criterion	Sub- criterion	Factor	Criterion Indicators	Impact	Desirable requirements
Physical	Soils	Drainage	Well drained, Seasonally wet, Permanently wet, open water	Construction costs	Well drained,
		Soil Type	Sandy, Clays, Sandy loams	Construction costs	Non-expansive soils
		Water table	High, low, very low	Structural stability	Very low water table
	Slope	Slope %	0-5%, 5 -10%, 10 - 15% ,15 - 20% , >20 %	Construction costs	0-15%, Flat, gentle to moderately steep
				Health risks	Non
	Hazards	Flooding Pollution	Non, occasional, rare, frequent, very frequent		
				Health risks	Non
	Land use	Vacant, non-agricultural land	Air, Ground water,		
					Vacant, non-agricultural
			Land for future expansion		

For each criterion, a suitability ranking was assigned for which seven-point scales to determine the qualitative rankings of the suitability criteria were transformed to be standard scores. The rankings range from 1 (Not suitable) to seven (Highly suitable). This "positive direction" Voogd, (1983) is chosen to keep the scores understandable since the higher the score, the more suitable the site is. The suitability scores in this model were based on personal knowledge about the area and limitations of factors to development.

For each of the site characteristics, a rating is given indicating the degree of suitability in relation to residential development. The following suitability levels are used in determining the model.

Not Suitable (1): This is attributed to sites with characteristics imposing certain constraints, which cannot be overcome or technically excluded for development e.g. ecological areas.

Least Suitable (3): A level for sites with characteristics imposing constraints, which can be overcome, but by massive investment.

Moderately Suitable (5): This is for factors with many criterion indicators. It denotes sites with constraints but where the investment is higher than in the suitable class.

Suitable (6): Sites with characteristics, which can be overcome by moderate investment.

Highly Suitable (7): Areas with characteristics imposing no significant constraints for development.

12.4.3 Determining Criterion Weights

Weights are applied when not all aspects have an equal importance in the evaluation. In this study, it is taken that the criteria used have unequal importance. Secondly the evaluation method adopted requires criteria to be assigned weights. Weights range from 9 (high importance) to 1 (not important). The parameters considered in this evaluation were assigned weights according to the opinion of the evaluator. Drainage assigned a weight of 9 is considered to be of high importance due to costs involved in making poorly drained sites workable. It is also due to the nature of the study area, which has wide valleys and swamps, (see map 3.5). These areas are considered to be of high ecological importance. Slope on the other hand is assigned a weight of 7 due to its importance on construction process. The study area has generally good soils with fewer limitations to construction. But there are also areas with soils that have severe limitations to construction. Thus this factor was assigned a weight of 5 being of less importance than slope or drainage. Accessibility to the network is assigned a weight of 3 because it may not significantly affect suitability, but it remains important. Although a long distance involves some extra costs and increase in travel time, areas may be opened if a new road is constructed or an existing one upgraded.

12.4.4 Environmental analysis using GIS

As noted earlier in the introductory section of this chapter, environmental analysis is vital for production of information necessary for advancement of humanity. The case study on environmental analysis is based on a study that assessed the environmental implication of urbanization and land use land cover change in the metropolitan area of Kampala (Nyakaana, Sengendo, et al 2005). The urbanization rate of Uganda is considerably high occurring at 5.5% p.a. while the process experienced has regional and urban specific disparities in respect to the speed of urban growth and expansion. Kampala city which is increasingly becoming a primate city is growing at an annual rate of 5.61% having more than 40% of the national urban population and 4.9% of the national population. The process of urbanization of Kampala and the city's rapid developments has created environment problems emanating from demographic dynamics as well as the existing development and planning problems experienced in the city. A spatial analysis of the city's growth was undertaken to determine the speed of growth in general and specifically for some of the land uses which exert pressure on the environmental resources of the city using Remote Sensing and GIS.

Figure 12.3: Built Up Area in Kampala

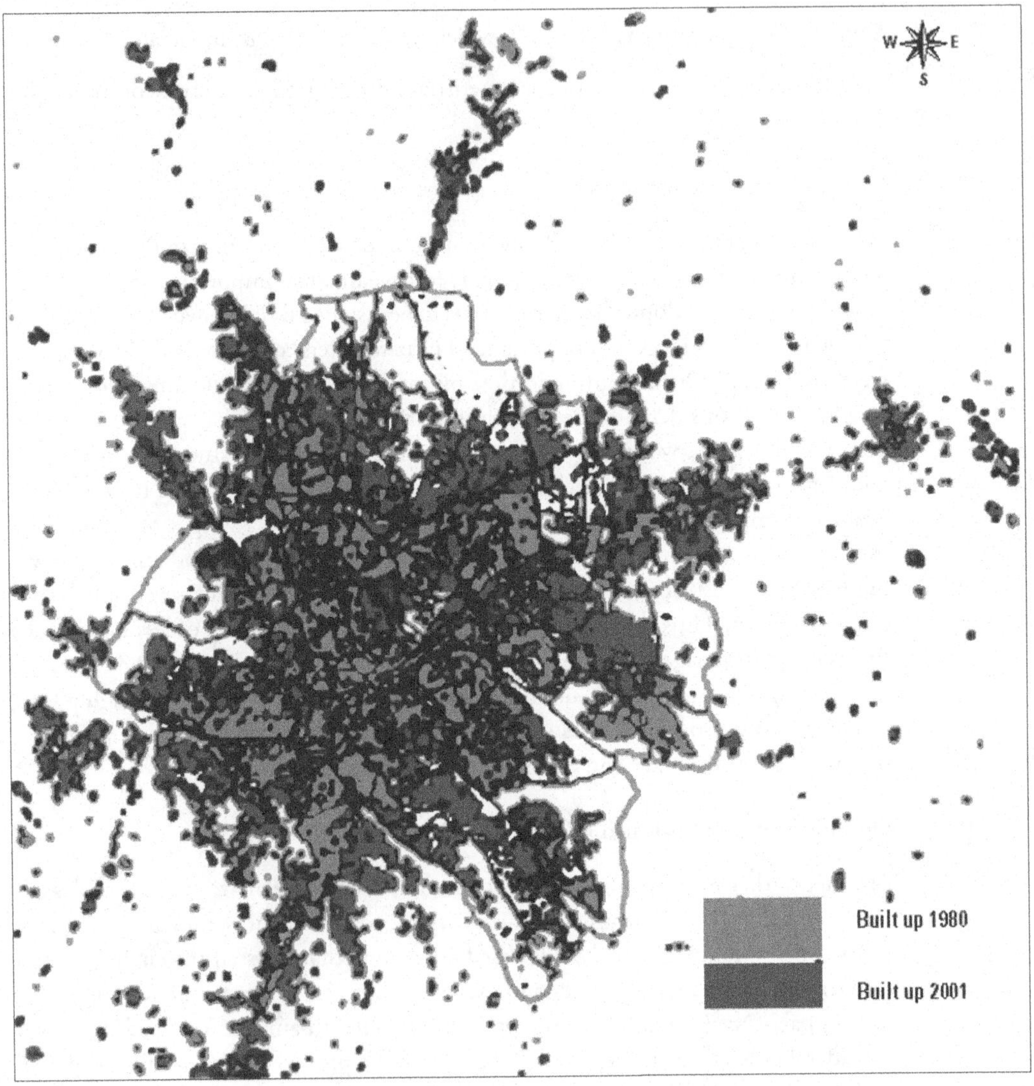

Source: Kampala City Planning Department

Using Remote Sensing and GIS techniques urban growth analysis was performed and related to environmental degradation in the metropolitan area. Two satellite imagery of Landsat, for 1980 and 2001 (resolution of 15 m after Pan Sharpening) were classified and results used in the urban spatial characterization of urban growth for the 20 year period. Since the resolution is low for urban land use mapping, classification of land use/cover was employed. Nine classes of land use and land cover were derived from an examination (with knowledge of the area characteristics and use of the QuickBird 2002 Imagery) of the spectral reflectance as shown in the table below.

Table 12.2: Classes of land use

Class	Remark
BG: Bare Ground	Significant rock outcrops and cleared plots for development
BO: Built Up Other	Characteristics of change cover to built up areas
F: Forest	The 1980 image still showed significant green areas mainly low land forests though mainly outside Kampala city boundaries
G: Grassland	This class combined low grass and tree areas with secondary vegetation but also re-grown grass after clearance of plots for development
ID: Industrial	With established industrial sites, the reflectance was found to be similar mainly due to material types of the roofs. This class however was found merging with built up and commercial (based on knowledge of area) areas especially in the civic area of the city
OP: Open Water	Represents the open water bodies in Kampala of Lake Victoria and the Kabaka's Lake
S: Swamp	A significant part of the wetland system in and around Kampala is predominated by papyrus type of vegetation from which this class derives
SA: Subsistence Agriculture	Areas of differentiated spectral reflectance with differing crops and other types of vegetation mixing with crops
SW: Swamp Forest	Wetland systems with significant tree cover that has not been disturbed that much

A supervised classification of the image analysis was undertaken after which, filtering was used to clean up the classified images in order to remove the unclassified pixels. Data was then input in the GIS for analysis of urban change, quantify the changes and relate to environmental variables.

As indicated in the table below, from the classification of the Landsat Images of 1980 and 2002, the analysis shows that in 1980, the land use land cover was predominated by agriculture activities occupying 62.2% of the total land area for the analysis area compared to 16% built up and 1.7% for industrial activity. The 2001 classification of the analysis area indicates significant changes for built up and industrial areas. For both built up and industrial uses/cover, the area occupied more than doubled in the period of 11 years while agriculture declined by a quarter mainly because it is easily convertible.

Table 12.3: Land - analysis of land use

Land Use/Landcover	Area Ha 1980	Percentage of Total 1980	Area Ha 2002	Percentage of Total 2002
BG : Bare Ground	0.0	0.0	362.2	0.9
BO: Built Up Other	6192.0	16.0	12269.6	31.7
Forest	458.6	1.2	1032.3	2.7
G: Grassland	1092.2	2.8	2155.4	5.6
ID: Industrial	669.4	1.7	1827.0	4.7
OP: Open Water	2193.6	5.7	2147.6	5.6
S: Swamp	1092.1	2.8	1112.6	2.9
SA: Subsistence Agriculture	24045.4	62.2	17622.6	45.6
SW: Swamp Forest	2921.5	7.6	135.4	0.4
TOTAL AREA	38664.7		38664.7	

A further analysis of the land use/land cover indicates that industrial land use changed by 172.9%, built up changed by 98.2% and forests by 125%. The increase in forest cover is attributed to the forest plantations which were also reported in (Norstrand, Development et 'al. 1994) the land use map prepared in 1991 using aerial photography. These forests were mainly in the north eastern part of Kampala which was still rural taking advantage of the demand for eucalyptus poles used for building in the city. For the other land use/cover classes of agriculture and swamp forest, the change was negative implying loss of swamp forests and agricultural land. Similarly swamps which are mainly covered by papyrus also reduced over the period of 11 years from 20.6% to 1.9% occupancy of the land area. When a rate of change was computed, industrial land use, forest land use and built up changed faster compared to other uses.

Built up cover changed at 8.9%, forest at 11.4% and industrial at 15.7%. The speed of change in industrial and built up is the main concern in terms of the relationship between industrialization and housing. The change rates of industrial and built up imply a possibility of the two classes of land use being related to each other. This is because industrial establishment imply increased demand for labour and housing for the laborers. It is also evident that industrial areas such as the main industrial complex that stretches from the civic area to the Nakivubo swamp is surrounded by high density residential areas including Kisugu, Namuwongo, Kibuli all of which provide residence to the many laborers for the industrial area and other industrial places in Kampala.

Table 12.4: Land use - change

Land Use/Landcover	Change Area Ha	Percentage Change	Percentage of Total 2002
BG : Bare Ground	362.2		
BO: Built Up Other	6077.6	98.2	8.9
Forest	573.8	125.1	11.4
G: Grassland	1063.2	97.3	8.8
ID: Industrial	1157.6	172.9.	15.7
OP: Open Water	-45.9	-2.1	-0.2
S: Swamp	20.6	1.9	0.2
SA: Subsistence Agriculture	-6422.8	-26.7	-2.4
SW: Swamp Forest	-2786.1	-95.4	-8.7

A spatial analysis of the two classified land use/cover maps through cross tabulation reveals that wetland areas have been degraded mainly by built up and industrial development. The built up land cover class combines the different uses of residential, commercial, infrastructural installations and utilities. Therefore wetlands have greatly been degraded by urban development in Kampala. The environmental impacts of the change described and analyzed indicate clearance of swamps, forests with the built up cover. Data on industrial establishment, waste management and pollution are required to supplement land cover change analysis as an indicator of environmental degradation in Kampala.

12.4.5 Opportunities and Challenges of GIS use

This section of the chapter focuses on the benefits of GIS. Whether defined from a functional point of view or defined according to what it is, GIS has proved to be an indispensable tool in the capture, analysis and presentation of geographic information. An attempt to describe the opportunities of GIS in current and future terms is rather a difficult task for, no one can tell in which direction GIS and IT related progress is heading. But below are some of the categories of opportunities from GIS.

From a scientific point of view, GIS offers powerful tools for the scientific investigation of spatial phenomena. As noted earlier, spatial scientists including geographers have always employed several techniques in deep investigation of spatial phenomena. Given the current literature on different applications as well as tested theories and models, from two dimensional to three dimensional models of reality, GIS offer great opportunities for deep analysis and investigation of spatial phenomena from description, explanation and prediction. For example natural resource managers are utilizing GIS to estimate bio-mass production and harvesting. Disaster risk assessment and reduction is also making use of Geo-Information tools while in-depth investigations are conducted in various fields using GIS.

Opportunities in GIS utilization can also be tapped into by the capability to support decision making. In many ways, decision making is an important component of all societal desires for advancement including resource utilization, human capital development to investments. Ideally decisions must be based on up to date and accurate information and one type of information required is spatial information. Due to the capability of GIS to have dynamics mechanisms of data updates, decision support is increasing drawing from the utility of GIS. Thus it is common to find in literature systems referred to as Decision Support Systems. This has emerged as an important field in GIS and a vital component of public and private systems of management information.

GIS is also increasingly being utilized as an ingredient of planning at various levels. From local project planning, sub-regional, regional, national to international levels, spatial planning in various fields is making use of GIS. It is equally common now to find terms in literature as Planning Support Systems that describe the information systems designed and managed to support various planning activities. Fields of urban and regional planning, nature conservation and natural resource management utilize the opportunities of timely information provided by GIS. These opportunities are yet to be fully exploited by the responsible organizations of managing natural resources and guiding urban and regional development.

In the context of cartography, GIS has an outstanding opportunity by having turned scaled geo-databases to seamless databases. Cartographic outputs have always been stored at various scales for reproduction but GIS can now store seamless databases for the entire globe. Even when it comes to cartographic querying and retrieval of data, GIS offers more flexible, visualization and modeling techniques of information representation.

One of the greatest opportunities that have developed with GIS progress is data sharing which has developed the concept of Spatial Data Infrastructure. By establishing SDI's data sharing, standardization and reduction of duplication will be a splendid opportunity in reduction of costs of data production and enhance spatial data utilization. SDI concern frameworks for data sharing pricing and involve clearing houses or portals and exchange modalities. SDI is now considered a vital ingredient of development in Africa and particularly Uganda where discrete units are engaged in data reproduction without any standards while duplicating efforts.

However though GIS offers opportunities it also has some limitations or challenges. Though capacity has been built in the field of Geo-Information to a substantive level, the demand for information requires even more personnel who are also up to date with the new techniques of GIS. This will require investment in education. The challenge of adapting Geo-Information tools for decision support or planning support in various organizations is already felt by the apparent threat of human resource with the GIS skills. This has and will continue slowing down the adoption of GIS in various organizations in Uganda.

12.5 Conclusion

In GIS has emerged as multifaceted field with capabilities of providing in-depth scientific analyses while offer easy to use, simple visualization and spatial information communication tools. It cannot be assumed that GIS is Geography or Geography is GIS given the disciplines from which GIS derives its theory, methodology and techniques. But there is assurance that GIS has transformed Geography by enhancing spatial analysis and representation techniques. GIS offers tools for Decision Support, Planning Support while it deepens scientific investigation with enhanced tools of spatial analysis. GIS based models are increasingly being utilized in various fields and replacing the analogue methods spatial representation and analysis. But for GIS to be fully utilized, the challenges of education, SDI and data sharing impediments have to be removed through collaborative efforts to reduce duplication, redundancy and quality of information produced.

Questions

References and further readings

Keith C. Clerk. 2003. *Getting Started with Geographic Information Systems*, Fourth Edition, Pearson Education Inc, New Jersey.

Aronoff S. 1989. *Geographic Information System: A management perspective*, Ottawa, WDL Publications.

Bonham-Carter, G. 1986. *Geographic Information System for Geoscientists: Modeling with GIS*, Tarrytown, NY: Pergamon Press.

Goodchild M. F. 2001. *Metrics of scale in remote sensing and GIS*, JAG, Vol 3, Issue 2, Enschede, Netherlands.

Voogd, et al,.1990. *Multicriteria Evaluation in physical planning*, North-Holland, Amsterdam.

Lwasa S. 2005. *A Geo-Information Approach for Urban Land Use Planning in Kampala*, FIG, Cairo, Egypt.